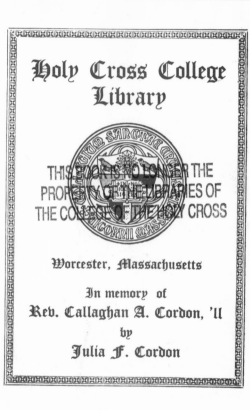

Multivariate Statistical Methods for Business and Economics

PRENTICE-HALL INTERNATIONAL SERIES IN MANAGEMENT

ATHOS AND COFFEY	*Behavior in Organizations: A Multidimensional View*
BALLOU	*Business Logistics Management*
BAUMOL	*Economic Theory and Operations Analysis, 3rd ed.*
BOLCH AND HUANG	*Multivariate Statistical Methods for Business and Economics*
BOOT	*Mathematical Reasoning in Economics and Management Science*
BROWN	*Smoothing, Forecasting and Prediction of Discrete Time Series*
CHAMBERS	*Accounting, Evaluation and Economic Behavior*
CHURCHMAN	*Prediction and Optimal Decision: Philosophical Issues of a Science of Values*
CLARKSON	*The Theory of Consumer Demand: A Critical Appraisal*
COHEN AND CYERT	*Theory of the Firm: Resource Allocation in a Market Economy*
CULLMAN AND KNUDSON	*Management Problems in International Environments*
CYERT AND MARCH	*A Behavioral Theory of the Firm*
FABRYCKY AND TORGERSEN	*Operations Economy: Industrial Applications of Operations Research*
FRANK, MASSY, AND WIND	*Market Segmentation*
GREEN AND TULL	*Research for Marketing Decisions, 2nd ed.*
GREENLAW, HERRON, AND RAWDON	*Business Simulation in Industrial and University Education*
HADLEY AND WHITIN	*Analysis of Inventory Systems*
HOLT, MODIGLIANI, MUTH, AND SIMON	*Planning Production, Inventories, and Work Force*
HYMANS	*Probability Theory with Applications to Econometrics and Decision-Making*
IJIRI	*The Foundations of Accounting Measurement: A Mathematical, Economic, and Behavioral Inquiry*
KAUFMANN	*Methods and Models of Operations Research*
LESOURNE	*Economic Analysis and Industrial Management*
MANTEL	*Cases in Managerial Decisions*
MASSÉ	*Optimal Investment Decisions: Rules for Action and Criteria for Choice*
MCGUIRE	*Theories of Business Behavior*
MILLER AND STARR	*Executive Decisions and Operations Research, 2nd ed.*
MONTGOMERY AND URBAN	*Management Science in Marketing*
MONTGOMERY AND URBAN	*Applications of Management Science in Marketing*
MORRIS	*Management Science: A Bayesian Introduction*
NICOSIA	*Consumer Decision Processes: Marketing and Advertising Decisions*
PETERS AND SUMMERS	*Statistical Analysis for Business Decisions*
PFIFFNER AND SHERWOOD	*Administrative Organization*
SIMONNARD	*Linear Programming*
SINGER	*Antitrust Economics: Selected Legal Cases and Economic Models*
VERNON	*Manager in the International Economy, 2nd ed.*
WAGNER	*Principles of Operations Research with Applications to Managerial Decisions*
WARD AND ROBERTSON	*Consumer Behavior: Theoretical Sources*
ZANGWILL	*Nonlinear Programming: A Unified Approach*
ZENOFF AND ZWICK	*International Financial Management*
ZIMMERMANN AND SOVEREIGN	*Quantitative Models for Production Management*

Multivariate Statistical Methods for Business and Economics

BEN W. BOLCH
Vanderbilt University

CLIFF J. HUANG
Vanderbilt University

PRENTICE-HALL, INC., ENGLEWOOD CLIFFS, NEW JERSEY

Library of Congress Cataloging in Publication Data

BOLCH, BEN W
Multivariate statistical methods for business and economics.

(Prentice-Hall international series in management)
Bibliography: p.
1. Multivariate analysis. I. Huang, Cliff J., joint author. II. Title.
QA278.B64 1973 519.5'3 72–13834
ISBN O–13–604819–6

Printed in the United States of America

10 9 8 7 6 5 4 3 2 1

PRENTICE-HALL INTERNATIONAL, INC., *London*
PRENTICE-HALL OF AUSTRALIA, PTY. LTD., *Sydney*
PRENTICE-HALL OF CANADA, LTD., *Toronto*
PRENTICE-HALL OF INDIA PRIVATE LIMITED, *New Delhi*
PRENTICE-HALL OF JAPAN, INC., *Tokyo*

PRENTICE-HALL, INC.
PRENTICE-HALL INTERNATIONAL, INC., *United Kingdom and Eire*
PRENTICE-HALL OF CANADA, LTD., *Canada*
DUNOD PRESS, *France*
MARUZEN COMPANY, LTD., *Far East*
HERRERO HERMANOS, SUCS., *Spain and Latin America*
R. OLDENBOURG VERLAG, *Germany*
ULRICO HOEPLI EDITORS, *Italy*

For
Suzanne *and* Tricia

Contents

chapter 2

The Normal Density and Statistical Inference 41

chapter 3

Tests Concerning Means 72

chapter 4

Linear Regression 105

chapter 5

Correlation, Analysis of Variance, and Covariance 154

chapter 6

Nonlinear Regression 208

chapter 7

Discrimination, Principal Components, and Canonical Correlations 229

chapter 8

Spectral Analysis 271

Appendix Tables 305

Bibliography 316

Index 323

Preface

This book grew out of a course which we first offered to advanced undergraduate and graduate students at Vanderbilt University in 1968. The course was designed to be taken prior to our econometrics sequence, and its purposes were (1) to present the standard matrix algebra approach to regression and thus free time for other matters in the econometrics courses and (2) to broaden our coverage of topics which were then usually not covered elsewhere. It was perfectly possible at that time for students to take all the courses that we offered in statistics and econometrics and yet have only vague ideas about discriminant analysis, principal component analysis, canonical correlation analysis, and spectral analysis. And, while we note that the newer econometrics texts are now including some or all of these techniques, we continue to believe that they more appropriately belong in a course which is taken prior to econometrics. This book is then biased toward topics that we believe will prove most useful to students in economics and business. It is heavy on standard continuous variate regression, is light on experimental design, and attempts to offer only an introductory exposition of many of the topics. This bias, plus the introduction to spectral analysis, sets the book apart from standard treatments of multivariate methods.

Another feature of the text is its computer programs. It is not hard to understand why many of the techniques presented here never came into much use until the day of wide computer availability. From one point of view the computer age has hurt economics by making data handling so easy that it has tended to draw some people away from analysis. But regardless of our feeling on this matter, the computer is with us and we consider ourselves obligated to teach our students how to use it. However, the text is so designed that it can be used without reference to the supportive materials dealing with the programs.

The manuscript for this book has been class-tested in one form or another for two years. Students are required to have one year of basic statistics and one year of basic calculus, but they are not required to know FORTRAN. The course makes much use of the problems given in the text, which are often done by use of the programs which are stored on a disc in our computer center. Except for a few minutes spent explaining how to punch the "magic" cards which access the disc, no instructional time is devoted to programming.

John Pilgrim has tested the programs on an IBM 360/40, and they ran without alteration. Marshall McMahon tested the programs on an 8K IBM 1130 and reported that some reductions in core requirements were necessary on three of the larger programs but that otherwise the programs ran without error. Our testing was done on an XDS Sigma 7. Source decks may be obtained at cost from the authors.

Many people have contributed their encouragement, patience, and advice to this project. Our friends and colleagues Dudley Cowden, Charles Federspiel, Yun Shung Ho, Marshall McMahon, John Pilgrim, William Steel, Fred Westfield, and Chi Lin Yen made helpful suggestions. Anne Bolch and Elizabeth Huang provided wonderful homes, and the students of EcBA 255 in 1970 and 1971 had an immeasurable influence on the final product. Mrs. Marie Cooper did an exceptional job in typing a difficult manuscript. Our thanks also go to the Prentice-Hall referees: Professor Jay E. Strum, Graduate School of Business Administration, New York University; Dr. John E. Freund; and Professor Lawrence L. Schkade, College of Business Administration, North Texas State University for reviewing the text.

We are indebted to the literary executor of the late Sir Ronald A. Fisher, F.R.S.; to Dr. Frank Yates, F.R.S.; and to Oliver & Boyd Ltd., Edinburgh, for permission to reprint from Table III and to abridge from Table IV from their book *Statistical Tables for Biological, Agricultural and Medical Research*, 6th ed., 1963. We also express our gratitude for permission to abridge from Tables 8 and 18 from E. S. Pearson and H. O. Hartley, ed., *Biometrika Tables for Statisticians* Vol. I (New York: Cambridge University Press, 1966).

We dedicate this book to our daughters, in the hope that they will grow to know the friendship that their fathers have found in working together.

BEN W. BOLCH
CLIFF J. HUANG
Nashville, Tennessee

Multivariate Statistical Methods for Business and Economics

chapter 1

Review of Selected Mathematical Topics

Intermediate and advanced statistical methods require a knowledge of certain topics in mathematics with which the reader may not be familiar. These topics are by no means specialized, and they appear quite often in the literature of economics and business. Therefore the student who makes the investment of a careful study of this chapter will derive as a payoff both a better understanding of statistical methods and a greater facility to understand the subject matter of economics and business.

1.1 elements of vector algebra

Consider the simultaneous equations

$$\begin{aligned} P &= 2.0 + 0.5Q \\ P &= 5.0 - 1.0Q \end{aligned} \tag{1-1}$$

As students of economics we might believe that the first of these equations describes a supply schedule which relates quantity supplied, Q, with price, P. The second of these equations might be thought of as a representation of a demand schedule. The two equations together imply that P and Q are *jointly determined* variables. Figure 1.1 gives a geometric representation of these

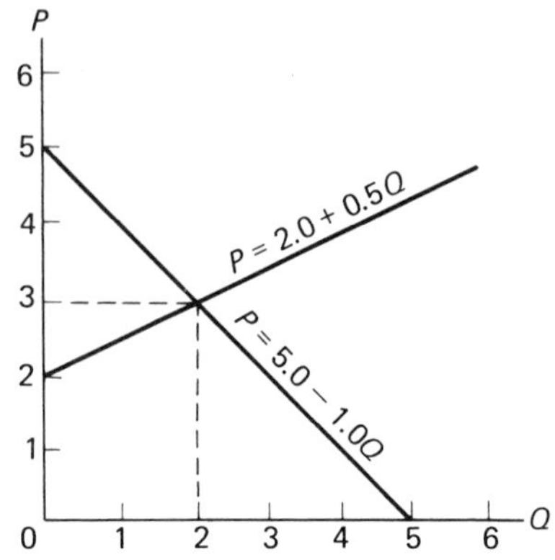

Figure 1.1. Geometric representation of eqs. (1-1) and their solution.

two equations for positive values of Q and P. By use of this figure we can easily find the simultaneous solution to these equations. It is the intersection of the two lines, that is, the point where $Q = 2.0$ and $P = 3.0$.

The solution could, of course, be obtained from the equations directly. Subtracting the second equation from the first gives

$$0 = -3.0 + 1.5Q$$

so that $Q = 2.0$. Substituting $Q = 2.0$ back into either of the original equations gives $P = 3.0$. For convenience we may write the solution of the equations as the point (2, 3).

Let us define the *directed line* segment from the origin to the point (2, 3) as $\mathbf{X} = \begin{bmatrix} 2 \\ 3 \end{bmatrix}$, where the first element represents a number on the horizontal axis and the second element a number on the vertical axis. Then **X** may be called a *vector*; it is depicted in Figure 1.2. The numbers 2 and 3 are called the *elements* of the vector **X**. Note carefully that in this book we shall distinguish a vector from a single number (a *scalar*) by use of boldface type for the former and italic type for the latter. Also notice that the elements of a vector are surrounded by brackets.[1]

If the elements of a vector are written one below the other, as above, the vector is called a *column vector*. A *row vector* is the *transpose* of a column vector, and the operation of transposition is indicated by a prime. Thus

$$\mathbf{X}' = [2 \quad 3]$$

is a row vector. Row and column vectors have the same geometric interpre-

[1]This same notation will hold for matrices, which are covered in later sections.

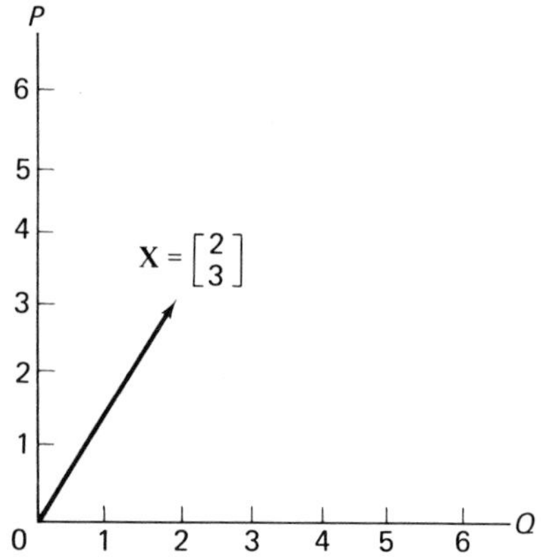

Figure 1.2. Geometric representation of the vector $\mathbf{X} = \begin{bmatrix} 2 \\ 3 \end{bmatrix}$.

tation. The advantage of a row vector is that it takes up less space on the printed page.

Three other important vectors for our coordinate system are worthy of mention at this point. First, the origin of the system is given by the *zero vector*:

$$\mathbf{0}' = [0 \quad 0]$$

Second, the two *unit* vectors

$$\mathbf{I}_1' = [1 \quad 0], \qquad \mathbf{I}_2' = [0 \quad 1]$$

are important because any point in the entire two-dimensional space may be represented by a *linear combination* of these two vectors.[2] Thus the point (2, 3) may be represented by the vector

$$\begin{bmatrix} 2 \\ 3 \end{bmatrix} = 2\begin{bmatrix} 1 \\ 0 \end{bmatrix} + 3\begin{bmatrix} 0 \\ 1 \end{bmatrix} = 2\mathbf{I}_1 + 3\mathbf{I}_2$$

Two concepts of vector algebra are illustrated by the above equation. The first concept is that of *scalar multiplication*, or multiplication of a vector by a single number. For an arbitrary vector, $\mathbf{A}'$, with elements A_1 and A_2, and an arbitrary scalar, k,

$$k\mathbf{A} = \mathbf{A}k = \begin{bmatrix} kA_1 \\ kA_2 \end{bmatrix}$$

[2]Formally, these are two possible *basis vectors* for the coordinated system.

That is, each of the elements of **A** is multiplied by the scalar quantity, and *premultiplication* of **A** by k (i.e., $k\mathbf{A}$) is the same as *postmultiplication* of **A** by k (i.e., $\mathbf{A}k$). To give two examples, let $\mathbf{A}' = [1 \quad 2]$ and $\mathbf{B}' = [-2 \quad 1]$. Then $2\mathbf{A}' = [2 \quad 4]$ and $(-1)\mathbf{B}' = [2 \quad -1]$.

The geometric explanation of scalar multiplication is that the vector is extended k times its original length in the same direction if k is positive. If k is negative, then the vector is extended k times its original length in the opposite direction. The vectors **A**, **B**, 2**A**, and (−1)**B** are shown in Figure 1.3.

A second concept in vector algebra is that of *vector addition*. Two vectors are added by adding their corresponding elements. Thus

$$\mathbf{A} + \mathbf{B} = \begin{bmatrix} 1 \\ 2 \end{bmatrix} + \begin{bmatrix} -2 \\ 1 \end{bmatrix} = \begin{bmatrix} -1 \\ 3 \end{bmatrix}$$

Geometrically, the sum $\mathbf{A} + \mathbf{B}$ may be viewed as the diagonal of a parallelogram with two sides given by **A** and **B**, as shown in Figure 1.3. We can reduce subtraction to addition by changing $\mathbf{A} - \mathbf{B}$ to $\mathbf{A} + (-1)\mathbf{B}$. Thus

$$\mathbf{A} - \mathbf{B} = \mathbf{A} + (-1)\mathbf{B} = \begin{bmatrix} 1 \\ 2 \end{bmatrix} + \begin{bmatrix} 2 \\ -1 \end{bmatrix} = \begin{bmatrix} 3 \\ 1 \end{bmatrix}$$

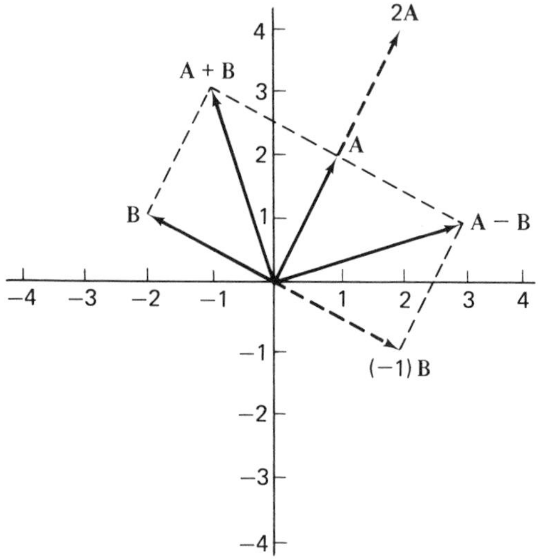

Figure 1.3. Multiplication of a vector by a scalar; addition and subtraction of vectors.

By changing $\mathbf{A} - \mathbf{B}$ to $\mathbf{A} + (-1)\mathbf{B}$, we have the sum of two vectors, and the vector $\mathbf{A} - \mathbf{B}$ is represented geometrically by the diagonal of a parallelogram with two sides given by $\mathbf{A}$ and $(-1)\mathbf{B}$ (see Figure 1.3).

One further rule is necessary to establish an algebraic system, and it concerns equality of vectors. Two vectors are *equal* if and only if their corresponding elements are all equal. This definition should be self-evident from the geometric interpretation that we have given of vectors.[3]

Linear Independence. In ordinary algebra, if $k_1 \neq 0$ in the linear equation

$$k_1 X + k_2 Y = 0 \tag{1-2}$$

the variable X may be said to be a linear function of the variable Y since[4]

$$X = \frac{-k_2}{k_1} Y$$

On the other hand, if Eq. (1-2) holds only when $k_1 = k_2 = 0$, the variable X cannot be said to be expressed as a linear function of the variable Y or vice versa. In this case we say that the two variables are *linearly independent*. This idea of linear independence can be extended to vectors.

The vectors $\mathbf{X}$ and $\mathbf{Y}$ are linearly independent if there do not exist scalars k_1 and k_2 such that

$$k_1 \mathbf{X} + k_2 \mathbf{Y} = \mathbf{0} \tag{1-2a}$$

unless both k_1 and k_2 are zero. Clearly $\mathbf{I}_1$ and $\mathbf{I}_2$ are linearly independent since

$$k_1 \begin{bmatrix} 1 \\ 0 \end{bmatrix} + k_2 \begin{bmatrix} 0 \\ 1 \end{bmatrix} = \begin{bmatrix} 0 \\ 0 \end{bmatrix}$$

can be satisfied only if both k_1 and k_2 are zero. When it is impossible to find nonzero k_1 and/or k_2 such that Eq. (1-2a) holds, the two vectors are linearly independent. The student should verify that $[2 \quad 3]$ and $[4 \quad 6]$ are *linearly dependent* since Eq. (1-2a) holds for $k_1 = 2$ and $k_2 = -1$.

All the vectors previously mentioned have been two-dimensional (i.e., they contained only two elements). In general, an n-dimensional vector contains n elements. An example of such a vector is $\mathbf{A}' = [A_1 \quad A_2 \quad \cdots \quad A_n]$.

---

[3]Technically, a real vector space is defined in terms of a set of vectors which follow certain rules of equality, addition, and scalar multiplication. See Paige and Swift (1961), p. 48 (in the Bibliography at the end of this book).

[4]Of course, we may reverse the argument and say that Y is a linear function of X if $k_2 \neq 0$. We assume that variables (or vectors) are nonzero.

Product of Vectors. The product of two vectors

$$\mathbf{A} = \begin{bmatrix} A_1 \\ A_2 \\ \cdot \\ \cdot \\ \cdot \\ A_n \end{bmatrix}, \qquad \mathbf{B} = \begin{bmatrix} B_1 \\ B_2 \\ \cdot \\ \cdot \\ \cdot \\ B_n \end{bmatrix}$$

is defined as

$$\mathbf{A}'\mathbf{B} = [A_1 \quad A_2 \quad \cdots \quad A_n] \begin{bmatrix} B_1 \\ B_2 \\ \cdot \\ \cdot \\ \cdot \\ B_n \end{bmatrix}$$

$$= A_1B_1 + A_2B_2 + \cdots + A_nB_n = \sum_{i=1}^{n} A_iB_i \tag{1-3}$$

For example if $\mathbf{A}' = [1 \quad 2]$ and $\mathbf{B}' = [-1 \quad 3]$, then $\mathbf{A}'\mathbf{B} = 1(-1) + 2(3) = 5$. Notice that the definition above extends to vectors with an arbitrary number of elements and that this same *inner product*, which is a scalar, could have been obtained by the multiplication $\mathbf{B}'\mathbf{A}$. Thus the inner product of two vectors is the sum of the products of the corresponding elements of the two vectors. The inner product of a vector with itself is clearly the sum of the squared elements of the vector:

$$\mathbf{A}'\mathbf{A} = A_1^2 + A_2^2 + \cdots + A_n^2 = \sum_{i=1}^{n} A_i^2 \tag{1-4}$$

A geometric representation of the inner product of two vectors is given in Figure 1.4 for vectors with two elements. From elementary geometry, the

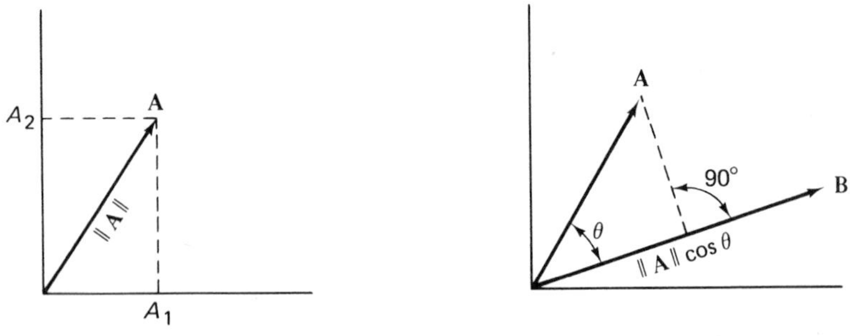

(a) Length, ‖**A**‖

(b) Inner product, ‖**A**‖ ‖**B**‖ cos θ

Figure 1.4. Length and inner product of vectors.

length of the vector **A** [denoted by $||\mathbf{A}||$ and shown in Figure 1.4(a)] is

$$||\mathbf{A}|| = \sqrt{A_1^2 + A_2^2} = (A_1^2 + A_2^2)^{1/2} = (\mathbf{A}'\mathbf{A})^{1/2} \tag{1-5}$$

Thus the length of a vector is the square root of the inner product of the vector with itself. In general, $||\mathbf{A}|| = (A_1^2 + A_2^2 + \cdots + A_n^2)^{1/2} = (\mathbf{A}'\mathbf{A})^{1/2}$.

The vectors **A** and **B** and the counterclockwise angle between them, θ, are shown in Figure 1.4(b). The projection of **A** onto **B** has a length of $||\mathbf{A}|| \cos \theta$, and the reader should prove that

$$\mathbf{A}'\mathbf{B} = ||\mathbf{A}||\,||\mathbf{B}|| \cos \theta$$

Thus the inner product is the product of the lengths of the two vectors and the cosine of the angle between them.

Suppose that two vectors are perpendicular (at right angles) to each other. Then the angle $\theta = 90°$, and $\cos \theta = 0$. Therefore the inner product of the vectors will be zero. Conversely, the inner product will be zero only when $\theta = 90°$; then the two vectors are said to be *orthogonal.* The reader should construct several examples to prove that all orthogonal vectors are linearly independent. One specific example is $\mathbf{A}' = [1 \quad 2]$ and $\mathbf{B}' = [-2 \quad 1]$. However, *not all* linearly independent vectors are orthogonal. An example is the pair $\mathbf{A}' = [1 \quad 2]$ and $\mathbf{B}' = [1 \quad 1]$.

A set of orthogonal vectors is called an *orthonormal* set if each vector has unit length. For example $\mathbf{I}_1' = [1 \quad 0]$ and $\mathbf{I}_2' = [0 \quad 1]$ are one such set. Another such orthonormal set is $\mathbf{P}_1' = [1/\sqrt{2} \quad 1/\sqrt{2}]$ and $\mathbf{P}_2' = [1/\sqrt{2} \quad -1/\sqrt{2}]$. These two sets of vectors, along with a vector **X**, are shown in Figure 1.5. Let the vector **X** have coordinates $X_1 = \frac{1}{2}$ and $X_2 = \frac{1}{4}$ in terms of $\mathbf{I}_1$ and $\mathbf{I}_2$. Thus

$$\mathbf{X} = \begin{bmatrix} X_1 \\ X_2 \end{bmatrix} = X_1\mathbf{I}_1 + X_2\mathbf{I}_2 = \frac{1}{2}\begin{bmatrix} 1 \\ 0 \end{bmatrix} + \frac{1}{4}\begin{bmatrix} 0 \\ 1 \end{bmatrix} = \begin{bmatrix} \frac{1}{2} \\ \frac{1}{4} \end{bmatrix}$$

In terms of $\mathbf{P}_1$ and $\mathbf{P}_2$ the vector **X** has coordinates, say Y_1 and Y_2, as yet unknown. To find these coordinates we need to evaluate

$$\mathbf{X} = Y_1\mathbf{P}_1 + Y_2\mathbf{P}_2$$

Premultiply both sides of the above equation by $\mathbf{P}_1'$:

$$\mathbf{P}_1'\mathbf{X} = Y_1\mathbf{P}_1'\mathbf{P}_1 + Y_2\mathbf{P}_1'\mathbf{P}_2$$

But $\mathbf{P}_1'\mathbf{P}_1 = 1$ and $\mathbf{P}_1'\mathbf{P}_2 = 0$ by virtue of the fact that $\mathbf{P}_1$ and $\mathbf{P}_2$ are orthonormal. Thus $\mathbf{P}_1'\mathbf{X} = Y_1$ or

$$Y_1 = \begin{bmatrix} \dfrac{1}{\sqrt{2}} & \dfrac{1}{\sqrt{2}} \end{bmatrix} \begin{bmatrix} \frac{1}{2} \\ \frac{1}{4} \end{bmatrix} = \frac{3}{4\sqrt{2}}$$

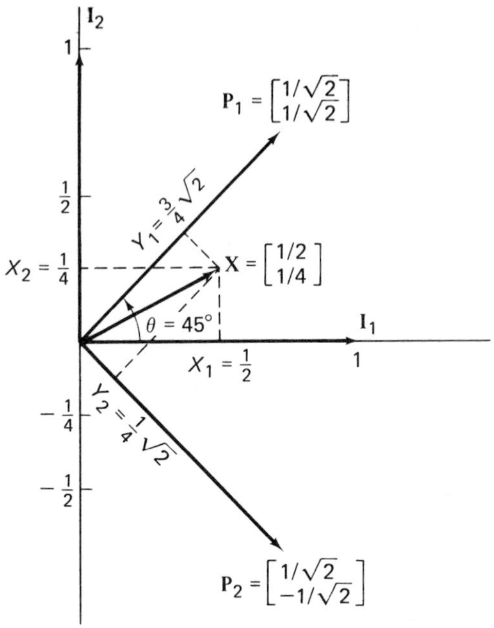

Figure 1.5. Two orthonormal sets and the vector **X**.

In a similar way, we find upon premultiplication of **X** by $\mathbf{P}_2'$ that $Y_2 = 1/4\sqrt{2}$.

Either of these orthonormal sets is suitable for a description of **X** in that we may pass from one coordinate system to the other without contradiction. Thus

$$\mathbf{X} = Y_1\mathbf{P}_1 + Y_2\mathbf{P}_2 = \frac{3}{4\sqrt{2}}\begin{bmatrix}\frac{1}{\sqrt{2}}\\ \frac{1}{\sqrt{2}}\end{bmatrix} + \frac{1}{4\sqrt{2}}\begin{bmatrix}\frac{1}{\sqrt{2}}\\ \frac{-1}{\sqrt{2}}\end{bmatrix} = \begin{bmatrix}\frac{1}{2}\\ \frac{1}{4}\end{bmatrix}$$

Therefore we may write

$$\mathbf{X} = X_1\mathbf{I}_1 + X_2\mathbf{I}_2$$

or

$$\mathbf{X} = Y_1\mathbf{P}_1 + Y_2\mathbf{P}_2$$

Furthermore, using the definition of vector multiplication we may find the angle between $\mathbf{P}_1$ and $\mathbf{I}_1$, which is

$$\cos\theta = \frac{\mathbf{I}_1'\mathbf{P}_1}{\|\mathbf{I}_1\|\|\mathbf{P}_1\|}$$

The reader may verify that $\cos\theta = 1/\sqrt{2}$ so that $\theta = 45°$. We shall generalize these concepts in later sections. This exercise of passing from one coordinate system to another will be important for our future presentations and should be mastered. We have attempted to illustrate that there is nothing sacred about using the coordinate system based upon the vectors $\mathbf{I}'_1 = [1 \quad 0]$ and $\mathbf{I}'_2 = [0 \quad 1]$. This system is merely the one with which we are accustomed to working, but in some cases other coordinate systems will not only be more convenient but will also be more meaningful. However, because the coordinate system based upon $\mathbf{I}_1$ and $\mathbf{I}_2$ is so ingrained in our thinking, it is comforting to know that we may pass from an arbitrary orthonormal coordinate system back to the one based upon $\mathbf{I}_1$ and $\mathbf{I}_2$.

Although it is useful to give some geometric meaning to vectors, a vector may be thought of as simply an ordered collection of numbers. This concept is the one that we shall make the most use of in this book. If the market price of a commodity is observed to be $1, $2, and $3 during three time periods, then we may summarize these data by the vector[5]

$$\mathbf{X}' = [1 \quad 2 \quad 3]$$

Clearly, $\mathbf{X}'$ could represent an unlimited number of observations. It is this capability to economize on notation that is one of the most attractive attributes of vectors.

1.2 some matrix operations

In the last section we noted that the solution values of Q and P for Eqs. (1-1) could be denoted by the vector $\mathbf{X}'_1 = [2 \quad 3]$. Suppose that there are two other sets of equations whose solutions are given by the vectors $\mathbf{X}'_2 = [1 \quad 2]$ and $\mathbf{X}'_3 = [2 \quad 1]$. We can summarize all these vectors in the following array, which is called a *matrix*:[6]

$$[\mathbf{X}_1 \quad \mathbf{X}_2 \quad \mathbf{X}_3] = \begin{bmatrix} 2 & 1 & 2 \\ 3 & 2 & 1 \end{bmatrix}$$

Just as a vector is an ordered collection of elements, so a matrix is an ordered collection of vectors. A matrix of *order* $n \times m$ (n by m) has n rows and m

[5]This vector may be thought of as a directed line segment in three-dimensional space.

[6]In this example the vectors are arranged columnwise. They could have been arranged rowwise if that had been more convenient.

columns. An arbitrary matrix, **X**, may be written as[7]

$$\mathbf{X} = \begin{bmatrix} X_{11} & X_{12} & \cdots & X_{1m} \\ X_{21} & X_{22} & \cdots & X_{2m} \\ \cdot & & & \cdot \\ \cdot & & & \cdot \\ \cdot & & & \cdot \\ X_{n1} & X_{n2} & \cdots & X_{nm} \end{bmatrix}$$

The first subscript on any element of **X** denotes the *row location* of the element. The second subscript on any element of **X** denotes the *column location* of the element. Thus X_{21} is located in the second row and the first column of the matrix **X**.

Since a matrix is a collection of vectors, we would expect that the rules for equality, addition, scalar multiplication, and transposition would carry over from vector algebra to matrix algebra. Indeed, as the examples below show, this is the case.

1. *Equality*. The matrix **X** and the matrix **Y** are *equal* if and only if their corresponding elements are equal. The following two matrices are equal:

$$\mathbf{X} = \begin{bmatrix} 2 & 3 \\ 4 & 5 \end{bmatrix}, \qquad \mathbf{Y} = \begin{bmatrix} 2 & 3 \\ 4 & 5 \end{bmatrix}$$

2. *Addition*. Matrices are added (subtracted) by adding (subtracting) their corresponding elements:

$$\begin{bmatrix} 1 & 2 \\ 3 & 4 \end{bmatrix} + \begin{bmatrix} -3 & 2 \\ 4 & 0 \end{bmatrix} = \begin{bmatrix} -2 & 4 \\ 7 & 4 \end{bmatrix}$$

3. *Scalar multiplication*. To multiply a matrix by a scalar, multiply each element of the matrix by the scalar:

$$2\begin{bmatrix} -3 & 2 \\ 4 & 0 \end{bmatrix} = \begin{bmatrix} -6 & 4 \\ 8 & 0 \end{bmatrix} = \begin{bmatrix} -3 & 2 \\ 4 & 0 \end{bmatrix}2$$

As for vectors, $k\mathbf{X} = \mathbf{X}k$. In the jargon of mathematics, scalar multiplication is *commutative*. That is, the sequence of multiplication does not matter. Later, we shall find that the sequence of multiplication will be important when two matrices are multiplied by each other.

4. *Transposition*. The transpose of the matrix **X**, written **X**′, is the matrix **X** with its rows and columns interchanged:

---

[7]A matrix of order $1 \times m$ is a row vector, and a matrix of order $n \times 1$ is a column vector. Thus vectors may be thought of as matrices.

$$\mathbf{X} = \begin{bmatrix} 1 & 2 & 3 \\ 4 & 5 & 6 \\ 7 & 8 & 9 \end{bmatrix}, \qquad \mathbf{X}' = \begin{bmatrix} 1 & 4 & 7 \\ 2 & 5 & 8 \\ 3 & 6 & 9 \end{bmatrix}$$

Before we continue with the elements of matrix algebra, let us discuss some important special matrices. The *identity* (or unit) matrix is a *square* matrix (one with the same number of rows as columns) which has 1 in every location of the main (northwest-southeast) diagonal and zero elsewhere. A 3×3 identity matrix is

$$\mathbf{I} = \begin{bmatrix} 1 & 0 & 0 \\ 0 & 1 & 0 \\ 0 & 0 & 1 \end{bmatrix}$$

The identity matrix is a special case of a *diagonal* matrix, which is one with nonzero elements only on its main diagonal. For example,

$$\mathbf{D} = \begin{bmatrix} 2 & 0 & 0 \\ 0 & 3 & 0 \\ 0 & 0 & 4 \end{bmatrix}$$

is a 3×3 diagonal matrix.

If a matrix has nonzero elements only on and above its main diagonal, it is called *upper-triangular*. If it has nonzero elements only on and below its main diagonal, it is called *lower-triangular*. Below, **X** is upper-triangular and **Y** is lower-triangular:

$$\mathbf{X} = \begin{bmatrix} 1 & 2 & 3 \\ 0 & 4 & 5 \\ 0 & 0 & 6 \end{bmatrix}, \qquad \mathbf{Y} = \begin{bmatrix} 3 & 0 & 0 \\ 4 & 7 & 0 \\ 8 & 9 & 11 \end{bmatrix}$$

A *symmetric matrix* is one which remains unchanged by transposition. That is, the elements below the main diagonal are the mirror image of the elements above the main diagonal. Since $\mathbf{X} = \mathbf{X}'$ in the following matrix, it is symmetric:

$$\mathbf{X} = \begin{bmatrix} 1 & 2 & 3 \\ 2 & 5 & 6 \\ 3 & 6 & 7 \end{bmatrix}$$

Many matrices found in statistics are symmetric.

Matrix Multiplication. Let us define the product matrix, A, as

$$\mathbf{A} = \mathbf{XY}$$

where **X** and **Y** are matrices. The ijth element of **A** is found by multiplying each element in the ith row of **X** by each element in the jth column of **Y** and summing the resulting terms. To illustrate,

$$\begin{bmatrix} 1 & 2 & 3 \\ 4 & 5 & 6 \end{bmatrix} \begin{bmatrix} 1 & 4 \\ 6 & 7 \\ 3 & 2 \end{bmatrix} = \begin{bmatrix} 1(1) + 2(6) + 3(3) & 1(4) + 2(7) + 3(2) \\ 4(1) + 5(6) + 6(3) & 4(4) + 5(7) + 6(2) \end{bmatrix}$$

$$= \begin{bmatrix} 22 & 24 \\ 52 & 63 \end{bmatrix}$$

It should be clear from this illustration that the product **XY** can be defined only if the number of columns in **X** is the same as the number of rows in **Y**. If this correspondence holds, **X** and **Y** are *conformable* under the desired sequence of multiplication. The following multiplication cannot be carried out because the two matrices are not conformable under the desired sequence of multiplication:

$$\begin{bmatrix} 1 & 2 & 3 \\ 4 & 5 & 6 \end{bmatrix} \begin{bmatrix} 1 & 2 \\ 2 & 1 \end{bmatrix}$$

However, if the sequence of multiplication is reversed, a product matrix can be obtained:

$$\begin{bmatrix} 1 & 2 \\ 2 & 1 \end{bmatrix} \begin{bmatrix} 1 & 2 & 3 \\ 4 & 5 & 6 \end{bmatrix} = \begin{bmatrix} 9 & 12 & 15 \\ 6 & 9 & 12 \end{bmatrix}$$

Clearly, the examples above show that the product **XY** is *not always* defined and that it is *not generally* true that **XY** = **YX**. A useful device for determining whether the product **XY** is possible—as well as for determining the order of the resulting product matrix if the multiplication is possible—is as follows. Write the order of the two matrices in the desired sequence of multiplication. For the two matrices to be multiplied directly above we have (2×2) and (2×3). If the inside numbers match (in this case, 2), the matrices are conformable under the desired sequence of multiplication since the first matrix has the same number of columns as the second matrix has rows. The outer numbers (in this case, 2 and 3) show the order of the resulting product matrix. The reader should try this rule with several matrices and should convince himself that the multiplication of vectors follows the same rule.

Important use will be made of the fact that multiplication using an identity matrix is commutative and that multiplication by the identity matrix does not change the matrix multiplied. Thus

$$\mathbf{IX} = \mathbf{XI} = \mathbf{X}$$

where **I** and **X** are of the same order.

Trace of a Matrix. The sum of the elements on the main (northwest-southeast) diagonal of a square matrix is called the *trace* of the matrix. Thus for the $n \times n$ matrix **X**

$$\text{tr}\,\mathbf{X} = X_{11} + X_{22} + \cdots + X_{nn} = \sum_{i=1}^{n} X_{ii}$$

where tr **X** stands for "the trace of **X**." The following points concerning the trace will be used in future presentations and may be verified by the reader:

$$\text{tr}\,(\mathbf{X} + \mathbf{Y}) = \text{tr}\,\mathbf{X} + \text{tr}\,\mathbf{Y}$$
$$\text{tr}\,(\mathbf{XY}) = \text{tr}\,(\mathbf{YX})$$
$$\text{tr}\,\mathbf{X}' = \text{tr}\,\mathbf{X}$$

1.3 matrix inversion

Using the matrix algebra that we have developed to this point, we can write Eqs. (1-1) in the following way:

$$\begin{bmatrix} -0.5 & 1.0 \\ 1.0 & 1.0 \end{bmatrix} \begin{bmatrix} Q \\ P \end{bmatrix} = \begin{bmatrix} 2.0 \\ 5.0 \end{bmatrix} \tag{1-1a}$$

If we call the 2×2 matrix of constants **A**, the 2×1 vector of unknowns **X**, and the 2×1 vector of constants **C**, we can write the system of equations as

$$\mathbf{AX} = \mathbf{C} \tag{1-1b}$$

Like vectors, matrices afford tremendous economy of notation. Equation (1-1b) could just as well stand for thousands of simultaneous linear equations rather than only two.

If **A**, **X**, and **C** were scalars, we would have the ordinary algebraic equation $AX = C$, and we could solve for X by evaluating

$$X = \frac{C}{A} = (A^{-1})C$$

provided that A was nonzero. In matrix algebra we solve for the column vector **X** by evaluating

$$\mathbf{X} = (\mathbf{A}^{-1})\mathbf{C}$$

if **A** is *nonsingular* (we shall explain the meaning of the word nonsingular shortly). The matrix $\mathbf{A}^{-1}$ is read "the *inverse* of **A**." Evidently the concept of the inverse matrix in matrix algebra is similar to the concept of division in ordinary algebra. Also, just as $AA^{-1} = 1$ in ordinary algebra, so $\mathbf{AA}^{-1} = \mathbf{I}$, the identity matrix of the same order as **A**, in matrix algebra. In fact, the inverse matrix $\mathbf{A}^{-1}$ is defined as the matrix such that $\mathbf{AA}^{-1} = \mathbf{I}$. The solution to our system of equations is

$$\mathbf{X} = \begin{bmatrix} Q \\ P \end{bmatrix} = (\mathbf{A}^{-1})\mathbf{C} = \begin{bmatrix} -\frac{2}{3} & \frac{2}{3} \\ \frac{2}{3} & \frac{1}{3} \end{bmatrix} \begin{bmatrix} 2.0 \\ 5.0 \end{bmatrix} = \begin{bmatrix} 2.0 \\ 3.0 \end{bmatrix}$$

which agrees with the results obtained by ordinary algebra. We now turn to a technique for finding the inverse matrix and leave it to the reader to verify that $\mathbf{AA}^{-1} = \mathbf{I}$ in the previous example.

Many techniques exist for matrix inversion.[8] We shall explain one method which is suitable for use with a desk calculator and which introduces some additional matrix concepts that will be needed in later chapters. A method of inversion that is suitable for use with a digital computer is presented in the Appendix to this chapter. However, before explaining a technique of inversion, we shall first introduce a function of a square matrix which is called the *determinant*.

Determinant. For every *square matrix* (one with the same number of rows and columns) there exists a scalar known as the determinant of the matrix. For example, the matrix

$$\begin{bmatrix} 4 & 6 \\ 1 & 2 \end{bmatrix}$$

has the determinant

$$\begin{vmatrix} 4 & 6 \\ 1 & 2 \end{vmatrix} = 4(2) - 1(6) = 2$$

We surround the determinant with *vertical lines* (rather than brackets) to avoid confusing it with a matrix. For a 2×2 matrix the determinant is the product of the elements on the main (northwest-southeast) diagonal minus

[8]See Faddeeva (1959).

the product of the elements on the secondary (northeast-southwest) diagonal. Thus the matrix

$$\mathbf{A} = \begin{bmatrix} A_{11} & A_{12} \\ A_{21} & A_{22} \end{bmatrix}$$

has the determinant

$$|\mathbf{A}| = \begin{vmatrix} A_{11} & A_{12} \\ A_{21} & A_{22} \end{vmatrix} = A_{11}A_{22} - A_{12}A_{21}$$

The problem of finding the determinant of matrices of higher order can be reduced to finding the determinant of successive 2 × 2 matrices by the use of *minor determinants* or *minors*. Consider the 3 × 3 determinant

$$|\mathbf{A}| = \begin{vmatrix} A_{11} & A_{12} & A_{13} \\ A_{21} & A_{22} & A_{23} \\ A_{31} & A_{32} & A_{33} \end{vmatrix} = \begin{vmatrix} 1 & 3 & 5 \\ 2 & 4 & 6 \\ 8 & 1 & 2 \end{vmatrix}$$

The minor of the element A_{ij}, denoted as M_{ij}, is the determinant found by crossing out the ith row and jth column of **A**. Thus the minor of $\mathbf{A}_{12}$ is

$$M_{12} = \begin{vmatrix} \cancel{1} & \cancel{3} & \cancel{5} \\ 2 & \cancel{4} & 6 \\ 8 & \cancel{1} & 2 \end{vmatrix} = \begin{vmatrix} 2 & 6 \\ 8 & 2 \end{vmatrix} = 2(2) - 6(8) = -44$$

When a positive or a negative sign is attached to a minor, the result is called a *cofactor*, denoted by C_{ij} and defined as

$$C_{ij} = (-1)^{i+j} M_{ij}$$

That is, if the sum of the subscripts is odd, the cofactor is the negative of the minor; if the sum of the subscripts is even, the cofactor is equal to the minor. In our example

$$C_{11} = (-1)^2 M_{11} = M_{11} = 2$$
$$C_{12} = (-1)^3 M_{12} = -M_{12} = 44$$
$$C_{13} = (-1)^4 M_{13} = M_{13} = -30$$

With these ideas of a minor and a cofactor, a 3 × 3 determinant can be expanded easily. Let us choose an arbitrary row, say the first row, of the matrix **A** and call it the *expansion row*. The determinant of **A** is the sum of the products of the elements on the expansion row and their corresponding cofactors. For our example

$$\begin{vmatrix} 1 & 3 & 5 \\ 2 & 4 & 6 \\ 8 & 1 & 2 \end{vmatrix} = A_{11}C_{11} + A_{12}C_{12} + A_{13}C_{13}$$

$$= 1(2) + 3(44) + 5(-30) = -16$$

In general the determinant of an $n \times n$ matrix can be expanded about its ith row as

$$|\mathbf{A}| = A_{i1}C_{i1} + A_{i2}C_{i2} + \cdots + A_{in}C_{in} \tag{1-6}$$

Each cofactor is an $(n-1) \times (n-1)$ determinant, but each can be reduced to a sequence of 2×2 determinants by repeated use of cofactors. The reader should convince himself that a determinant can be expanded about any row or column and lead to the same result.[9]

A square matrix is called *singular* if its determinant is zero. If the determinant is nonzero, the matrix is called *nonsingular*, or of *full rank*. The rank of a matrix, whether square or not, is the order of the largest nonzero determinant that can be calculated from the matrix. The following matrix is of rank 2. The determinant of the entire matrix is zero, but if one row and column are removed, the determinant of the remaining 2×2 matrix is nonzero. Thus

$$\begin{vmatrix} 1 & 2 & 3 \\ 4 & 5 & 6 \\ 7 & 8 & 9 \end{vmatrix} = 0 \quad \text{but} \quad \begin{vmatrix} 1 & 2 \\ 7 & 8 \end{vmatrix} = -6$$

upon removing the second row and the third column of the original matrix. For this 3×3 matrix the columns are linearly dependent, for upon treating the columns as vectors we can define constants $k_1 = 1$, $k_2 = -2$, and $k_3 = 1$ to obtain

$$1\begin{bmatrix} 1 \\ 4 \\ 7 \end{bmatrix} - 2\begin{bmatrix} 2 \\ 5 \\ 8 \end{bmatrix} + 1\begin{bmatrix} 3 \\ 6 \\ 9 \end{bmatrix} = \begin{bmatrix} 0 \\ 0 \\ 0 \end{bmatrix}$$

In a similar way the 3×4 matrix

$$\begin{bmatrix} 2 & 1 & 3 & 1 \\ 0 & 0 & 1 & 2 \\ 1 & 1 & 0 & 0 \end{bmatrix}$$

[9] The row or column used for the expansion is a matter of convenience. If the row or column with the most zeros is used, the computation will be shortened.

has rank of at most 3 since there is no nonzero determinant of order 4×4 that can be calculated from this matrix. However, if we remove the third column, the remaining matrix has a nonzero determinant (in this case, -2) and the original matrix has rank 3. Again, the columns of the full matrix are linearly dependent, as the reader may wish to verify. Thus the rank of a matrix can also be defined as the maximum number of linearly independent vectors (either row or column) in the matrix.

Finally, the determinant of the product of two square matrices is equal to the product of the determinants of the matrices. Thus

$$|\mathbf{AB}| = |\mathbf{A}||\mathbf{B}|$$

As an exercise, the reader is asked to illustrate this contention.

Cofactor and Adjoint Matrix. The cofactor matrix, denoted as cof **A**, is the matrix whose elements are the cofactors of the original matrix. Returning to our original 3×3 matrix **A**, we see that

$$\text{cof } \mathbf{A} = \text{cof} \begin{bmatrix} 1 & 3 & 5 \\ 2 & 4 & 6 \\ 8 & 1 & 2 \end{bmatrix} = \begin{bmatrix} C_{11} & C_{12} & C_{13} \\ C_{21} & C_{22} & C_{23} \\ C_{31} & C_{32} & C_{33} \end{bmatrix} = \begin{bmatrix} 2 & 44 & -30 \\ -1 & -38 & 23 \\ -2 & 4 & -2 \end{bmatrix} \tag{1-7}$$

The *adjoint matrix*, denoted as adj **A**, is the transpose of the cofactor matrix:

$$\text{adj } \mathbf{A} = (\text{cof } \mathbf{A})' = \begin{bmatrix} 2 & -1 & -2 \\ 44 & -38 & 4 \\ -30 & 23 & -2 \end{bmatrix} \tag{1-8}$$

Inverse Matrix. The inverse matrix is the adjoint matrix multiplied by the reciprocal of the determinant of the original matrix:

$$\mathbf{A}^{-1} = \frac{1}{|\mathbf{A}|} \text{adj } \mathbf{A} \tag{1-9}$$

For our example

$$\mathbf{A}^{-1} = \frac{-1}{16} \begin{bmatrix} 2 & -1 & -2 \\ 44 & -38 & 4 \\ -30 & 23 & -2 \end{bmatrix} = \begin{bmatrix} -0.1250 & 0.0625 & 0.1250 \\ -2.7500 & 2.3750 & -0.2500 \\ 1.8750 & -1.4375 & 0.1250 \end{bmatrix}$$

Three points should be clear from this example. First, it is obvious that if $|\mathbf{A}|$ is zero, the matrix cannot be inverted since $1/|\mathbf{A}|$ is undefined. Second, if $|\mathbf{A}|$ is close to zero, other things the same, the numerical size of the elements of the inverse will be large and may give trouble if the calculation is done on a machine which will store only a few digits. When $|\mathbf{A}|$ is close to zero, the matrix is termed *near-singular* and inversion on a digital computer can be troublesome. Finally, it is always a good idea to check the inverse by checking the equality of $\mathbf{AA}^{-1} = \mathbf{I}$. In our example the equality is exact, but depending on the matrix the equality may end up only as an approximation because of rounding errors.[10] The reader is urged to check the inverse matrix given above.

Diagonal Matrix. The inverse of a diagonal matrix is another diagonal matrix with the reciprocal of the corresponding elements of the original matrix on its main diagonal. Thus

$$\mathbf{D} = \begin{bmatrix} 2 & 0 & 0 \\ 0 & 3 & 0 \\ 0 & 0 & 4 \end{bmatrix} \quad \text{and} \quad \mathbf{D}^{-1} = \begin{bmatrix} \frac{1}{2} & 0 & 0 \\ 0 & \frac{1}{3} & 0 \\ 0 & 0 & \frac{1}{4} \end{bmatrix}$$

Partitioned Inverse. It is often advantageous to invert a matrix by *partitioning*. Partitioning involves grouping certain parts of the matrix together. Suppose that (for convenience) we partition $\mathbf{A}$ as follows:

$$\mathbf{A} = \left[\begin{array}{cc:c} 1 & 3 & 5 \\ 2 & 4 & 6 \\ \hdashline 8 & 1 & 2 \end{array}\right] = \left[\begin{array}{c:c} \mathbf{A}_{11} & \mathbf{A}_{12} \\ \hdashline \mathbf{A}_{21} & \mathbf{A}_{22} \end{array}\right]$$

The dashed lines indicate that $\mathbf{A}$ has been partitioned into four parts, which may be regarded as *submatrices*. If the partitioning is done so that $\mathbf{A}_{22}$ is nonsingular, as it is in this case, then the elements of the inverse are given by

$$\mathbf{A}^{-1} = \left[\begin{array}{c:c} \mathbf{Q}^{-1} & -\mathbf{Q}^{-1}\mathbf{A}_{12}\mathbf{A}_{22}^{-1} \\ \hdashline -\mathbf{A}_{22}^{-1}\mathbf{A}_{21}\mathbf{Q}^{-1} & \mathbf{A}_{22}^{-1} + \mathbf{A}_{22}^{-1}\mathbf{A}_{21}\mathbf{Q}^{-1}\mathbf{A}_{12}\mathbf{A}_{22}^{-1} \end{array}\right] \tag{1-10}$$

where

$$\mathbf{Q} = \mathbf{A}_{11} - (\mathbf{A}_{12}\mathbf{A}_{22}^{-1}\mathbf{A}_{21})$$

From our previous example, $A_{22}^{-1} = \frac{1}{2}$ and[11]

$$\mathbf{Q} = \begin{bmatrix} 1 & 3 \\ 2 & 4 \end{bmatrix} - \frac{1}{2}\begin{bmatrix} 5 \\ 6 \end{bmatrix}[8 \quad 1] = \begin{bmatrix} -19 & 0.5 \\ -22 & 1.0 \end{bmatrix}$$

---

[10] If the initial inverse is not suitably exact, it can generally be improved by a method given by Hotelling. For a computing routine, see McCalla (1967), pp. 162–163.

[11] To invert a 2×2 matrix, divide each element by the determinant; then interchange the elements on the main diagonal and change the signs of the elements on the secondary diagonal. If A is 1×1, a scalar, then its inverse is simply the reciprocal of the scalar. Almon (1967), pp. 20–21, gives the partitioned inverse formulas when $\mathbf{A}_{11}$ is nonsingular.

Upon inversion we find that

$$\mathbf{Q}^{-1} = \begin{bmatrix} -0.1250 & 0.0625 \\ -2.7500 & 2.3750 \end{bmatrix}$$

which agrees with our previous result for the matrix **A**. The rest of the inverse is left as an exercise.

Inversion by partitioning has important theoretical implications and is useful when the size of the matrix is so large that it exceeds the capacity of the available computer and must be inverted by stages. Also, if the matrix can be partitioned so that $\mathbf{A}_{12}$ and $\mathbf{A}_{21}$ are *zero matrices*,[12] then from Eq. (1-10)

$$\left[\begin{array}{c|c} \mathbf{A}_{11} & \mathbf{0} \\ \hline \mathbf{0} & \mathbf{A}_{22} \end{array}\right]^{-1} = \left[\begin{array}{c|c} \mathbf{A}_{11}^{-1} & \mathbf{0} \\ \hline \mathbf{0} & \mathbf{A}_{22}^{-1} \end{array}\right]$$

Calculation is also shortened when one or more of the submatrices that must be inverted are identity matrices since $\mathbf{I}^{-1} = \mathbf{I}$.

Determinants may also be calculated from partitioned matrices. If $\mathbf{A}_{22}$ is nonsingular, then

$$|\mathbf{A}| = |\mathbf{A}_{22}|\,|\mathbf{Q}| \tag{1-11}$$

1.4 quadratic forms

If **A** is an $n \times n$ *symmetric* matrix and **X** is an $n \times 1$ column vector, then[13]

$$\mathbf{X}'\mathbf{A}\mathbf{X} = [X_1 \quad X_2 \quad \cdots \quad X_n] \begin{bmatrix} A_{11} & A_{12} & \cdots & A_{1n} \\ A_{21} & A_{22} & \cdots & A_{2n} \\ \vdots & & & \vdots \\ A_{n1} & A_{n2} & \cdots & A_{nn} \end{bmatrix} \begin{bmatrix} X_1 \\ X_2 \\ \vdots \\ X_n \end{bmatrix}$$

is called a *quadratic form*. In particular, we may write the quadratic equation

$$z = 2X_1^2 + 2X_1X_2 + 3X_2^2 \tag{1-12}$$

as

$$z = [X_1 \quad X_2] \begin{bmatrix} 2 & 1 \\ 1 & 3 \end{bmatrix} \begin{bmatrix} X_1 \\ X_2 \end{bmatrix} \tag{1-12a}$$

---

[12] A zero matrix is one whose elements are all zero.

[13] We restrict ourselves to real matrices in this book, that is, matrices whose elements are all real numbers.

A quadratic form is called *positive definite* and the matrix **A** is called a positive definite matrix if all leading principal minor determinants of **A** are positive. Thus if the minors shown below are positive, i.e.,

$$A_{11} > 0, \quad \begin{vmatrix} A_{11} & A_{12} \\ A_{21} & A_{22} \end{vmatrix} > 0, \quad \begin{vmatrix} A_{11} & A_{12} & A_{13} \\ A_{21} & A_{22} & A_{23} \\ A_{31} & A_{32} & A_{33} \end{vmatrix} > 0, \quad \cdots, \quad |\mathbf{A}| > 0$$

then $\mathbf{X}'\mathbf{A}\mathbf{X}$ and **A** are positive definite. Notice that the principal minor determinants are formed from submatrices which start with the northwest element of the matrix and increase by the addition of one row and column until the determinant of the entire matrix is formed. Clearly, all positive definite matrices are of full rank since the determinant of the entire matrix is greater than zero.

The positive definite quadratic form is most important for future presentation and we shall explore its properties in some detail. One special case of this form is

$$\begin{aligned} z &= [X_1 \quad X_2] \begin{bmatrix} A_{11} & 0 \\ 0 & A_{22} \end{bmatrix} \begin{bmatrix} X_1 \\ X_2 \end{bmatrix} \\ &= A_{11}X_1^2 + A_{22}X_2^2 \end{aligned} \tag{1-13}$$

For this case both A_{11} and A_{22} are greater than zero so that the value of z must always be greater than zero.

The topmost diagrams in Figure 1.6 illustrate the case where $A_{11} = A_{22}$. For a given z, Eq. (1-13) will generate a circle. If z is allowed to vary from zero through a continuous range of positive values, then a *quadratic surface* like the one in the top left diagram in Figure 1.6 will be generated. The figure resembles an ice-cream cone and is technically called an *elliptic paraboloid.* Any slice through this figure which is parallel to the X_1, X_2 axes will generate a circle like the one drawn in the top right diagram in Figure 1.6. If the slice is taken at $z = 1.0$, then the radius of the circle is $1/\sqrt{A_{11}} = 1/\sqrt{A_{22}}$.

If $A_{11} \neq A_{22}$, then the figure generated by varying z resembles a "squashed" ice-cream cone and any slice taken parallel to the X_1, X_2 axes will trace out an *ellipse* whose axes correspond to segments of the X_1, X_2 axes. The left center diagram in Figure 1.6 shows this type of quadratic surface, where $A_{11} < A_{22}$, and the right center diagram shows the corresponding ellipse generated by slicing the quadratic surface at $z = 1.0$. The *major* (longer) *axis* of the ellipse corresponds to a segment of the X_1 axis in this case, and the *minor* (shorter) axis corresponds to a segment of the X_2 axis. For a slice taken at $z = 1.0$, the major axis has a *semilength* (half-length) of $1/\sqrt{A_{11}}$ and the minor axis has a semilength of $1/\sqrt{A_{22}}$. The ellipses and their axes will

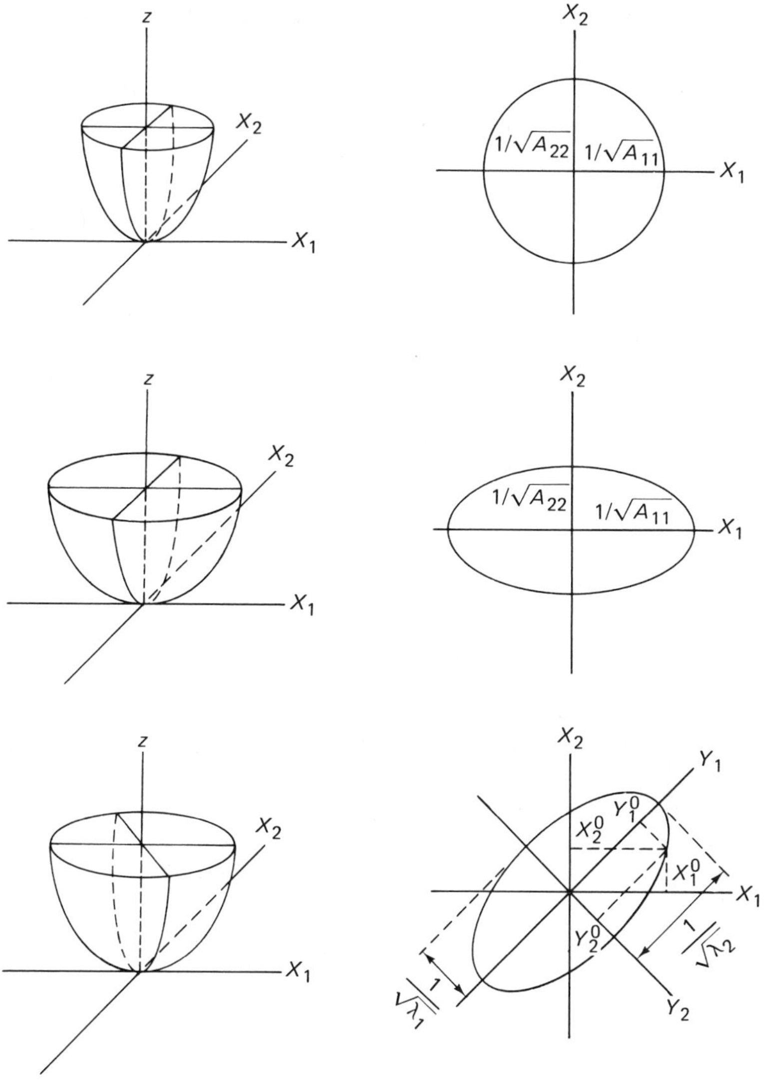

Figure 1.6. Quadratic forms and slices taken at $z = 1.0$.

increase in size as slices are taken for larger values of z. In general, the major axis will have a semilength of $\sqrt{z/A_{11}}$ and the minor axis will have a semilength of $\sqrt{z/A_{22}}$ for $A_{11} < A_{22}$.

If the positive definite matrix $\mathbf{A}$ is not a diagonal matrix, then the major and minor axes of the ellipse generated by slicing the quadratic surface will no longer be on the X_1, X_2 axes. For example (since $A_{12} = A_{21}$),

$$z = [X_1 \quad X_2]\begin{bmatrix} A_{11} & A_{12} \\ A_{21} & A_{22} \end{bmatrix}\begin{bmatrix} X_1 \\ X_2 \end{bmatrix}$$
$$= A_{11}X_1^2 + 2A_{12}X_1X_2 + A_{22}X_2^2 \tag{1-14}$$

will generate a surface such as the one shown at the lower left in Figure 1.6. The cross-product term will cause the quadratic surface to rotate in such a way that the major and minor axes of the ellipse will no longer be on the X_1, X_2 axes. If $A_{12} < 0$, the ellipse will have its major axis pointing in a southwest-northeast direction, as in the bottom right diagram in Figure 1.6. If $A_{12} > 0$, the major axis will point in a northwest-southeast direction.

Since this ellipse is a contour of the quadratic surface, every point on the ellipse corresponds to the same value of z. The point $\mathbf{X}^0$, say, has coordinates

$$\mathbf{X}^0 = \begin{bmatrix} X_1^0 \\ X_2^0 \end{bmatrix}$$

on the X_1, X_2 axes with

$$z_1 = (\mathbf{X}^0)'\mathbf{A}(\mathbf{X}^0) \tag{1-15}$$

However, this same point has coordinates $(\mathbf{Y}^0)' = [Y_1^0 \; Y_2^0]$ in terms of its major and minor axes. If the major and minor axes were on the X_1, X_2 axes, then the ellipse would resemble the one given in the center right diagram in Figure 1.6, and the coefficient matrix of its quadratic form would be diagonal. We can reason, therefore, as if the ellipse had been rotated into a position such that its major and minor axes were on the X_1, X_2 axes and express its coordinates in terms of its major and minor axes by use of a diagonal matrix. Thus

$$z_1 = (\mathbf{Y}^0)'\mathbf{D}(\mathbf{Y}^0) \tag{1-15a}$$

where $\mathbf{D}$ is some diagonal matrix

$$\mathbf{D} = \begin{bmatrix} \lambda_1 & 0 \\ 0 & \lambda_2 \end{bmatrix}$$

Suppose also that the semilengths of the major and minor axes are given by

$$\text{Semilength of major axis} = \sqrt{\frac{z_1}{\lambda_2}}$$

$$\text{Semilength of minor axis} = \sqrt{\frac{z_1}{\lambda_1}}$$

where $\lambda_1 > \lambda_2$.

From the discussion at the end of Sec. 1.1, we see that the relationship between these two sets of coordinates in Eqs. (1-15) and (1-15a) is

$$\mathbf{X}^0 = [\mathbf{P}_1 \quad \mathbf{P}_2]\mathbf{Y}^0 = \mathbf{P}\mathbf{Y}^0 \tag{1-16}$$

where $\mathbf{P}_1$ and $\mathbf{P}_2$ are orthonormal vectors directed along the major and minor axes of the ellipse, respectively. The matrix $\mathbf{P}$ is called the *modal* matrix of $\mathbf{A}$.

Substitute Eq. (1-16) into Eq. (1-15) and get

$$z_1 = (\mathbf{P}\mathbf{Y}^0)'\mathbf{A}(\mathbf{P}\mathbf{Y}^0) = (\mathbf{Y}^0)'\mathbf{P}'\mathbf{A}\mathbf{P}(\mathbf{Y}^0)$$

But since

$$z_1 = (\mathbf{Y}^0)'\mathbf{D}(\mathbf{Y}^0)$$

we have

$$\mathbf{P}'\mathbf{A}\mathbf{P} = \mathbf{D}$$

Therefore any ellipse expressed in terms of the X coordinate system may be expressed in terms of the Y coordinate system, where the coordinate systems are related as

$$z = \mathbf{X}'\mathbf{A}\mathbf{X} = \mathbf{Y}'\mathbf{D}\mathbf{Y}$$

and the elements on the main diagonal of $\mathbf{D}$ determine the length of the major and minor axes. In Sec. 1.6 we shall call these elements of $\mathbf{D}$ (λ_1 and λ_2) the *characteristic roots* of $\mathbf{A}$. We shall call the vectors $\mathbf{P}_1$ and $\mathbf{P}_2$ the *normalized characteristic vectors* associated with these characteristic roots.

1.5 differentiation

In this section we shall give some results which apply to differentiation of vectors and quadratic forms of matrices. It is assumed that the reader has some knowledge of calculus, and parallels to ordinary scalar calculus will be cited.

Let the scalar Y be defined as

$$Y = \mathbf{A}'\mathbf{X}$$

where $\mathbf{A}$ and $\mathbf{X}$ are both $n \times 1$ vectors. Then the vector of partial derivatives with respect to $\mathbf{X}$ is defined as

$$\frac{\partial Y}{\partial \mathbf{X}} = \begin{bmatrix} \dfrac{\partial Y}{\partial X_1} & \dfrac{\partial Y}{\partial X_2} & \cdots & \dfrac{\partial Y}{\partial X_n} \end{bmatrix}'$$

For example, the function

$$Y = 2X_1 + 3X_2 + 5X_3$$

may be written as

$$Y = [2 \quad 3 \quad 5]\begin{bmatrix} X_1 \\ X_2 \\ X_3 \end{bmatrix} = \mathbf{A}'\mathbf{X}$$

and the vector of partial derivatives is

$$\left[\frac{\partial Y}{\partial X_1} \quad \frac{\partial Y}{\partial X_2} \quad \frac{\partial Y}{\partial X_3}\right]' = [2 \quad 3 \quad 5]'$$

1. If Y is a constant for all X, then

$$\frac{\partial Y}{\partial \mathbf{X}} = \mathbf{0} \tag{1-17}$$

   where $\mathbf{0}$ is $n \times 1$. This result is strictly analogous to ordinary calculus.
2. If Y is not constant for all X, then, as in ordinary calculus,

$$\frac{\partial Y}{\partial \mathbf{X}} = \mathbf{A} \tag{1-18}$$

3. The vector of partial derivatives of the quadratic form $z = \mathbf{X}'\mathbf{A}\mathbf{X}$ is given by

$$\frac{\partial z}{\partial \mathbf{X}} = 2\mathbf{A}\mathbf{X} \tag{1-19}$$

For example, if

$$\begin{aligned} z &= 2X_1^2 + 2X_1X_2 + 3X_2^2 \\ &= [X_1 \quad X_2]\begin{bmatrix} 2 & 1 \\ 1 & 3 \end{bmatrix}\begin{bmatrix} X_1 \\ X_2 \end{bmatrix} \end{aligned} \tag{1-20}$$

then

$$\frac{\partial z}{\partial \mathbf{X}} = \begin{bmatrix} \dfrac{\partial z}{\partial X_1} \\ \dfrac{\partial z}{\partial X_2} \end{bmatrix} = \begin{bmatrix} 4X_1 + 2X_2 \\ 2X_1 + 6X_2 \end{bmatrix} = 2\begin{bmatrix} 2 & 1 \\ 1 & 3 \end{bmatrix}\begin{bmatrix} X_1 \\ X_2 \end{bmatrix}$$

$\mathbf{A}$ may be an identity matrix, in which case $z = \mathbf{X}'\mathbf{X}$ and $\partial z/\partial \mathbf{X} = 2\mathbf{X}$.

Maxima and Minima. We shall find it necessary on many occasions to maximize or minimize functions which are expressed using vectors and matrices. As in ordinary calculus unconstrained maximization or minimization is accomplished under suitable conditions by setting the first partial derivatives equal to zero. The second derivatives determine whether the function has been maximized or minimized.

Very often we shall be concerned with constrained maxima or minima of quadratic forms. Suppose that we were interested in maximizing Eq. (1-20) subject to the restriction that $X_1^2 + X_2^2 = 1$ (the sum of the squared values of X is unity). Geometrically, the restriction $X_1^2 + X_2^2 = 1$ defines a circle of unit radius (a *unit circle*). Referring to Figure 1.7, we find that the minimum of the quadratic form subject to this unit circle restriction will be that value z_2 where the quadratic form will just fit into the unit circle. The maximum will be that value z_1 where the unit circle will just fit into the quadratic form. In Figure 1.7 we see that the smaller ellipse tangent to the unit circle at the extreme points of its major axis will represent the value of z_2. The larger ellipse tangent to the unit circle at the extreme points of its minor axis will represent z_1.

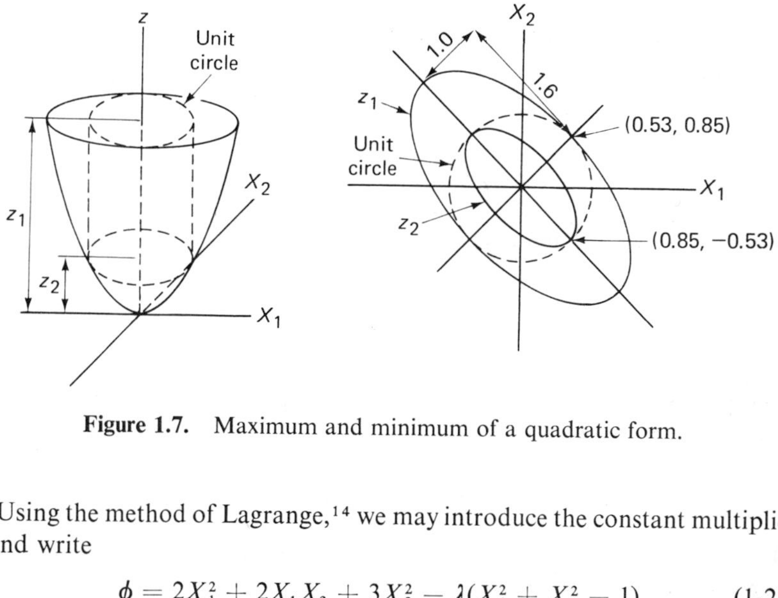

Figure 1.7. Maximum and minimum of a quadratic form.

Using the method of Lagrange,[14] we may introduce the constant multiplier λ and write

$$\phi = 2X_1^2 + 2X_1X_2 + 3X_2^2 - \lambda(X_1^2 + X_2^2 - 1) \tag{1-21}$$

Next, we set the partial derivatives of ϕ with respect to X_1 and X_2 equal to zero,

$$\frac{\partial \phi}{\partial X_1} = 4X_1 + 2X_2 - 2\lambda X_1 = 0 \tag{1-22}$$

$$\frac{\partial \phi}{\partial X_2} = 2X_1 + 6X_2 - 2\lambda X_2 = 0 \tag{1-23}$$

[14]See Taylor (1955), p. 198.

and solve these two equations along with the restraint

$$X_1^2 + X_2^2 = 1 \tag{1-24}$$

Dividing Eq. (1-22) by X_1, we have

$$\frac{X_2}{X_1} = \lambda - 2$$

and dividing Eq. (1-23) by X_2, we find that

$$\frac{X_2}{X_1} = \frac{1}{\lambda - 3}$$

Upon setting these two expressions equal to each other, we obtain the polynomial

$$\lambda^2 - 5\lambda + 5 = 0$$

which may be solved for the two roots[15]

$$\lambda_1 = \frac{5 + \sqrt{5}}{2} = 3.618034$$

$$\lambda_2 = \frac{5 - \sqrt{5}}{2} = 1.381966$$

Write Eqs. (1-22) and (1-23) as

$$2X_1 + X_2 = \lambda X_1$$
$$X_1 + 3X_2 = \lambda X_2$$

Multiply the first equation by X_1, the second equation by X_2, and sum the two resulting equations (using the restriction that $X_1^2 + X_2^2 = 1$):

$$2X_1^2 + 2X_1X_2 + 3X_2^2 = \lambda \tag{1-25}$$

The left-hand side of Eq. (1-25) is the original quadratic form. It will be a maximum when the larger root λ_1 replaces λ and a minimum when the smaller root λ_2 replaces λ. Inserting the larger root into Eq. (1-22),

$$X_2 = (1.618034)X_1$$

and using Eq. (1-24),

$$X_1^2 + (1.618034)^2 X_1^2 = 1$$

---

[15]Solutions to an equation of the type $a\lambda^2 + b\lambda + c = 0$ are given by $\lambda = (-b \pm \sqrt{b^2 - 4ac})/2a$.

we find that[16] $X_1 = 0.52573$. Inserting $X_1^2 = 0.2764$ into Eq. (1-23), we find that $X_2 = 0.85065$. Thus the quadratic equation is maximized (see Figure 1.7) at

$$\mathbf{P}_1' = [X_1 \quad X_2] = [0.52573 \quad 0.85065]$$

with $z_1 = 2(0.2764) + 2(0.52573)(0.85065) + 3(0.7236) = 3.62$. Proceeding in the same way, using the smaller root, we see that the quadratic equation is minimized at

$$\mathbf{P}_2' = [X_1 \quad X_2] = [0.85065 \quad -0.52573]$$

with $z_2 = 1.38$.

Notice that $\mathbf{P}_1$ and $\mathbf{P}_2$ are orthonormal vectors. Therefore we may make an overall check on our calculations by observing that the semilength of the minor axis is, after rounding, $\sqrt{z_1/\lambda_1} = 1$, as it should be. The semilength of the major axis is $\sqrt{z_1/\lambda_2} = 1.619$.

If we wish to plot the ellipse, we form the diagonal matrix

$$\mathbf{D} = \begin{bmatrix} \lambda_1 & 0 \\ 0 & \lambda_2 \end{bmatrix} = \begin{bmatrix} 3.62 & 0 \\ 0 & 1.38 \end{bmatrix}$$

as discussed in Sec. 1.4. In terms of the Y coordinates, $z_1 = [Y_1 \ Y_2]\mathbf{D}[Y_1 \ Y_2]'$, and upon setting $z_1 = 3.62$, we have $Y_1 = \pm\sqrt{1.0 - 0.381Y_2^2}$. As an exercise the student should choose several values for Y_2 which lie between zero and 1.619 in absolute value and calculate the resulting values for Y_1. Then, knowing the location of the major and minor axes of the ellipse, he should sketch the ellipse itself.

1.6 characteristic roots and vectors

In the last section we showed how Eq. (1-20) could be maximized or minimized subject to the restriction that $X_1^2 + X_2^2 = 1$ by use af ordinary scalar algebra and calculus. In this section we shall show how the solution to this problem may be obtained by matrix algebra and calculus. The matrix presentation will enable us to generalize the problem to more than two dimensions.

Write Eq. (1-21) as

$$\phi = \mathbf{X}'\mathbf{A}\mathbf{X} - \lambda(\mathbf{X}'\mathbf{X} - 1)$$

---

[16] X_1 should be ± 0.52573. However, it will not alter the value of the quadratic equation if the negative value is used.

where

$$\mathbf{A} = \begin{bmatrix} 2 & 1 \\ 1 & 3 \end{bmatrix}$$

Then, using Eq. (1-19), we wish to solve

$$\frac{\partial \phi}{\partial \mathbf{X}} = 2\mathbf{AX} - 2\lambda\mathbf{X} = \mathbf{0} \tag{1-26}$$

Equation (1-26) may be written as

$$(\mathbf{A} - \lambda\mathbf{I})\mathbf{X} = \mathbf{0} \tag{1-27}$$

and it will have *nonzero* solutions if and only if $(\mathbf{A} - \lambda\mathbf{I})$ is singular, that is, if and only if

$$|\mathbf{A} - \lambda\mathbf{I}| = \mathbf{0} \tag{1-28}$$

Equation (1-28) is a polynomial in λ. We shall call this equation the *characteristic equation* of the matrix **A**. We wish to find values for λ, the *characteristic roots*,[17] such that Eq. (1-28) holds. To do this, we write the polynomial implied by Eq. (1-28) explicitly,

$$|\mathbf{A} - \lambda\mathbf{I}| = \begin{vmatrix} 2-\lambda & 1 \\ 1 & 3-\lambda \end{vmatrix} = \lambda^2 - 5\lambda + 5 = 0$$

and the solutions to this polynomial, given in the last section, are 3.618034 and 1.381966. As in the last section, let us call the larger of these roots λ_1 and the smaller λ_2. These roots correspond to the Lagrange multipliers of the last section.

Associated with each characteristic root is a *characteristic vector*, which is the vector **X** of Eq. (1-27) for a given value of λ. Using $\lambda_1 = 3.618034$ and Eq. (1-27), solve

$$(\mathbf{A} - \lambda_1\mathbf{I})\mathbf{X} = \begin{bmatrix} -1.618034 & 1.0 \\ 1.0 & -0.618034 \end{bmatrix} \begin{bmatrix} X_1 \\ X_2 \end{bmatrix} = \begin{bmatrix} 0 \\ 0 \end{bmatrix}$$

which reduces to the simultaneous equations

$$-1.618034X_1 + X_2 = 0$$
$$X_1 - 0.618034X_2 = 0$$

that is,

$$X_1 = 0.618034X_2 \tag{1-29}$$

---

[17]These are also called *latent*, *eigen*, and *proper* roots.

There are infinitely many values of X_1 and X_2 which satisfy Eq. (1-29). This fact can be deduced from Eq. (1-27) directly since if some vector **X** satisfies that equation, so will any scalar multiple of the vector **X**. To reduce this ambiguity,[18] we require in addition that $\mathbf{X}'\mathbf{X} = 1$. That is, we require that the sum of the squared elements of **X** be unity. This is the same restriction of the last section. A vector which satisfies Eq. (1-27) and the restriction

$$\mathbf{X}'\mathbf{X} = 1 \tag{1-30}$$

is called a *normalized characteristic vector*. Proceeding in the same manner as that explained in the last section, we find that the normalized characteristic vector associated with λ_1 is

$$\mathbf{P}_1' = [X_1 \quad X_2] = [0.52573 \quad 0.85065]$$

Similarly, using λ_2, we find that the associated normalized characteristic vector is

$$\mathbf{P}_2' = [X_1 \quad X_2] = [0.85065 \quad -0.52573]$$

These findings, of course, agree with those of the last section. The characteristic vectors form an orthonormal set and the maximization or minimization of Eq. (1-20) can be carried out using the characteristic roots and vectors.

The relationship between the coordinate systems X and Y, as described in previous sections, can also be generalized. From Eq. (1-16) we can say that in general

$$\mathbf{X} = [\mathbf{P}_1 \quad \mathbf{P}_2]\mathbf{Y} = \mathbf{P}'\mathbf{Y}$$

and that

$$z = \mathbf{X}'\mathbf{A}\mathbf{X} = \mathbf{Y}'\mathbf{D}\mathbf{Y}$$

where

$$\mathbf{D} = \mathbf{P}'\mathbf{A}\mathbf{P} = \begin{bmatrix} \lambda_1 & 0 \\ 0 & \lambda_2 \end{bmatrix} \tag{1-31}$$

Thus, in general, **P** is the matrix of characteristic vectors associated with the characteristic roots, and **D** is a diagonal matrix with the characteristic roots on its main diagonal. In our particular example we may express z in terms of X: $z = 2X_1^2 + 2X_1X_2 + 3X_2^2$—or in terms of Y: $z = 3.618034\,Y_1^2 + 1.381966\,Y_2^2$.

For a 3×3 matrix there will be three characteristic roots, for a 4×4 matrix there will be four characteristic roots, and so forth. For matrices

---

[18]The ambiguity that remains is inconsequential for our purposes, since the remaining indeterminancy is that if **X** satisfies Eq. (1-27), so does $(-1)\mathbf{X}$. See footnote 16.

larger than 4×4 the roots cannot, in general, be obtained in a finite number of steps and must be approximated.[19] The computer program given in the Appendix to this chapter will carry out such calculations for the kinds of matrices used in this book.

Four properties of characteristic roots are important for our future work. The student may verify these properties using the example just discussed.

1. The characteristic roots of a positive definite symmetric matrix are all positive. In the previous example λ_1 and λ_2 are positive since **A** is positive definite. That is, the principal minors of **A** are positive.

$$A_{11} = 2, \qquad \begin{vmatrix} A_{11} & A_{12} \\ A_{21} & A_{22} \end{vmatrix} = 5$$

2. The product of the characteristic roots of a symmetric matrix equals the determinant of the matrix. From Eq. (1-31)

$$|\mathbf{D}| = |\mathbf{P'AP}| = |\mathbf{P'P}||\mathbf{A}| = |\mathbf{A}|$$

since $\mathbf{P'P} = \mathbf{I}$.

3. The sum of the characteristic roots of a symmetric matrix equals the trace of the matrix. Again from Eq. (1-31)

$$\text{tr } \mathbf{D} = \text{tr}(\mathbf{P'AP}) = \text{tr}(\mathbf{P'PA}) = \text{tr } \mathbf{A}$$

since $\mathbf{P'P} = \mathbf{I}$.

4. For any positive definite symmetric matrix **A**, the characteristic roots of $\mathbf{A}^{-1}$ are the reciprocal of the characteristic roots of **A**. The normalized characteristic vectors of **A** and $\mathbf{A}^{-1}$ are identical. Using Eq. (1-27), where λ is a characteristic root and **X** the associated characteristic vector,

$$\mathbf{AX} = \lambda\mathbf{X}$$

Premultiply both sides by $\mathbf{A}^{-1}$:

$$\mathbf{X} = \lambda\mathbf{A}^{-1}\mathbf{X}$$

or

$$\mathbf{A}^{-1}\mathbf{X} = \frac{1}{\lambda}\mathbf{X}$$

Thus if λ is a characteristic root of **A**, then $1/\lambda$ is the corresponding characteristic root of $\mathbf{A}^{-1}$. The vector **X** remains unchanged. This point is important because we shall sometimes calculate the characteristic roots of the quadratic form $z = \mathbf{X'AX}$ using $\mathbf{A}^{-1}$ rather than **A**.

---

[19]From a theorem first rigorously proved by Niels Abel that states that polynomials of higher than fourth degree are generally incapable of algebraic solution.

1.7 some statistical quantities expressed in matrix algebra

As we have mentioned before, one of the main reasons matrix algebra is so important in statistics is that it affords a very convenient notation system. The examples below illustrate this fact.

Arithmetic Means. The simple[20] arithmetic mean of a series of n observations on the variable X is defined by[21]

$$\bar{X} = \sum_{i=1}^{n} \frac{X_i}{n} \tag{1-32}$$

If we define $\mathbf{A}$ as the $n \times 1$ vector containing only 1's and $\mathbf{X}$ as the $n \times 1$ vector of observations, then

$$\bar{X} = \frac{1}{n}\mathbf{A}'\mathbf{X} = \frac{1}{n}[1 \quad 1 \quad \cdots \quad 1]\begin{bmatrix} X_1 \\ X_2 \\ \cdot \\ \cdot \\ \cdot \\ X_n \end{bmatrix}$$

Very often we shall have p variables and shall wish to compute the means of each of them. Define $\mathbf{A}$ as the $n \times 1$ vector containing only 1's and $\mathbf{X}$ as the

---

[20]By simple, we mean that all observations are given the weight 1. The general formula for an arithmetic mean is

$$\bar{X} = \frac{\sum_{i=1}^{n} W_i X_i}{\sum_{i=1}^{n} W_i}$$

where W_i represents a series of weights.

[21]$\sum_{i=1}^{n} X_i$ is often written $\sum X$, where the end points of summation are understood. The following points of review concerning the summation operator, $\sum$, may be useful:

1. $\sum_{i=1}^{n} k = nk$, where k is a constant.
2. $\sum kX = k \sum X$.
3. $\sum X^2 = X_1^2 + X_2^2 + \cdots + X_n^2$.
4. $(\sum X)^2 = (X_1 + X_2 + \cdots + X_n)^2$.
5. $\sum (X + Y) = \sum X + \sum Y$.
6. $\sum XY = X_1 Y_1 + X_2 Y_2 + \cdots + X_n Y_n$.
7. $\sum X \sum Y = (\sum X)(\sum Y)$.

Care should be taken not to confuse points 3 and 4 or 6 and 7.

$n \times p$ matrix of observations where any column of $\mathbf{X}$ represents the observations on one variable. This convention will represent a slight change in our notation, since the first subscript on $\mathbf{X}$ will now represent the variable number and the second subscript will represent the item number for that variable. The vector of means is given by

$$\bar{\mathbf{X}}' = \frac{1}{n}\mathbf{A}'\mathbf{X} = \frac{1}{n}[1 \quad 1 \quad \cdots \quad 1]\begin{bmatrix} X_{11} & X_{21} & \cdots & X_{p1} \\ X_{12} & X_{22} & \cdots & X_{p2} \\ \cdot & & & \cdot \\ \cdot & & & \cdot \\ \cdot & & & \cdot \\ X_{1n} & X_{2n} & \cdots & X_{pn} \end{bmatrix}$$

$$= [\bar{X}_1 \quad \bar{X}_2 \quad \cdots \quad \bar{X}_p] \tag{1-33}$$

Sums of Squares and Cross Products. The sum of squares of a variable X is defined as

$$\sum X^2 = X_1^2 + X_2^2 + \cdots + X_n^2$$

and the sum of cross products of the two variables X and Y is given by

$$\sum XY = X_1Y_1 + X_2Y_2 + \cdots + X_nY_n$$

For p variables the matrix of sums of squares and cross products is given by

$$\mathbf{X}'\mathbf{X} = \begin{bmatrix} X_{11} & X_{12} & \cdots & X_{1n} \\ X_{21} & X_{22} & \cdots & X_{2n} \\ \cdot & & & \cdot \\ \cdot & & & \cdot \\ \cdot & & & \cdot \\ X_{p1} & X_{p2} & \cdots & X_{pn} \end{bmatrix}\begin{bmatrix} X_{11} & X_{21} & \cdots & X_{p1} \\ X_{12} & X_{22} & \cdots & X_{p2} \\ \cdot & & & \cdot \\ \cdot & & & \cdot \\ \cdot & & & \cdot \\ X_{1n} & X_{2n} & \cdots & X_{pn} \end{bmatrix}$$

$$= \begin{bmatrix} \sum X_1^2 & \sum X_1X_2 & \cdots & \sum X_1X_p \\ \sum X_2X_1 & \sum X_2^2 & \cdots & \sum X_2X_p \\ \cdot & & & \cdot \\ \cdot & & & \cdot \\ \cdot & & & \cdot \\ \sum X_pX_1 & \sum X_pX_2 & \cdots & \sum X_p^2 \end{bmatrix} \tag{1-34}$$

Since $\sum X_iX_j = \sum X_jX_i$, this matrix will be symmetric, with the sums of squares on the main diagonal and the sums of cross products elsewhere.

Questions and Problems

Sec. 1.1

1. Given the vectors $\mathbf{A}'_1 = [2 \quad 3]$ and $\mathbf{A}'_2 = [-1 \quad -2]$,
   a. Sketch the vectors.
   b. Add the vectors and sketch the sum.
   c. Are the vectors linearly independent? Orthogonal? Orthonormal?
   d. Form the inner product of the vectors.
   e. Find the lengths of $\mathbf{A}_1$ and $\mathbf{A}_2$.

2. Give an example other than the one provided in the text of two vectors which are linearly independent but not orthogonal.

3. Let $\mathbf{A}' = [A_1 \; A_2]$ and $\mathbf{B}' = [B_1 \; B_2]$. Prove that $\mathbf{A}'\mathbf{B} = ||\mathbf{A}|| \; ||\mathbf{B}|| \cos \theta$. (*Hint:* Let a, b, and c represent the lengths of the three sides of a triangle. Also, let the angle between the sides with length a and b be θ. Then by the law of cosines $c^2 = a^2 + b^2 - 2ab \cos \theta$. Also, $||\mathbf{A} - \mathbf{B}||^2 = (\mathbf{A} - \mathbf{B})'(\mathbf{A} - \mathbf{B}) = \mathbf{A}'\mathbf{A} - \mathbf{B}'\mathbf{A} - \mathbf{A}'\mathbf{B} + \mathbf{B}'\mathbf{B}$.)

4. Let two orthonormal vectors be

$$\mathbf{P}'_1 = [4/\sqrt{20} \quad 2/\sqrt{20}] \quad \text{and} \quad \mathbf{P}'_2 = [-2/\sqrt{20} \quad 4/\sqrt{20}]$$

Express $\mathbf{A}'_1 = [3 \quad 4]$ in terms of these orthonormal vectors. Check your calculations by converting these coordinates back to the $\mathbf{I}_1$ and $\mathbf{I}_2$ axes.

Sec. 1.2

1. If $\mathbf{X}$ is 2×2 and $\mathbf{Y}$ is 2×10, what will be the dimension of $\mathbf{XY}$? Is $\mathbf{YX}$ possible?

2. Let

$$\mathbf{X} = \begin{bmatrix} 1 & 3 & 5 \\ 2 & 4 & -1 \end{bmatrix} \quad \text{and} \quad \mathbf{Y}' = \begin{bmatrix} 0 & 1 & 1 \\ 2 & 0 & 0 \end{bmatrix}$$

Find $\mathbf{XY}$.

3. Give an algebraic proof of $\text{tr}(\mathbf{XY}) = \text{tr}(\mathbf{YX})$.

4. Let the matrices $\mathbf{A}$, $\mathbf{B}$, and $\mathbf{C}$ all be of order $n \times n$. Show that the following properties hold:
   a. $(\mathbf{AB})\mathbf{C} = \mathbf{A}(\mathbf{BC})$.
   b. $\mathbf{A}(\mathbf{B} + \mathbf{C}) = \mathbf{AB} + \mathbf{AC}$.
   c. $(\mathbf{A} + \mathbf{B})\mathbf{C} = \mathbf{AC} + \mathbf{BC}$.
   d. $(\mathbf{A}')' = \mathbf{A}$.
   e. $(\mathbf{A} + \mathbf{B})' = \mathbf{A}' + \mathbf{B}'$.
   f. $(\mathbf{AB})' = \mathbf{B}'\mathbf{A}'$.

Sec. 1.3

1. Verify that the following matrix has a rank of 3:

$$\begin{bmatrix} 2 & 1 & 3 & 1 \\ 0 & 0 & 1 & 2 \\ 1 & 1 & 0 & 0 \end{bmatrix}$$

2. Solve the following system of equations by matrix inversion:

$$\begin{aligned} -0.5X_1 - X_2 + 0.25X_3 + X_4 &= 4 \\ 0.5X_2 - X_3 + X_4 &= 5 \\ 0.5(X_1 + X_3) &= 6 \\ 0.5(X_1 + X_4) &= 7 \end{aligned}$$

Solve the same system using a partitioned inverse scheme.

3. Using 2×2 matrices, verify numerically that the following properties hold:

a. $(\mathbf{ABC})^{-1} = \mathbf{C}^{-1}\mathbf{B}^{-1}\mathbf{A}^{-1}$. b. $(\mathbf{A}')^{-1} = (\mathbf{A}^{-1})'$.

4. Suppose that **A** and **B** are two square matrices of order $n \times n$. It can be shown that the determinant of the product of the two matrices is equal to the product of the determinants of the matrices. That is, $|\mathbf{AB}| = |\mathbf{A}||\mathbf{B}|$. Give a numerical example of this theorem. Also give a numerical example of the theorem $|\mathbf{A}| = |\mathbf{A}'|$.

Sec. 1.4

1. Let $z = X_1^2 + 2X_2^2 + 3X_3^2 - 2X_1X_2 + 2X_1X_3 - 4X_2X_3$. Write a matrix representation of this quadratic form similar to that of Eq. (1-12a). Show that the quadratic form is positive definite.

2. Let

$$z = [X_1 \quad X_2]\begin{bmatrix} 2 & 0 \\ 0 & 3 \end{bmatrix}\begin{bmatrix} X_1 \\ X_2 \end{bmatrix}$$

When $z = 1$, sketch the ellipse that will be generated.

Sec. 1.5

1. Write $\partial z/\partial \mathbf{X}$ for the quadratic forms for Problems 1 and 2 of Sec. 1.4.
2. Sketch the ellipse mentioned at the end of Sec. 1.5 in the text.

Sec. 1.6

1. Find the normalized characteristic vectors and the characteristic roots of the following matrix:

$$\mathbf{A} = \begin{bmatrix} 2 & 3 \\ 3 & 5 \end{bmatrix}$$

If this matrix represents the coefficients of a quadratic form $z = \mathbf{X}'\mathbf{A}\mathbf{X}$, maximize the quadratic form subject to a unit circle. Give the semilengths of the major and minor axes of the ellipse that results from the maximization.

2. Since the product of the characteristic roots of a symmetric matrix equals the determinant of the matrix, an $n \times n$ nonsingular matrix contains no zero characteristic roots. Generally, if there are k zero characteristic roots of an $n \times n$ matrix, the rank of the matrix will not be less than $n - k$. Show that the following matrix is rank 2:

$$\begin{bmatrix} 1 & -2 & 0 \\ -2 & 4 & 0 \\ 0 & 0 & 3 \end{bmatrix}$$

Sec. 1.7

1. Let $\mathbf{X}$ be an $n \times p$ matrix of observations. Devise a matrix $\mathbf{A}$ such that $\mathbf{AX}$ will produce a $2 \times p$ matrix which contains only the first and last observation for any X variable in its columns.

APPENDIX

Throughout this book computer programs which will do most of the calculations discussed are given in appendices. These programs are made up of subroutines, which carry out a specific mathematical chore, and main programs, which call the subroutines. The subroutines are all written using standard FORTRAN statements and should work with only minor alterations, if any, on almost any computer that compiles some level of FORTRAN.

The main programs contain all READ and WRITE statements. In this appendix the main programs are simply suggestive ways to use the two subroutines given, and they may be altered quite easily to suit the taste of the user.

The two subroutines in this appendix carry out matrix inversion and characteristic root and vector calculation. There must be hundreds of programs which exist to do these calculations, and the methods that we have chosen are not necessarily the most efficient available. The programs were chosen mainly because they can be understood from first principles by the student interested in checking into the references given. Furthermore, the programs are short, easy to use, and in the authors' experience gave good results. The programs are *not intended* for serious research work but rather as aids for class exercises. However, if double precision arithmetic is used, the programs have been found to work well with typical research problems encountered in economics.

A1.1 subroutine INVS

*A. **Description.*** This subroutine inverts a matrix by *sweeping.*[1] The matrix to be inverted need *not* be symmetric, but, of course, it must be square.

B. Limitations. The matrix to be inverted must be nonsingular and cannot have *any* zeros on its main diagonal, and the matrix is destroyed upon inversion. Therefore, if it is desired to retain the original matrix, it must be stored prior to calling INVS. The maximum dimension of the matrix to be inverted is 10 × 10. This dimension may be increased or reduced by altering the DIMENSION statement in the subroutine.

C. Use. Assuming that the matrix A and the scalar M are in storage, the statement

CALL INVS (A, M)

in the main program will cause entry into the subroutine. **A** is the matrix

```
      SUBROUTINE INVS(A,M)
      DIMENSION A(10,10)
      DO 20 K=1,M
      A(K,K)=-1./A(K,K)
      DO 5  I=1,M
      IF(I-K)3,5,3
3     A(I,K)=-A(I,K)*A(K,K)
5     CONTINUE
      DO 10 I=1,M
      DO 10 J=1,M
      IF((I-K)*(J-K))9,10,9
9     A(I,J)=A(I,J)-A(I,K)*A(K,J)
10    CONTINUE
      DO 20 J=1,M
      IF(J-K)18,20,18
18    A(K,J)=-A(K,J)*A(K,K)
20    CONTINUE
      DO 25 I=1,M
      DO 25 J=1,M
25    A(I,J)=-A(I,J)
      RETURN
      END
```

[1]See Dempster (1969), pp. 62–65, for an explanation of this technique.

to be inverted upon entry into the subroutine, and it is the inverse matrix upon exit from the subroutine. M is the number of rows in **A** (or columns, since **A** must be square). The main program of the next section gives specific details for the use of INVS.

A1.2 main for INVS

The first data card supplied to this calling program has the numerical value of M punched in columns 1–3. This numerical value must be punched without a decimal point and placed as far to the right in this field as possible.[2] If $M = 2$, punch 2 in column 3. If $M = 10$, punch 10 in columns 2 and 3.

The remaining data cards contain the matrix **A**, rowwise. The elements of **A** are punched, with the decimal point, in eight fields of width 10, that is, in columns 1–10, 11–20, 21–30, . . . , 71–80. If **A** is less than 9×9, the first card contains the first row of **A**, the second card the second row of **A**, and so forth. If **A** is larger than 8×8, the rows are continued from one card to the next. For example, if **A** is 10×10, the first card contains $\mathbf{A}_{11}$ through A_{18}, the second card contains A_{19} through $A_{1\ 10}$. The next two cards contain the second row of **A**, and so forth. The following is an example of the data card arrangement for the matrix associated with Eq. (1-1a), followed by the main program itself. Again we warn the reader that while this routine seems to do well on small matrices which are not badly conditioned, it may not hold up under the stress of some actual research problems. Furthermore, it is advisable to check the accuracy of an inverse matrix.

1	2	3	4	5	6	7	8	9	10	11	12	13	14	15	16	17	18	19	20	21
		2																		
							-	.	5								1	.	0	
							1	.	0								1	.	0	

```
C     INVS NEEDED
      DIMENSION A(10,10)
      READ(5,10) M
10    FORMAT(I3)
      DO 20 I=1,M
20    READ(5,25)(A(I,J),J=1,M)
25    FORMAT(8F10.0)
      CALL INVS(A,M)
      DO 30 I=1,M
30    WRITE(6,40)(A(I,J),I,J,J=1,M)
40    FORMAT(4(E17.7,'(',2I2,')'))
      CALL EXIT
      END
```

[2] M is an integer variable.

The output from the program is shown below. Notice that the elements of the inverse are given in the exponential format[3] and that the row and column location of each element of the inverse is printed to the right of the element itself (in parentheses) for ease of reference.

```
-.6666670E+00( 1 1)     .6666666E+00( 1 2)
 .6666666E+00( 2 1)     .3333333E+00( 2 2)
```

A1.3 subroutine CHAR

A. Description. This subroutine calculates all characteristic roots and their associated normalized characteristic vectors from a matrix.[4] The matrix from which the roots and vectors are calculated is **R**. The vector **ROOT** contains the characteristic roots. The matrix **VEC** contains the normalized characteristic vectors columnwise. That is, the first column of **VEC** contains the characteristic vector associated with the first root, the second column contains the vector associated with the second root, and so forth. The roots are not necessarily sorted in order of magnitude.

B. Limitations. ***R*** *must be symmetric.* It is destroyed by the subroutine. The maximum size of **R** is 10×10, but this size may be altered by changing the DIMENSION statement.

C. Use. Assuming that the matrix $\boldsymbol{R}$ and the scalar M are in storage, the statement

CALL CHAR (R, M, VEC, ROOT)

will cause entry into the subroutine. **VEC** and **ROOT** are returned to the main and may be printed out. M is the number of rows in **R**. The main program below furnishes further details.

---

[3]The two digits following the E tell where to move the decimal point. Thus .6666667E + 02 indicates that the decimal point should be moved two places to the *right*, giving the number 66.66667. The number .6666667E − 02 indicates that the decimal point should be moved two places to the left, giving the number .006666667. If the numbers following E are both zero (as they are in this example), the decimal point is in the right location. A minus sign preceding the entire number indicates that the number is negative.

[4]See Dempster (1969), pp. 89ff., for an explanation of this technique, which is the method of Jacobi. However, there are some obvious errors in Dempster's exercise on p. 102. Anderson (1958) illustrates the *exhaustion method.* The authors have found this method to produce serious rounding errors on a digital computer. However, this method is useful if only the largest root is to be calculated from a nonsymmetric matrix.

```
      SUBROUTINE CHAR(R,M,VEC,ROOT)
      DIMENSION R(10,10),VEC(10,10),ROOT(10)
      DO 10 I=1,M
      DO 5 J=1,M
5     VEC(I,J)=0.0
10    VEC(I,I)=1.0
      M1=M-1
12    DO 40 I=1,M1
      I1=I+1
      DO 40 J=I1,M
      IF(ABS(R(I,J))-0.000001) 40,40,13
13    XL=-R(I,J)
      XM=0.5*(R(I,I)-R(J,J))
      XO = XL/ SQRT(XL*XL+XM*XM)
      IF(XM) 14,15,15
14    XO=-XO
15    ST=XO/( SQRT(2.0*(1.0+ SQRT(1.0-XO*XO))))
      CT= SQRT(1.0-ST*ST)
      DO 17 KK=1,M
      ROOT(KK)= CT*VEC(KK,I)-ST*VEC(KK,J)
      VEC(KK,J)= ST*VEC(KK,I)+CT*VEC(KK,J)
17    VEC(KK,I)=ROOT(KK)
      ROOT(1)= R(I,J)
      ROOT(2)=R(I,I)
      R(I,J)=(R(I,I)-R(J,J))*ST*CT +ROOT(1)*(CT*CT-ST*ST)
      R(J,I)=R(I,J)
      R(I,I)=CT*CT*ROOT(2)-2.0*CT*ST*ROOT(1)+ST*ST*R(J,J)
      R(J,J)=ST*ST*ROOT(2)+CT*CT*R(J,J)+2.0*CT*ST*ROOT(1)
      DO 30 K=1,M
      IF((I-K)*(J-K)) 25,30,25
25    ROOT(1)= R(I,K)
      R(I,K)=CT*R(I,K)-ST*R(J,K)
      R(J,K) =ST*ROOT(1)+CT*R(J,K)
      R(K,J)=R(J,K)
      R(K,I)=R(I,K)
30    CONTINUE
40    CONTINUE
      KOUNT=0
      DO 60 I=1,M1
      I1=I+1
      DO 60 J=I1,M
      IF( ABS(R(I,J))-0.000001)60,60,55
55    KOUNT =KOUNT+1
60    CONTINUE
      IF(KOUNT) 70,70,12
70    DO 80 I=1,M
80    ROOT(I)=R(I,I)
      RETURN
      END
```

A1.4 main for CHAR

The following main program reads M and the matrix $\mathbf{R}$ in the same way as M and $\mathbf{A}$ were read by the main for INVS. An example of the output for the matrix associated with Eq. (1-20) is shown on page 40.

```
C       CHAR NEEDED
        DIMENSION R(10,10),VEC(10,10),ROOT(10)
        READ(5,10) M
  10    FORMAT(I3)
        DO 20 I=1,M
  20    READ(5,25)(R(I,J),J=1,M)
  25    FORMAT(8F10.0)
        CALL CHAR(R,M,VEC,ROOT)
        DO 40 K=1,M
        WRITE(6,30) ROOT(K)
  30    FORMAT(/,/,' CHARACTERISTIC ROOT=',E15.7,/,' CHARACTERISTIC VECTOR
       1')
  40    WRITE(6,45)(VEC(J,K),J=1,M)
  45    FORMAT(10F11.7)
        CALL EXIT
        END
```

```
CHARACTERISTIC ROOT=    .1381965E+01
CHARACTERISTIC VECTOR
  .8506507   -.5257310


CHARACTERISTIC ROOT=    .3618032E+01
CHARACTERISTIC VECTOR
  .5257310    .8506507
```

chapter 2

The Normal Density and Statistical Inference

Statistical inference is concerned with the making of decisions about populations on the basis of information obtained from a sample. These decisions are based upon probability theory, and as a result they are invariably based not only upon the information obtained from the sample but also upon probability densities. In this chapter we shall review some properties of one of the most important probability densities in applied statistics—the univariate (or single-variate) normal density. We shall then discuss the bivariate (or two-variate) normal density and finally the multivariate (or many-variate) normal density. In addition, we shall discuss the estimation of the parameters of these densities and shall review some elements of statistical inference.

2.1 the univariate normal density

The univariate normal density is a fundamental one for statistical inference for several interrelated reasons. First, it is a member of a class of probability densities that depend only on two parameters: the mean and the variance. Therefore the density is relatively simple to deal with analytically. Second, the density has been found to afford a close representation of a wide variety of random phenomena. As a result, the density is justified on practical grounds. Third, the density is associated with the central limit theorem. This theorem tells us that the distribution of sample means that have been randomly drawn from a population with known finite variance will be asymptotically normal as the sample size increases. Thus the normal density is widely used in the analysis of sample means.

The normal probability density function of the random variable X can be written as[1]

$$f(X) = \frac{1}{(2\pi\sigma^2)^{1/2}} \exp\left[-\frac{1}{2}\frac{(X-\mu)^2}{\sigma^2}\right], \qquad -\infty < X < \infty \qquad \text{(2-1)}$$

Aside from the constants in Eq. (2-1), which include $\pi \doteq 3.1416$ and $e \doteq 2.7183$, the function depends only on μ, the *mean*, and σ^2, the *variance*. The mean gives a measure of the location of the density, and since the density is symmetric about this point, the mean is also the median and the mode of the density.[2] Figure 2.1 shows three normal densities with the same variance but with different means.

We mention here that some authors use capital letters to refer to a random variable and lowercase letters to refer to the value of a random variable. We have not adopted that rule in an attempt to keep the notation as simple as possible. In a similar way, we do not distinguish between the terms *random variable*, *variable*, and *variate*. In applied statistical analysis at the level presented in this book, no confusion is likely to result from failure to make these distinctions.

The variance σ^2, or, alternatively, its square root σ, the *standard deviation*, gives a measure of the dispersion of the density. The variance has a mini-

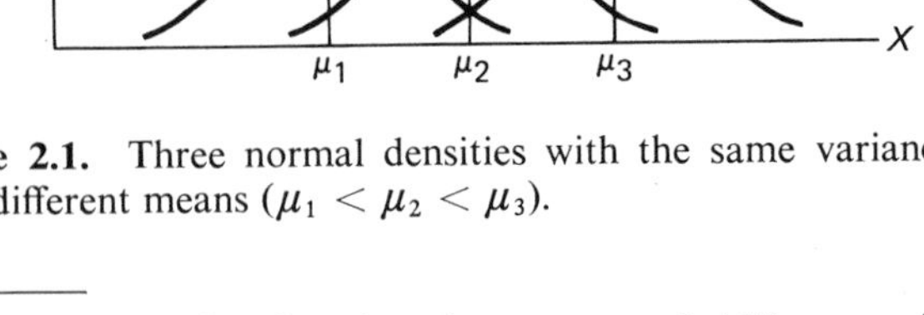

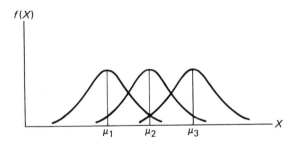

Figure 2.1. Three normal densities with the same variance but with different means ($\mu_1 < \mu_2 < \mu_3$).

---

[1]A probability density function gives the average probability per unit length of the interval ΔX. The function does not give probabilities directly. Recall also that $\exp(Y) = e^Y$.

[2]The mode is the value of X that gives the maximum value to $f(X)$. The median is the value of X that divides the density into equal areas.

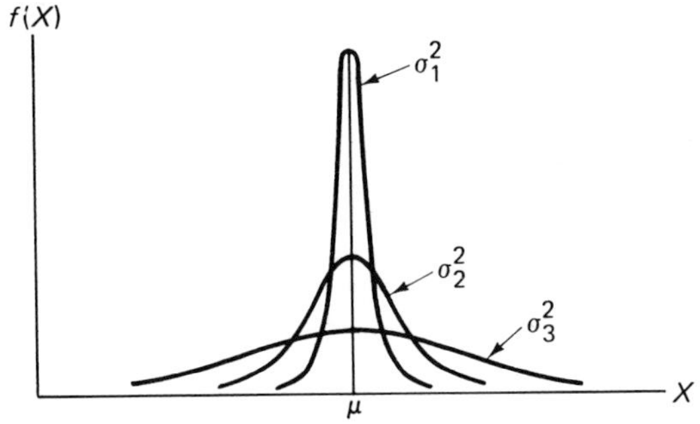

Figure 2.2. Three normal densities with the same mean but with different variances ($\sigma_1^2 < \sigma_2^2 < \sigma_3^2$).

mum value of zero but may be indefinitely large. Figure 2.2 shows three normal densities with the same mean but with different variances.

Notice that in both of the previous figures the density does not touch the horizontal axis. The normal density is asymptotic to the X axis; it is never zero no matter how large or small X becomes.

Obviously there are infinitely many normal densities depending on various combinations of μ and σ^2. Fortunately we may express the normal density in *standard* form by writing it as a function of z, the standard measure, rather than X. The standard measure is defined as

$$z = \frac{X - \mu}{\sigma} \tag{2-2}$$

and we may now write the normal density function in standard form as[3]

$$f(z) = \frac{1}{(2\pi)^{1/2}} \exp\left[-\frac{1}{2}z^2\right] \tag{2-3}$$

The function $f(z)$ is called the *standardized normal density function.* In this form the density has a mean of zero and a variance of 1. Thus in this form the density is unique, a fact that allows a single table to be computed which gives areas (probabilities) associated with the density. Such a table is given in the Appendix Tables at the end of the book (Table 1), and it relates to the *right tail* of the standardized normal density function.

A numerical example of the use of this table may be helpful for purposes of review. Suppose that it is known that a variable X is distributed normally

[3]It may seem obvious why the exponent of e in Eq. (2-3) becomes $-z^2/2$ by virtue of the fact that the exponent on e in Eq. (2-1) is $-z^2/2$. However, it may not seem at all obvious why σ vanishes from the denominator of Eq. (2-1). Actually, the formal relationship between $f(X)$ and $f(z)$ is $f(z) = f(X)\sigma$, where σ is the absolute value of dX/dz, the Jacobian.

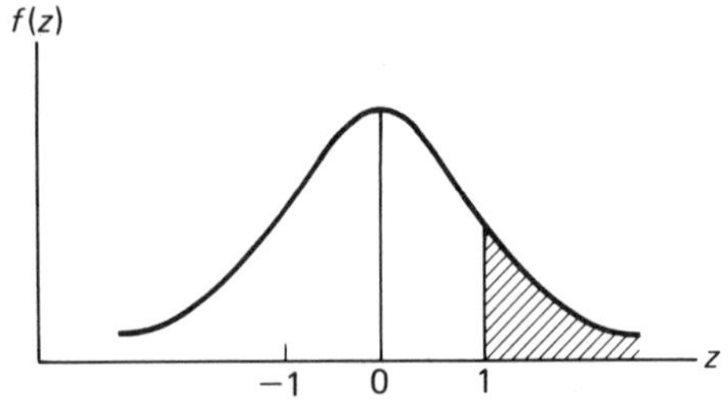

Figure 2.3. Normal density.

with a mean of 10 units and a variance of 4 (standard deviation of 2 units). What is the probability that a random selection from this density will produce a value of X as large as or larger than 12? Figure 2.3 shows the relevant density expressed as a function of z, and the shaded area is the probability in question.

The standard score for the X value is

$$z = \frac{X - \mu}{\sigma} = \frac{12 - 10}{2} = 1$$

as indicated in Figure 2.3. In Table 1 in the Appendix Tables we find in the left and top margins $z = 1.00$, and we locate in the body of the table the number 0.15866 or 15.866 percent. The probability that X will exceed 12 is 0.15866 and the probability that it will be less than 12 is $1.0 - 0.15866 = 0.84134$, since the total area under the curve is unity. Since the density is symmetric, the probability of obtaining an X value as small as or smaller than 8 (i.e., $z = -1.0$) is also 0.15866.

2.2 expectation and variance

Up to this point we have viewed the mean and the variance as parameters of the normal density function. We shall now give a more general definition of these statistical quantities.

Mathematical Expectation. If X is a continuous random variable and if the integral converges, the expected value of X is defined as

$$E(X) = \int_{-\infty}^{\infty} X f(X)\, dX \tag{2-4}$$

where $f(X)$ is the probability density function of X. If X is a discrete variable, then $f(X)$ represents probability, and if the sums converge,

$$E(X) = \frac{\sum [Xf(X)]}{\sum f(X)} = \sum [Xf(X)] \tag{2-5}$$

Equation (2-5) is an arithmetic mean of the X values with $f(X)$ forming the weights. The denominator of Eq. (2-5) vanishes since $\sum f(X) = 1$ for any probability distribution.

To illustrate, suppose that we define the *uniform* density function as

$$\begin{aligned} f(X) &= \tfrac{1}{3}, \qquad 0 \leq X \leq 3 \\ &= 0, \qquad \text{elsewhere} \end{aligned}$$

This function is graphed in Figure 2.4.

Let us first verify that $f(X)$ is a density function. The two necessary and sufficient conditions for a continuous density function are that (1) $f(X)$ is greater than or equal to zero for all X and (2) $\int_{-\infty}^{\infty} f(X)\,dX = 1.0$. Clearly, $f(X)$ given above satisfies these two conditions, and for this density the expected value is

$$E(X) = \int_{-\infty}^{\infty} \frac{1}{3} X\,dX = \frac{1}{3}\int_{0}^{3} X\,dX = \frac{3}{2}$$

Variance. If X is a continuous variable, the variance of X is

$$\mathrm{var}(X) = \int_{-\infty}^{\infty} (X - E(X))^2 f(X)\,dX \tag{2-6}$$

If X is a discrete variable, then

$$\begin{aligned} \mathrm{var}(X) &= \sum [(X - E(X))^2 f(X)] \\ &= E[X - E(X)]^2 \end{aligned} \tag{2-7}$$

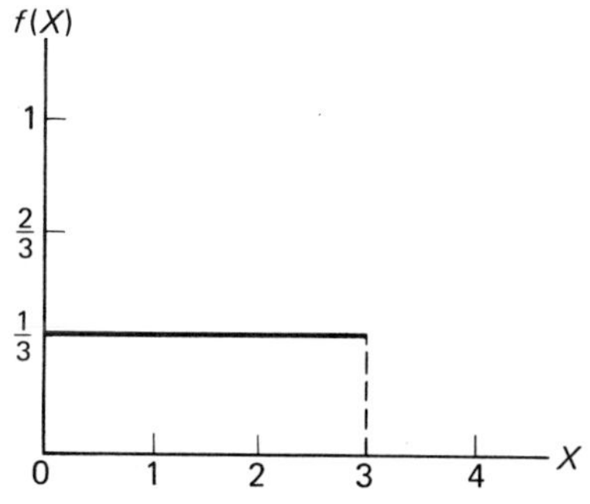

Figure 2.4. Uniform density.

The second form of Eq. (2-7) holds for both discrete and continuous variables. For the uniform density given above,

$$\operatorname{var}(X) = \frac{1}{3}\int_0^3 \left(X - \frac{3}{2}\right)^2 dX = \frac{3}{4}$$

Notice that while the expected value can be calculated without a knowledge of the variance, the variance cannot be calculated without a knowledge of the expected value.

Properties of the Expected Value and Variance. In this book we shall use the symbols μ and $E(X)$ interchangeably; both denote the population mean, and in certain cases it is more convenient to use $E(X)$ to avoid subscripts and the like. The following properties of the expectation operator are useful and should be verified by the reader:

1. The expected value of a constant is the constant itself. Thus, if c is a constant, $E(c) = c$.
2. The expected value of a constant times a variable is the constant multiplied by the expected value of the variable. Thus, $E(cX) = cE(X)$, where c is a constant and X is a variable.
3. The expected value of a sum is the sum of the expected values. That is $E(\sum X) = \sum (E(X))$.

We shall use the symbol σ^2 and $\operatorname{var}(X)$ interchangeably in this book; both denote the population variance. The following properties of the variance are useful and should be verified:

1. The variance of a constant is zero. Hence, if c is any constant, $\operatorname{var}(c) = 0$.
2. The variance of a constant times a variable is the constant squared times the variance of the variable. For a constant c and a variable X, $\operatorname{var}(cX) = c^2 \operatorname{var}(X)$.
3. The variance of a variable is unchanged if a constant is added to each value of the variable. Thus, $\operatorname{var}(c + X) = \operatorname{var}(X)$.

Instead of dealing with a single variable, we are often interested in dealing with two or more variables. In the case of two variables, say X_1 and X_2, we may represent them with the vector $\mathbf{X}$, where

$$\mathbf{X}' = [X_1 \quad X_2]$$

Let us call the density function that applies to both variables jointly the *joint density function*, $f(\mathbf{X})$, where[4]

$$f(\mathbf{X}) = f(X_1, X_2)$$

In this case both variables will have means and both will have variances. In addition, the two variables will have joint variation measured by *covariance*. Figure 2.5 shows items taken from a hypothetical scatter of points relating to the variables X_1 and X_2. The scatter of points is located, for the most part, in the first and third quadrants and thus indicates some positive joint relationship between the two variables. Suppose that we pick some particular point and measure it in terms of its deviation from the center of the scatter, μ_1 and μ_2. Since most of the points lie in the first and third quadrants, the product of these deviations for any point, $(X_1 - \mu_1)(X_2 - \mu_2)$, is more likely to be positive than negative. Thus, if we define the covariance, σ_{12}, as

$$\sigma_{12} = E[(X_1 - \mu_1)(X_2 - \mu_2)] \tag{2-8}$$

this quantity will be positive for our example. In general, if the two variables are positively related, their covariance will be positive. If the two variables are negatively related, their covariance will be negative. If the two variables are statistically independent, their covariance will be zero.[5] We shall often

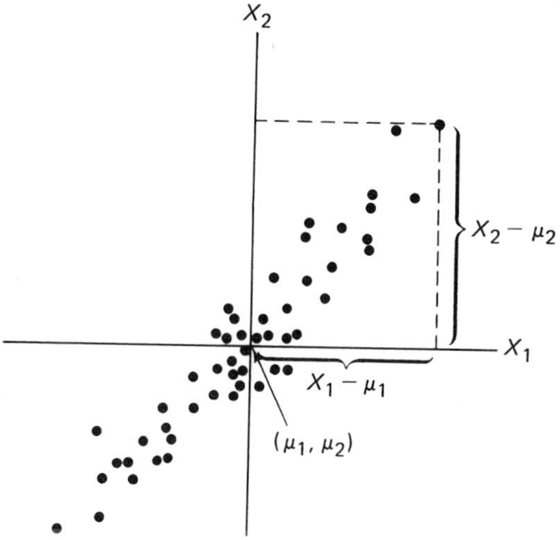

Figure 2.5. Scatter showing positive covariance.

[4]The joint density function gives the average probability per square unit area, $\Delta X_1 \Delta X_2$.

[5]But the reverse is not true. Two variables may even be functionally related and have zero covariance.

use the notation $\text{cov}(X_1, X_2)$ to indicate the covariance between X_1 and X_2. Note that if either μ_1 or μ_2 is zero, then $\text{cov}(X_1, X_2) = E(X_1X_2)$.

If we arrange the two means into a vector, called the *mean vector*, we have

$$E(\mathbf{X}) = \begin{bmatrix} E(X_1) \\ E(X_2) \end{bmatrix} = \begin{bmatrix} \mu_1 \\ \mu_2 \end{bmatrix} = \boldsymbol{\mu} \tag{2-9}$$

We can also arrange all variances and covariances into a *covariance matrix*, $\mathbf{\Sigma}$. Thus

$$\mathbf{\Sigma} = \begin{bmatrix} E(X_1 - \mu_1)^2 & E(X_1 - \mu_1)(X_2 - \mu_2) \\ E(X_2 - \mu_2)(X_1 - \mu_1) & E(X_2 - \mu_2)^2 \end{bmatrix} = \begin{bmatrix} \sigma_{11} & \sigma_{12} \\ \sigma_{21} & \sigma_{22} \end{bmatrix} \tag{2-10}$$

In the covariance matrix, the variances of the two variables are located on the main diagonal and the covariance is located off the main diagonal. The matrix is symmetric, since $\sigma_{12} = \sigma_{21}$, and we shall assume that it is positive definite. No confusion should exist between the matrix $\mathbf{\Sigma}$ and the summation operator Σ since the former will always be set in boldface type. Notice also that we shall use double subscripts rather than the "square" to denote the variance of a variable. This practice seems now to be a standard one for matrix representation. Therefore, for a single variable, σ^2 is the population variance. In the covariance matrix this variance is represented by, say, σ_{11}.

The *correlation* between X_1 and X_2 is indicated by the symbol ρ_{12} (rho). The correlation coefficient is closely connected with the covariance matrix. Just as σ_{12} measures the association between X_1 and X_2, so ρ_{12} measures the association between these variables. One advantage of using ρ_{12} instead of σ_{12} is that ρ_{12} is independent of the unit of measurement of the variables. The correlation coefficient is defined as

$$\rho_{12} = \frac{\sigma_{12}}{\sqrt{\sigma_{11}\sigma_{22}}} \tag{2-11}$$

Equation (2-11) indicates that ρ_{12} must take the same sign as σ_{12} since σ_{11} and σ_{22} are both positive.[6] If σ_{12} is zero, so is ρ_{12}. Furthermore,

$$-1 < \rho_{12} < 1 \tag{2-12}$$

which follows from the assumption that $\mathbf{\Sigma}$ is positive definite and the fact that the determinant of $\mathbf{\Sigma}$ can be written as $\sigma_{11}\sigma_{22}(1 - \rho_{12}^2)$. If ρ_{12} is equal to ± 1, the bivariate density degenerates into a univariate density.

Again, we may arrange the correlation coefficients into a *correlation*

[6]If either σ_{11} or σ_{22} is zero, then ρ_{12} is undefined.

matrix:

$$\boldsymbol{\rho} = \begin{bmatrix} \rho_{11} & \rho_{12} \\ \rho_{21} & \rho_{22} \end{bmatrix} = \begin{bmatrix} 1.0 & \rho_{12} \\ \rho_{21} & 1.0 \end{bmatrix} \tag{2-13}$$

The coefficients ρ_{11} and ρ_{22} are unity since they represent the correlation of a variable with itself. Also, $\rho_{12} = \rho_{21}$ since the correlation of X_1 with X_2 is the same as the correlation of X_2 with X_1. The correlation matrix is, therefore, symmetric, and it will be positive definite. It also follows that if the variables are statistically independent, then both $\boldsymbol{\rho}$ and $\boldsymbol{\Sigma}$ will be diagonal matrices since $\sigma_{12} = 0$.

The covariance matrix can be written in terms of variances and correlation coefficients. By Eqs. (2-10) and (2-11)

$$\boldsymbol{\Sigma} = \begin{bmatrix} \sigma_{11} & \rho_{12}\sqrt{\sigma_{11}\sigma_{22}} \\ \rho_{12}\sqrt{\sigma_{11}\sigma_{22}} & \sigma_{22} \end{bmatrix} \tag{2-14}$$

The inverse of the covariance matrix in this form is

$$\boldsymbol{\Sigma}^{-1} = \frac{1}{1 - \rho_{12}^2} \begin{bmatrix} \dfrac{1}{\sigma_{11}} & \dfrac{-\rho_{12}}{\sqrt{\sigma_{11}\sigma_{22}}} \\ \dfrac{-\rho_{12}}{\sqrt{\sigma_{11}\sigma_{22}}} & \dfrac{1}{\sigma_{22}} \end{bmatrix} \tag{2-15}$$

We shall make use of this inverse in the next section.

2.3 the bivariate normal density

The bivariate normal density function, which gives the *joint* probability density of two random variables X_1 and X_2, can be written for finite values of X_1 and X_2 as

$$f(X_1, X_2) = \frac{1}{2\pi|\boldsymbol{\Sigma}|^{1/2}} \exp\left[-\frac{1}{2}(\mathbf{X} - \boldsymbol{\mu})'\boldsymbol{\Sigma}^{-1}(\mathbf{X} - \boldsymbol{\mu})\right] \tag{2-16}$$

where

$$(\mathbf{X} - \boldsymbol{\mu})' = [X_1 - \mu_1 \quad X_2 - \mu_2]$$

and $|\boldsymbol{\Sigma}|$ is the determinant of $\boldsymbol{\Sigma}$. The function can be represented by a three-dimensional surface, such as the one shown in Figure 2.6. The surface is centered at the point (μ_1, μ_2), and it has the property that any vertical slice made through the surface parallel to the X_1 axis or the X_2 axis will produce a univariate normal density function. Figure 2.7 shows such a slice. The density produced in this manner is called a *conditional density*. In Figure 2.7 the conditional density of X_2 for a given value of X_1 is depicted. We shall use the

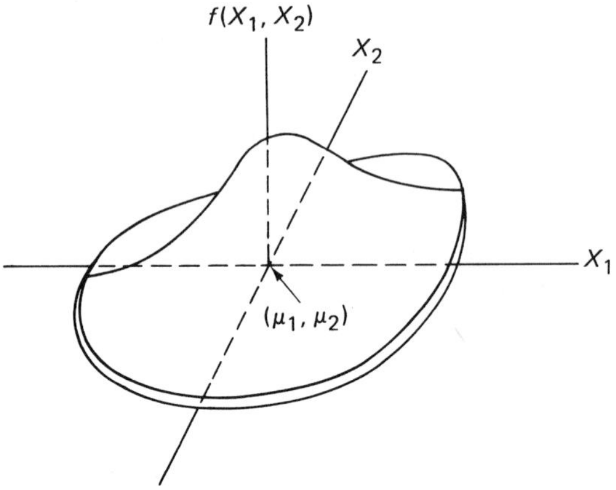

Figure 2.6. Bivariate normal density.

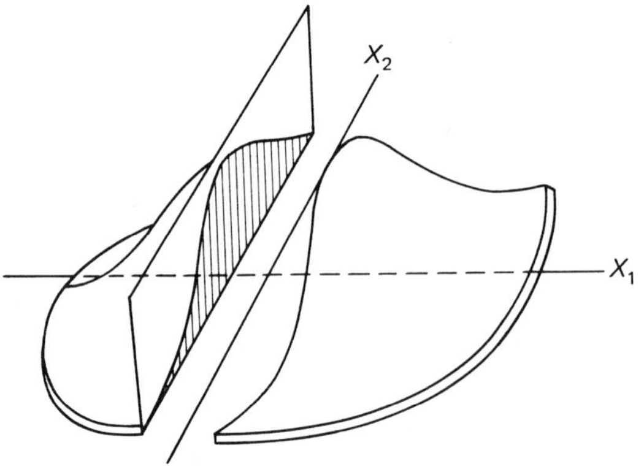

Figure 2.7. Conditional normal density, $f(X_2 \mid X_1)$.

notation $f(X_2 \mid X_1)$, where | is read "given," to indicate the conditional density of X_2 for a given value of X_1. The conditional density $f(X_1 \mid X_2)$, i.e., the density of X_1 for a given value of X_2, would be shown by a slice parallel to the X_1 axis.

The density function described by Eq. (2-16) will be constant when

$$[\mathbf{X} - \boldsymbol{\mu}]'\boldsymbol{\Sigma}^{-1}[\mathbf{X} - \boldsymbol{\mu}] = c \tag{2-17}$$

where c is a constant. That is, Eq. (2-16) will contain only constants. The contour generated by setting the exponent on e equal to a constant is equivalent to a slice through the density function which is made parallel to the X_1, X_2 axes. The left-hand side of Figure 2.8 shows such a slice, and the right-hand side shows the ellipse generated by the slice.

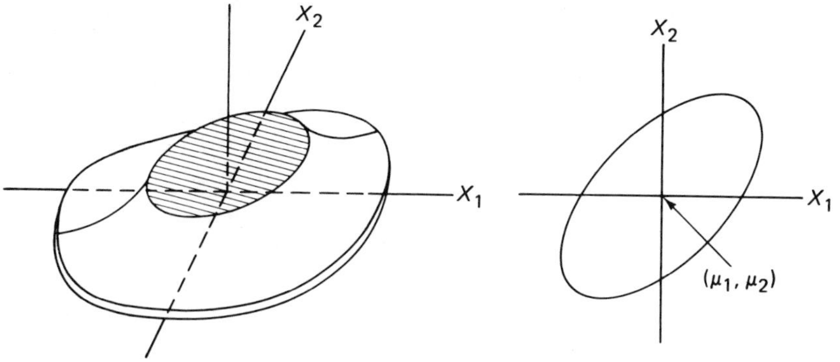

Figure 2.8. Constant density slice through a bivariate normal density and the associated ellipse.

Equation (2-17) is a quadratic equation that describes the ellipse generated by taking such a constant density slice. The ellipse is centered at (μ_1, μ_2) and its orientation is given by $\boldsymbol{\Sigma}^{-1}$. From Eq. (2-15) the orientation will depend on the two variances and the correlation coefficient.

If we standardize both X_1 and X_2, forming z_1 and z_2, then the bivariate density function will be given by

$$f(z_1, z_2) = \frac{1}{2\pi |\boldsymbol{\rho}|^{1/2}} \exp\left[-\frac{1}{2}\mathbf{z}'\boldsymbol{\rho}^{-1}\mathbf{z}\right] \tag{2-18}$$

where

$$\mathbf{z}' = [z_1 \quad z_2]$$

Since the means of both z variates are zero and the variances of both z variates are unity, the joint density function will be centered at the point (0, 0), and the covariance and correlation matrices computed from the standardized variables will be identical.

The ellipse generated by taking a constant density slice from this standardized bivariate normal density function will be defined by[7]

$$\mathbf{z}'\boldsymbol{\rho}^{-1}\mathbf{z} = c$$

Since the bivariate density is now only a function of the correlation coefficient, the ellipse generated is now functionally dependent only on the correlation coefficient for its orientation. If the correlation coefficient is

[7]The correlation coefficient is unaffected by whether it is calculated from the original X variates or the standardized X variates.

positive, the major axis of the ellipse will lie along a 45° line. If the correlation coefficient is negative, the major axis of the ellipse will lie along a 135° line. If the correlation coefficient is zero, the ellipse will degenerate into a circle. Furthermore, the major axis of the ellipse will become longer relative to the minor axis as the correlation coefficient increases in absolute value.

These remarks are easily verified. The major axis of the constant density ellipse will have the greatest squared semilength of any vector passing through the origin. To find the major axis, we need to find the vector $\mathbf{z}' = [z_1 \ z_2]$ with the maximum squared semilength subject to the ellipse, $\mathbf{z}'\boldsymbol{\rho}^{-1}\mathbf{z} = c$. Using the method of Lagrange described in Sec. 1.5, we write

$$\phi = \mathbf{z}'\mathbf{z} - \lambda(\mathbf{z}'\boldsymbol{\rho}^{-1}\mathbf{z} - c)$$

where $\mathbf{z}'\mathbf{z}$ is the squared semilength. Thus we consider

$$\frac{\partial \phi}{\partial \mathbf{z}} = 2\mathbf{z} - 2\lambda\boldsymbol{\rho}^{-1}\mathbf{z} = \mathbf{0}$$

Simplifying and multiplying both sides by $\boldsymbol{\rho}$, we obtain

$$(\boldsymbol{\rho} - \lambda\mathbf{I})\mathbf{z} = \mathbf{0} \tag{2-19}$$

Equation (2-19) is in the same form as Eq. (1-27) and hence the λs are the characteristic roots of $\boldsymbol{\rho}$ with associated characteristic vectors $\mathbf{z}$. The characteristic roots of the $\boldsymbol{\rho}$ matrix will both be positive since $\boldsymbol{\rho}$ is assumed to be positive definite, and they are $\lambda_1 = 1 + \rho_{12}$ and $\lambda_2 = 1 - \rho_{12}$. λ_1 will be the larger root if ρ_{12} is positive; otherwise λ_2 will be the larger root unless ρ_{12} is zero. Since we are calculating the characteristic roots from the inverse of the matrix associated with the quadratic form describing the ellipse (i.e., $\mathbf{z}'\boldsymbol{\rho}^{-1}\mathbf{z} = c$), if ρ_{12} is positive, the semilength of the major axis will be $\sqrt{\lambda_1 c}$ and will increase as ρ_{12} increases. The semilength of the minor axis will be $\sqrt{\lambda_2 c}$ and will decrease as ρ_{12} increases. The definition of the major and minor axes is reversed if ρ_{12} is negative, but the argument concerning the elongation of the major axis relative to the minor axis as ρ_{12} increases in absolute value remains in force. As a special case, if ρ_{12} is zero, then the two axes have equal length.[8] Two normalized characteristic vectors associated with these characteristic roots are $[1/\sqrt{2} \quad 1/\sqrt{2}]$ associated with λ_1 and $[1/\sqrt{2} \quad -1/\sqrt{2}]$ associated with λ_2. If λ_1 is the larger root (ρ_{12} is positive), then the major axis of the ellipse will lie along the line $z_1 = z_2$. If λ_2 is the larger root (ρ_{12} is negative), then the major axis will lie along the line $z_1 = -z_2$.

[8] There are infinitely many axes that may be called major and minor.

2.4 the multivariate normal density

The multivariate joint normal density function is a straightforward extension of the bivariate case. For variates $X_1, X_2, \ldots, X_p$, all assumed to lie in the interval $(-\infty, \infty)$, the multivariate normal density is given by

$$f(X_1, X_2, \ldots, X_p) = \frac{1}{(2\pi)^{P/2}|\boldsymbol{\Sigma}|^{1/2}} \exp\left[-\frac{1}{2}(\mathbf{X}-\boldsymbol{\mu})'\boldsymbol{\Sigma}^{-1}(\mathbf{X}-\boldsymbol{\mu})\right] \quad (2\text{-}20)$$

where

$$[\mathbf{X}-\boldsymbol{\mu}]' = [X_1 - \mu_1 \quad X_2 - \mu_2 \quad \cdots \quad X_p - \mu_p] \quad (2\text{-}21)$$

Again, if

$$\mathbf{X}' = [X_1 \quad X_2 \quad \cdots \quad X_p]$$

the mean vector is given by

$$\begin{aligned} E(\mathbf{X}') &= [E(X_1) \quad E(X_2) \quad \cdots \quad E(X_p)] \\ &= [\mu_1 \quad \mu_2 \quad \cdots \quad \mu_p] = \boldsymbol{\mu}' \end{aligned} \quad (2\text{-}22)$$

and the covariance matrix, assumed to be positive definite, is defined as

$$\boldsymbol{\Sigma} = \begin{bmatrix} \sigma_{11} & \sigma_{12} & \cdots & \sigma_{1p} \\ \sigma_{21} & \sigma_{22} & \cdots & \sigma_{2p} \\ \cdot & & & \cdot \\ \cdot & & & \cdot \\ \cdot & & & \cdot \\ \sigma_{p1} & \sigma_{p2} & \cdots & \sigma_{pp} \end{bmatrix} \quad (2\text{-}23)$$

In the covariance matrix the elements on the main diagonal $(\sigma_{11}, \sigma_{22}, \ldots, \sigma_{pp})$ are the variances of each of the p variables. The elements off the main diagonal are the covariances of the variables. The matrix is symmetric since $\sigma_{ij} = \sigma_{ji}$.

The correlation matrix, also assumed to be positive definite, is defined as

$$\boldsymbol{\rho} = \begin{bmatrix} 1 & \rho_{12} & \cdots & \rho_{1p} \\ \rho_{21} & 1 & \cdots & \rho_{2p} \\ \cdot & & & \cdot \\ \cdot & & & \cdot \\ \cdot & & & \cdot \\ \rho_{p1} & \rho_{p2} & \cdots & 1 \end{bmatrix} = \begin{bmatrix} \frac{\sigma_{11}}{\sqrt{\sigma_{11}\sigma_{11}}} & \frac{\sigma_{12}}{\sqrt{\sigma_{11}\sigma_{22}}} & \cdots & \frac{\sigma_{1p}}{\sqrt{\sigma_{11}\sigma_{pp}}} \\ \frac{\sigma_{21}}{\sqrt{\sigma_{22}\sigma_{11}}} & \frac{\sigma_{22}}{\sqrt{\sigma_{22}\sigma_{22}}} & \cdots & \frac{\sigma_{2p}}{\sqrt{\sigma_{22}\sigma_{pp}}} \\ \cdot & & & \cdot \\ \cdot & & & \cdot \\ \cdot & & & \cdot \\ \frac{\sigma_{p1}}{\sqrt{\sigma_{pp}\sigma_{11}}} & \frac{\sigma_{p2}}{\sqrt{\sigma_{pp}\sigma_{22}}} & \cdots & \frac{\sigma_{pp}}{\sqrt{\sigma_{pp}\sigma_{pp}}} \end{bmatrix} \quad (2\text{-}24)$$

Here ρ_{12} is the correlation between X_1 and X_2, ρ_{1p} is the correlation between X_1 and X_p, and so forth.

If we standardize each variate by forming $z_i = (X_i - \mu_i)/\sqrt{\sigma_{ii}}$, then the density is given by

$$f(z_1, z_2, \ldots, z_p) = \frac{1}{(2\pi)^{p/2} |\boldsymbol{\rho}|^{1/2}} \exp\left[-\frac{1}{2}\mathbf{z}'\boldsymbol{\rho}^{-1}\mathbf{z}\right] \tag{2-25}$$

The p-variate normal density function with mean vector $\boldsymbol{\mu}$ and covariance matrix $\boldsymbol{\Sigma}$ will be abbreviated by the symbols $N_p(\boldsymbol{\mu}, \boldsymbol{\Sigma})$. Notice that if p is set equal to 1, then Eq. (2-20) becomes the univariate density function, and if p is set equal to 2, then Eq. (2-20) becomes the bivariate joint normal density function. Thus Eq. (2-20) is the general expression for the normal density function. In a similar way, Eq. (2-25) is a general expression for the standardized normal density function.

A constant density *ellipsoid* of p dimensions may be calculated from the p-variate normal density function. The semilengths of the axes of this ellipsoid may be calculated from the characteristic roots of the $\boldsymbol{\Sigma}$ matrix if the original X variates are used, or, equivalently, from the characteristic roots of the $\boldsymbol{\rho}$ matrix if the standardized variates are used. The orientation of the ellipsoid will be given by the characteristic vectors associated with these characteristic roots. We shall illustrate these calculations in later sections.

2.5 estimation

In practical statistical work we are almost never given the parameters of a population, and these parameters are usually estimated on the basis of sample information. Two types of estimates are generally discussed in the literature: point estimation and interval estimation. With *point estimation*, we make a single guess about the value of the population parameter. This guess is "best" in some sense, and we shall discuss the meaning of the word *best* below. For example, we might guess that the average height of all students in a university is 5′ 10″. The guess is often made that the population parameter, say μ, is identical to a sample statistic, say $\bar{X}$. With *interval estimation* we guess a range of values for the population parameter. For example, we may estimate that the average height of all students in a certain university is between 5′ 8″ and 6′ 0″. Before drawing the sample, we may state that we are confident to some degree that the population parameter will be located within a certain interval set about the sample statistic. If we took a random sample of university students in a given university and found that the average height was $\bar{X} = 5' \, 10''$, we might state with 95 percent confidence that upon repeated sampling the population mean would be included in the interval $\bar{X} \pm k$, where k might be 2 inches. The number k may be a constant from sample to sample or it may vary, depending on the problem.

Two estimation *methods* will prove useful for our further discussions. The first of these methods is the *method of least-squares*. This method, associated with the names of Legendre and Gauss, is widely applied.[9]

Suppose that we have n sample observations $X_1, X_2, \ldots, X_n$ which have been drawn at random from a population with mean μ. We assume that each X_i observation can be expressed as

$$X_i = \mu + \epsilon_i$$

where ϵ_i is a deviation about the true population mean. Suppose that we have some estimate of μ, say $\hat{\mu}$. For $\hat{\mu}$ there is a corresponding deviation of X_i about this estimated mean. One of these deviations,

$$e_i = X_i - \hat{\mu}$$

is shown in Figure 2.9. One objective method of estimating the mean would be to choose the $\hat{\mu}$ that minimizes the sum of the squared values of e_i. That is, we wish to minimize $\sum e^2 = \sum (X - \hat{\mu})^2$ over all n sample observations. Setting the first derivative with respect to $\hat{\mu}$ equal to zero, we have

$$\frac{d \sum e^2}{d\hat{\mu}} = -2 \sum (X - \hat{\mu}) = 0$$

which reduces to

$$\hat{\mu} = \sum \frac{X}{n}$$

and is the ordinary arithmetic mean, $\bar{X}$.

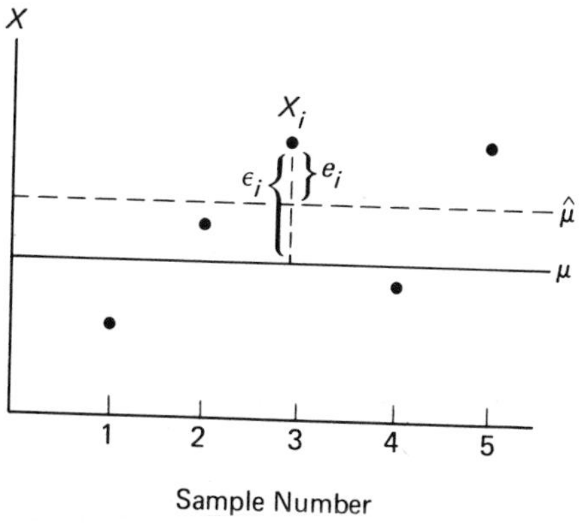

Figure 2.9. Least-squares estimation.

---

[9]For a history of estimation methods, see Deutsch (1965). The approach followed here is based upon Deming (1938). Some other estimation methods include the method of moments and the method of Bayes.

The method of maximum likelihood, associated mainly with the statistician R. A. Fisher, is a second important estimation technique. Let us assume that each X_i, $i = 1, 2, \ldots, n$ is drawn from a normal population with density

$$f(X_i) = \frac{1}{(2\pi\sigma^2)^{1/2}} \exp\left[-\frac{1}{2}\frac{(X_i - \mu)^2}{\sigma^2}\right]$$

That is, each X_i is assumed to be drawn from an identical normal distribution with mean μ and variance σ^2. For each X_i the ordinate of its normal density is given by $f(X_i)$. Furthermore, given X_i, the ordinate $f(X_i)$ will be a function of μ.

Since the mean of a normal density function is also its mode, the ordinate associated with the mean will be the maximum ordinate of the density. Thus, if all X_i sample values happen to have the same value as the mean, the product of these n ordinates is

$$\begin{aligned} L &= f(X_1)f(X_2)\cdots f(X_n) \\ &= \frac{1}{(2\pi\sigma^2)^{n/2}} \exp\left[-\frac{1}{2}\frac{\sum (X - \mu)^2}{\sigma^2}\right] \end{aligned} \tag{2-26}$$

and the product will be a maximum since the exponent will vanish. However, it is very improbable that all X_i values will have exactly the same value as μ. Thus we choose a value of $\hat{\mu}$ which will maximize the function L, called the likelihood function. That is, we estimate the mean in such a way that L will be as large as or larger than it will be for any other estimate of the mean. Figure 2.10 shows that L is maximized when $\mu = \hat{\mu}$.

To evaluate the maximum, we set the first partial derivatives of ln L (to ease the taking of derivatives) equal to zero. The natural logarithm of L is

$$\ln L = -\frac{n}{2}\ln(2\pi) - \frac{n}{2}\ln\sigma^2 - \frac{1}{2\sigma^2}\sum (X - \mu)^2$$

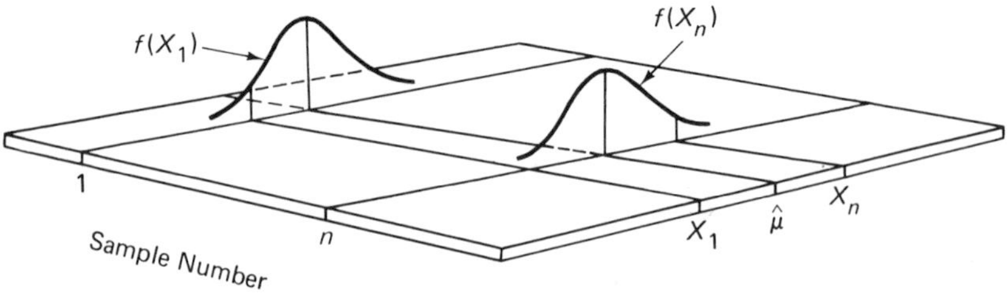

Figure 2.10. Maximum likelihood estimation.

To evaluate the maximum, we set the first partial derivatives of $\ln L$ with respect to μ and σ^2 equal to zero:

$$\frac{\partial \ln L}{\partial \mu} = \frac{1}{\sigma^2}\sum (X - \mu) = 0$$

$$\frac{\partial \ln L}{\partial \sigma^2} = \frac{-n}{2\sigma^2} + \frac{1}{2\sigma^4}\sum (X - \mu)^2 = 0$$

Upon simplification and upon replacing μ and σ^2 by their estimates $\hat{\mu}$ and $\hat{\sigma}^2$, we find that

$$\hat{\mu} = \sum \frac{X}{n} \tag{2-27}$$

$$\hat{\sigma}^2 = \sum \frac{(X - \hat{\mu})^2}{n} \tag{2-28}$$

The maximum likelihood estimate of μ is identical to the least-squares estimate under the assumption of normality. Thus we generally estimate the population mean by the sample mean, $\hat{\mu} = \bar{X} = \sum X/n$. Clearly, from Figure 2.10, maximizing the ordinate of X_i is the same as minimizing the squared deviation of X_i about $\hat{\mu}$. It is for this reason that under the assumption of normality the least-squares and maximum likelihood estimates of μ are the same. Furthermore, the maximum likelihood estimate of the population variance is also the familiar sample variance $SD^2 = \sum x^2/n = \sum (X - \bar{X})^2/n$.

"Good" Estimators. How good are the estimators $\bar{X}$ and SD^2? This question can be answered only with regard to some set of norms, and statisticians have set forth various criteria that are worthy of consideration. No list of these criteria can be exhaustive, and like most ethical structures, in certain situations some criteria can be achieved only at the expense of others. Situation ethics therefore plays a role in the choice of an estimator.

1. *Ease of calculation.* Other things the same, most people will pick an estimator that is easier to calculate over one that is more difficult. For very small samples, the sample range can be made to perform about as well as the computationally more difficult estimator SD (as an estimator of the population standard deviation). As the sample size increases, however, the range does badly in comparison to SD and is not clearly easier to compute on a digital computer.

2. *Lack of bias.* Other things the same, we would like an estimator that is "on the average correct." An estimator is *unbiased* if its expected value is the same as the parameter it is being used to estimate. The estimator $\bar{X}$ is unbiased since

$$E(\bar{X}) = E\left(\sum \frac{X}{n}\right) = \frac{1}{n}\sum E(X) = \frac{n}{n}\mu = \mu$$

However, the estimator SD^2 is a biased estimator of the population variance σ^2 since[10]

$$E(SD^2) = \frac{n-1}{n}\sigma^2$$

In this case the bias of SD^2 is known and can be corrected for. If we define

$$s^2 = \frac{n}{n-1}SD^2$$

then

$$s^2 = \frac{\sum (X - \bar{X})^2}{n-1} = \frac{\sum x^2}{n-1} \tag{2-29}$$

and the expected value of s^2 is σ^2. Since s^2 is an unbiased estimator of the population variance,[11] it is generally used in preference to SD^2.

3. *Consistency*. An estimator is consistent if as the sample size increases, the estimator converges in probability to the value of the parameter. Let $\hat{\theta}_n$ be an estimator of a population parameter θ. The estimator is based upon a random sample of n observations. If for some small positive ϵ and δ there is some N such that the probability of $|\hat{\theta}_n - \theta| < \epsilon$ is greater than $1 - \delta$ for all $n > N$, then $\hat{\theta}_n$ is called a consistent estimator of θ. That is, $\text{Prob}[|\hat{\theta}_n - \theta| < \epsilon] > 1 - \delta$ for all $n > N$. For brevity we say that the probability limit of $\hat{\theta}_n$ is θ, and we write this statement as $\text{plim}\ \hat{\theta}_n = \theta$, where *plim* stands for probability limit.[12] Both $\bar{X}$ and s^2 are consistent estimators of μ and σ^2, respectively.

4. *Efficiency*. Even if an estimator is unbiased, its sampling distribution may have such great variance that little reliability can be placed on the expected value of the estimator. If two competing estimators $\hat{\theta}_1$ and $\hat{\theta}_2$ are both unbiased estimators of θ, then $\hat{\theta}_1$ is said to be relatively efficient if $\text{var}(\hat{\theta}_1)$ is less than $\text{var}(\hat{\theta}_2)$. A measure of relative efficiency is $\text{var}(\hat{\theta}_1)/\text{var}(\hat{\theta}_2)$. An unbiased estimator is particularly desirable if its distribution is closely compacted about its expected value. One problem is that the variance of an estimator may change with the sample size, and in this case it is sometimes necessary to choose between two estimators on the basis of their asymptotic

---

[10]The proof of this contention is rather lengthy although not difficult. We refer the interested reader to Goldberger (1964), p. 97, or to Morrison (1967), p. 16.

[11]However, both $s = \sqrt{s^2}$ and $SD = \sqrt{SD^2}$ are biased estimators of the population standard deviation but the bias can be corrected. See Bolch (1968) for a table of correction factors.

[12]For additional remarks, see Dhrymes (1970), pp. 88ff. Notice that here the vertical bars stand for the absolute value rather than for the determinant.

variances. However, these asymptotic variances may be misleading if in practice one works with small samples. Both $\bar{X}$ and s^2 are relatively efficient estimators of μ and σ^2, respectively.

5. *Minimum mean square error.* Sometimes one is faced with a choice between two estimators which are both biased and which have different variances. In such a case, a logical method of choosing between them would be to do so on the basis of some combined index of their bias and variance. Define the bias of an estimator as $E(\hat{\theta}) - \theta$. Then the *mean square error* is defined as

$$E(\hat{\theta} - \theta)^2 = E[\hat{\theta} - E(\hat{\theta})]^2 + [E(\hat{\theta}) - \theta]^2 = \text{var}(\hat{\theta}) + (\text{bias})^2$$

While the concept of minimum mean square error makes good sense, it is unfortunately not operational in many practical situations because of unknown bias. If we consider only unbiased estimators, then

$$E(\hat{\theta} - \theta)^2 = \text{var}(\hat{\theta})$$

so that the mean square error is the same as the variance of the estimator. Moreover, if $\text{var}(\hat{\theta})$ is less than the variance of any other unbiased estimator, then $\hat{\theta}$ is called the minimum variance unbiased estimator of θ.

To estimate the multivariate normal density function as given by Eq. (2-20), it is necessary to estimate the mean vector $\boldsymbol{\mu} = [\mu_1 \; \mu_2 \; \cdots \; \mu_p]$ and the covariance matrix $\boldsymbol{\Sigma}$. Suppose then that we have p variates with n observations on each variate. Let the observations be arranged in the matrix $\mathbf{X}$ as given in Eq. (1-33). The least-squares and maximum likelihood estimate of the mean vector $\boldsymbol{\mu}$ is given by the vector of sample means, $\bar{\mathbf{X}}$, as defined by Eq. (1-33). For the covariance matrix we transform the sample observations into deviations from their respective means. Thus transform $\mathbf{X}$ into $\mathbf{x}$, where

$$\mathbf{x} = \begin{bmatrix} X_{11} & X_{21} & \cdots & X_{p1} \\ X_{12} & X_{22} & \cdots & X_{p2} \\ \cdot & & & \cdot \\ \cdot & & & \cdot \\ \cdot & & & \cdot \\ X_{1n} & X_{2n} & \cdots & X_{pn} \end{bmatrix} - \begin{bmatrix} \bar{X}_1 & \bar{X}_2 & \cdots & \bar{X}_p \\ \bar{X}_1 & \bar{X}_2 & \cdots & \bar{X}_p \\ \cdot & & & \cdot \\ \cdot & & & \cdot \\ \cdot & & & \cdot \\ \bar{X}_1 & \bar{X}_2 & \cdots & \bar{X}_p \end{bmatrix} = \begin{bmatrix} x_{11} & x_{21} & \cdots & x_{p1} \\ x_{12} & x_{22} & \cdots & x_{p2} \\ \cdot & & & \cdot \\ \cdot & & & \cdot \\ \cdot & & & \cdot \\ x_{1n} & x_{2n} & \cdots & x_{pn} \end{bmatrix} \tag{2-30}$$

The *covariation matrix* is defined [see Eq. (1-34)] for the sample as

$$\mathbf{x}'\mathbf{x} = \begin{bmatrix} \sum x_1^2 & \sum x_1 x_2 & \cdots & \sum x_1 x_p \\ \sum x_2 x_1 & \sum x_2^2 & \cdots & \sum x_2 x_p \\ \cdot & & & \cdot \\ \cdot & & & \cdot \\ \cdot & & & \cdot \\ \sum x_p x_1 & \sum x_p x_2 & \cdots & \sum x_p^2 \end{bmatrix} \tag{2-31}$$

and an estimate of the covariance matrix is

$$\mathbf{S} = \frac{1}{n-1}\mathbf{x}'\mathbf{x} = \begin{bmatrix} s_{11} & s_{12} & \cdots & s_{1p} \\ s_{21} & s_{22} & \cdots & s_{2p} \\ \cdot & & & \cdot \\ \cdot & & & \cdot \\ \cdot & & & \cdot \\ s_{p1} & s_{p2} & \cdots & s_{pp} \end{bmatrix} \tag{2-32}$$

If only one variate is involved, the single variance σ^2 is estimated by s^2 according to Eq. (2-29). Therefore the **S** matrix corresponds to s^2 when only one variate is involved, and the **S** matrix enjoys the same properties as s^2.

The correlation matrix is estimated by appealing to Eq. (2-11) and replacing the population variances and covariances by their estimates as defined by Eq. (2-32). If we call r_{ij} the estimator of ρ_{ij}, then

$$r_{ij} = \frac{s_{ij}}{\sqrt{s_{ii}s_{jj}}} = \frac{\sum x_i x_j}{\sqrt{\sum x_i^2 \sum x_j^2}} \tag{2-33}$$

and the sample correlation matrix is

$$\mathbf{R} = \begin{bmatrix} 1 & r_{12} & r_{13} & \cdots & r_{1p} \\ r_{21} & 1 & r_{23} & \cdots & r_{2p} \\ \cdot & & & & \cdot \\ \cdot & & & & \cdot \\ \cdot & & & & \cdot \\ r_{p1} & r_{p2} & r_{p3} & \cdots & 1 \end{bmatrix} \tag{2-34}$$

Programs are given in the Appendix to this chapter to calculate $\bar{\mathbf{X}}$, **S**, and **R**.

2.6 hypothesis testing

A statistical hypothesis is usually a statement about one or more population parameters that is open to doubt. We do not make the hypothesis that all mothers are women, but we might make the hypothesis that the accident rate for workers in some factory on the day shift is the same as that for workers on the night shift. Since hypotheses are open to doubt, they need to be tested, and in statistical inference we usually test hypotheses on the basis of sample information and on some statement about the probability that we shall reject the hypothesis under test when it is true. In most practical statistical work we deal with two competing hypotheses: the null hypothesis, denoted H_0, and the alternative hypothesis, denoted H_1. The null hypothesis is the hypothesis under test, and if we reject the null hypothesis as untenable in some statistical sense, then we accept the alternative hypothesis. Since we are dealing with an unknown population parameter and making our judgment on the basis of sample information, we may not reach the correct conclusion concerning that parameter. We shall be correct in our conclusion if we reject the null hypothesis when it is false or if we accept the null hypothesis when it is true. We shall be incorrect in our conclusion if we reject the null hypothesis when it is true (a *type I error*) or if we accept the null hypothesis when it is false (a *type II error*). Most of the time, when conducting classic hypothesis tests in business and economics, we set an acceptable level of probability of making a type I error and conduct the test on the basis of sample information and this stated level of probability.[13] In this text we shall call the probability of making an error of the first kind *alpha*, α. The value of alpha must lie between zero and 1 since it is a probability. In classic statistical inference two general rules are given for the determination of the size of alpha. First, the greater the degree of belief in the null hypothesis, the smaller the value of alpha should be. Second, the greater the cost of rejecting a true null hypothesis, the smaller the value of alpha should be.

Let us illustrate these ideas with a simple example. Suppose that we form the null hypothesis that the population mean is not different from some hypothetical value μ_0. The alternative hypothesis is that it is different from this hypothetical value. We can state the two hypotheses as follows:

$$H_0\colon \quad \mu = \mu_0$$
$$H_1\colon \quad \mu \neq \mu_0$$

We decide in advance of the test that we are willing to allow a probability $\alpha = 0.05$ of making a type I error. Assume that we also know the value of the population variance, σ^2. This is an unrealistic assumption in most cases and it will be dropped in the next chapter.

Since we are interested in a hypothesis concerning the population mean, it would seem natural to test that hypothesis by use of the sample mean, $\bar{X}$. We already know one important fact about $\bar{X}$, and that is that $\bar{X}$ is an

[13] We can generally control the probability of making a type I and a type II error if we are free to use a large enough sample. In quality control the sample size is often determined by such considerations.

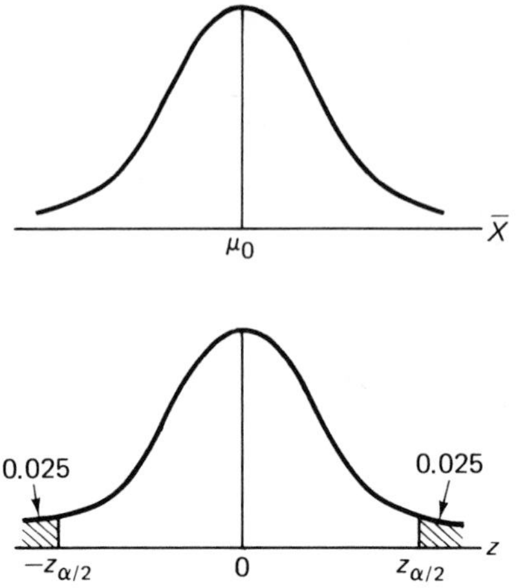

Figure 2.11. Distribution of $\bar{X}$ and $z = (\bar{X} - \mu_0)/\sigma_{\bar{X}}$ under the null hypothesis for large n.

unbiased estimator of the population mean. Therefore the expected value of the sampling distribution of the sample mean will be the same as the population mean. In fact, the sampling distribution of the sample means will look something like the distribution shown in the upper part of Figure 2.11 for a large sample size.

Figure 2.11 indicates that the distribution of sample means is approximately normal. Indeed, this is the case, as the following theorem (a special case of the central limit theorem) states:

> Under general conditions if $\bar{X}$ is the mean of a sample of size n drawn from a population with mean μ and finite variance σ^2, then the distribution of the sample means drawn at random from this population will approach $N_1(\mu, \sigma^2/n)$ for large (but finite) samples.[14]

This theorem does much to account for the central place of the normal density in the theory of statistical inference, and it generalizes to vectors of means and the multivariate normal density function. Notice that this theorem does not require that the population from which the means were drawn be normal.

Knowing the mean, variance, and density function of the distribution of sample means, we are now in a position to conduct a test of the hypothesis. Let us standardize the distribution of means by forming

$$z = \frac{\bar{X} - \mu_0}{\sigma_{\bar{X}}} \tag{2-35}$$

[14]See Wilks (1962), pp. 256ff., for a discussion of the proof of this theorem.

where $\sigma_{\bar{X}} = \sigma/\sqrt{n}$. This standardization assumes that the mean of the sampling distribution is given by the null hypothesis and that we know the value of the population variance. The standardization is equivalent to that given by Eq. (2-2) since the standard deviation of the sampling distribution is given by $\sigma_{\bar{X}}$.

If the null hypothesis is true, we should expect that relatively few sample means would generate z values that deviate a great distance from zero, although occasionally we shall encounter such extreme deviations. Suppose that we bound $\alpha/2 = 0.025$, or 2.5 percent of the distribution, in each tail of the standardized normal density. The values of z which bound these *critical* regions, $\pm z_{\alpha/2}$, are easily found. Look at the body of Table 1 in the Appendix Tables and locate 0.025. The corresponding value of z in the margin of that table is 1.96. Since the normal density is symmetric, the values of z which bound $\alpha/2$ percent of the density in each tail are ± 1.96. In sampling we would expect to encounter values of z greater than 1.96 or smaller than -1.96 only 5 percent of the time if the null hypothesis is true. Therefore, if we reject the null hypothesis whenever z falls into the critical region, we can expect to make a type I error 5 percent of the time. In terms of the sample mean, our rule is to reject the null hypothesis when we encounter a sample mean outside $\mu_0 \pm z_{\alpha/2}(\sigma_{\bar{X}})$. These critical points are found, of course, by solving Eq. (2-35) for $\bar{X}$.

An equivalent method of testing *this* null hypothesis is to set central *confidence limits*. If the confidence limits cover the hypothetical value of the population mean, we accept the null hypothesis; if they do not cover the hypothetical value of the population mean, we reject the null hypothesis. $100(1 - \alpha)$ percent confidence limits are determined by use of Eq. (2-35). In that equation we replace the constant μ_0 with the variable μ, we replace the variable z with the constant $z_{\alpha/2}$, and then we solve for μ. In general we shall call this process of solving for the parameter in terms of the test statistic an *inversion* of the test statistic equation. The confidence limits are, by an inversion of Eq. (2-35), given by

$$\bar{X} \pm z_{\alpha/2}(\sigma_{\bar{X}}) \tag{2-36}$$

Given the value of alpha, we would expect upon repeated sampling that these limits would cover the value of the population mean $100(1 - \alpha)$ percent of the time.

The hypothesis test discussed above is *two-sided*. That is, we are interested in both positive and negative deviations about the hypothetical population mean. If we are interested only in positive *or* negative deviations about the assumed population mean, we may perform a *one-sided* hypothesis test. For positive deviations we would test

$$H_0: \quad \mu = \mu_0$$
$$H_1: \quad \mu > \mu_0$$

and for negative deviations we would test

$$H_0: \quad \mu = \mu_0$$
$$H_1: \quad \mu < \mu_0$$

For the hypotheses tests of this section the only principal difference in conducting a one-sided hypothesis test rather than a two-sided test is that a single value, z_α, is taken to be critical. Thus we bound alpha percent of the distribution in one tail of the distribution. If the alternative hypothesis is $\mu > \mu_0$, then the critical value of z is set in the right tail of the distribution. If the alternative hypothesis is $\mu < \mu_0$, then the critical value of z is set in the left tail of the distribution.

2.7 intelligent use of the normal distribution

Most of the hypothesis tests which are conducted in this book are based upon the assumption that the population distribution from which the observations are drawn at random is normal. This assumption is often violated in practice and it should be made with caution.

There are two broad classes of hypothesis test procedures in statistical inference: *parametric* and *nonparametric* (or distribution-free). Parametric statistical inference is generally the more "powerful" of the two when the assumptions which underpin the model are satisfied. In parametric statistical inference we generally need to assume some specific population distribution. In nonparametric inference we usually need to make less strict assumptions about the specific configuration of the population distribution. The price that we pay for the additional generality of nonparametric methods is usually some loss of power to reject the null hypothesis when it is actually false.

A clear-cut advantage of some nonparametric techniques is their ability to handle problems when the measurement level is only on a nominal or an ordinal scale, When we are able to classify an item, we measure it on a nominal scale. Thus, saying that an individual works for firm A is a nominal measurement. If we can say that one measurement is greater (or smaller) than another, we are able to measure on an ordinal scale. Most modern utility theory is based upon the assumption that people can rank (or order) their preferences.

The methods in this book are based upon the assumption that measurements are made on a higher scale, which is sometimes called ratio. That is, we not only know that A is greater than B but we also know by how many units A exceeds B. Furthermore, there is a point on the measurement scale which represents zero.

Admittedly, the assumption of a ratio scale and the assumption of a normal population (together with other assumptions concerning independence and the like) place severe restrictions on the applicability of the methods to be presented. We justify their presentation on several grounds:

1. A broad class of statistical problems in economics and business is associated with measurement on a ratio scale. Economics is indeed fortunate among the social sciences to be able to measure so many things (prices, quantities, and so forth) at this level of measurement.
2. Methods like those explained in this text are used (rightly or wrongly) in the majority of the quantitative research reported in the literature of economics and business. Even though the reader may never do any quantitative research in these fields, he should have a reading knowledge of these statistical techniques to aid him in understanding and evaluating modern research.
3. It is the opinion of the authors that the parametric methods presented in this text should be mastered before their nonparametric counterparts. The theory of nonparametric inference is generally much more mathematically complicated than that of parametric inference, and a full appreciation of nonparametric methods can be had, in our opinion, only after some exposure to parametric methods.
4. Some of the techniques presented in this book can be used to describe a phenomenon without the normality assumption. Regression analysis can be used as a descriptive technique by assuming only that a certain matrix has an inverse. We sometimes tend to forget that descriptive statistics are often very useful in their own right apart from any inferences made from them.

These points, however, do not free us of the need for careful examination and justification of our statistical model, and the rich literature of nonparametric statistical methods should not be overlooked. For an introductory treatment of nonparametric methods, the reader may consult Siegel (1956). A more advanced treatment is given by Puri and Sen (1971).

Questions and Problems

Sec. 2.1

1. Let X be normally distributed with $\mu = 100$ and $\sigma^2 = 9$. Find the probability that a random selection will produce

   a. A value greater than 110.

b. A value between 100 and 110.
c. A value less than 98.
d. A value greater than 98.

2. The lifetime of a certain kind of light bulb is a random variable having a normal distribution with $\mu = 430$ hours and $\sigma = 32$ hours. Eighty percent of the bulbs can be expected to last more than what number of hours?

Sec. 2.2

1. Prove each of the three properties for the mean and the variance given in Sec. 2.2.
2. Prove that if either μ_1 or μ_2 is zero, then $\text{cov}(X_1, X_2) = E(X_1X_2)$.
3. Prove that $-1 \leq \rho_{12} \leq 1$.

Sec. 2.3

1. Let a bivariate normal distribution have $\boldsymbol{\mu}' = [1\ 2]$ and

$$\boldsymbol{\Sigma} = \begin{bmatrix} 2 & 0 \\ 0 & 4 \end{bmatrix}$$

Sketch the constant density ellipse using $c = 0.2$.

2. Prove that the correlation coefficient r_{12} will be the same whether calculated from X_1 and X_2 or z_1 and z_2.

Sec. 2.4

1. Let $\boldsymbol{\Sigma}_1$ be a $p \times p$ diagonal matrix which contains on its main diagonal the square roots of the elements that are on the main diagonal of $\boldsymbol{\Sigma}$. Show that $\boldsymbol{\Sigma}_1^{-1}\boldsymbol{\Sigma}\boldsymbol{\Sigma}_1^{-1} = \boldsymbol{\rho}$.

2. Suppose that the random variables $\mathbf{X}$ of Eq. (2-20) can be partitioned into two subsets $\mathbf{X}' = [\mathbf{X}_1'\ \mathbf{X}_2']$, where $\mathbf{X}_1' = [X_1 \cdots X_q]$ and $\mathbf{X}_2' = [X_{q+1} \cdots X_p]$. Let $\mathbf{X}_1'$ and $\mathbf{X}_2'$ both be normally distributed and independent. Then $\boldsymbol{\Sigma}$ of Eq. (2-33) can be written as

$$\boldsymbol{\Sigma} = \left[\begin{array}{c|c} \boldsymbol{\Sigma}_{11} & \boldsymbol{\Sigma}_{12} \\ \hline \boldsymbol{\Sigma}_{21} & \boldsymbol{\Sigma}_{22} \end{array}\right] = \left[\begin{array}{c|c} \boldsymbol{\Sigma}_{11} & \mathbf{0} \\ \hline \mathbf{0} & \boldsymbol{\Sigma}_{22} \end{array}\right]$$

where the $\boldsymbol{\Sigma}_{ii}$s are the covariance matrices of the subset $\mathbf{X}_i$. Clearly, $|\boldsymbol{\Sigma}| = |\boldsymbol{\Sigma}_{11}||\boldsymbol{\Sigma}_{22}|$. Show that the joint density function of Eq. (2-20) can be written as the product of two density functions. That is,

$$f(X_1, \ldots, X_p) = f_1(X_1, \ldots, X_q)f_2(X_{q+1}, \ldots, X_p)$$

where

$$f_1(X_1, \ldots, X_q) = \frac{1}{(2\pi)^{q/2}|\mathbf{\Sigma}_{11}|^{1/2}} \exp\left[-\frac{1}{2}(\mathbf{X}_1 - \boldsymbol{\mu}_1)'\mathbf{\Sigma}_{11}^{-1}(\mathbf{X}_1 - \boldsymbol{\mu}_1)\right]$$

$$f_2(X_{q+1}, \ldots, X_p) = \frac{1}{(2\pi)^{(p-q)/2}|\mathbf{\Sigma}_{22}|^{1/2}} \exp\left[-\frac{1}{2}(\mathbf{X}_2 - \boldsymbol{\mu}_2)'\mathbf{\Sigma}_{22}^{-1}(\mathbf{X}_2 - \boldsymbol{\mu}_2)\right]$$

and $\boldsymbol{\mu}_1$ and $\boldsymbol{\mu}_2$ are the mean vectors of $\mathbf{X}_1$ and $\mathbf{X}_2$, respectively.

Sec. 2.5

1. "Bias is really no problem if it is known. Who would not rather question a man who always lies rather than a man who lies at random?" Comment.
2. "Likelihood does not mean probability." Comment.

Sec. 2.6

1. Fill in the type I and type II errors in the following table:

Statistician's conclusion	*State of nature*	
	θ_1	θ_2
θ_1		
θ_2		

2. As a purchasing agent you desire copper tubing 20 millimeters in diameter. Lots whose tubes average larger or smaller than 20 millimeters should be rejected. A large lot of tubes arrives and a random sample of 100 tubes gives a sample mean of 21 millimeters. If σ is known to be 8 millimeters and $\alpha = 0.05$, should the lot be rejected? How would your conclusions change if tubes smaller than 20 millimeters were acceptable but those larger were not? How would your conclusions change if $\alpha = 0.10$?

APPENDIX

In this appendix we shall give three subroutines which calculate the sample mean vector, $\bar{\mathbf{X}}$, of Eq. (1-33); the sample covariation matrix, $\mathbf{x}'\mathbf{x}$, of Eq. (2-31); and the sample correlation matrix, $\mathbf{R}$, of Eq. (2-34). We shall also present a main program which is a suggested way of using these three subroutines and which calculates the covariance matrix, S, of Eq. (2-32), from **x′x**.

A2.1 subroutine MEAN

A. Description. This subroutine computes a vector of sample means, $\bar{\mathbf{X}}$, from a matrix of observations, **X**.

B. Limitation. The maximum number of variables is 10 and the maximum number of observations on each variable is 100. This dimension may be increased or reduced by altering the DIMENSION statement in the subroutine.

C. Use. Assuming that the matrix of observations X and the scalars M and N are in storage, the statement

CALL MEAN (M, N, X, XBAR)

will cause entry into the subroutine. M is the number of variables and N is the number of observations on each variable. None of the entering variables is affected by the subroutine, and the vector **XBAR** contains the M means upon exit from the subroutine. The main program below gives further details and MEAN is shown below.

```
      SUBROUTINE MEAN (M,N,X,XBAR)
      DIMENSION X(10,100),XBAR(10)
      DO 20 I=1,M
      SUM=0.0
      DO 10 J=1,N
10    SUM=SUM+X(I,J)
20    XBAR(I)=SUM/FLOAT(N)
      RETURN
      END
```

A2.2 subroutine COVAR

A. Description. This subroutine computes the matrix **x′x** from a matrix of observations **X**.

B. Limitations. Same as for MEAN.

C. Use. Assuming that the matrix of observations X, the scalars M and N, and the vector **XBAR** are in storage, the subroutine computes **x′x**. **X**, M, N, and **XBAR** have the same meaning as they do for MEAN. None of the entering variables is affected by the subroutine and the statement

CALL COVAR (M, N, X, S, XBAR)

will cause entry into the subroutine. The $M \times M$ matrix **S** is **x′x** upon exit. To convert **x′x** into the sample covariance matrix, we need only to divide each element of **S** by $n - 1$. The subroutine is shown below.

```
      SUBROUTINE COVAR(M,N,X,S,XBAR)
      DIMENSION X(10,100),S(10,10),XBAR(10)
      DO 20 I=1,M
      DO 20 K=I,M
      SIK=0.0
      DO 10 J=1,N
10    SIK=SIK+(X(I,J)-XBAR(I))*(X(K,J)-XBAR(K))
      S(I,K)=SIK
20    S(K,I)=SIK
      RETURN
      END
```

A2.3 subroutine CORR

A. Description. This subroutine computes the correlation matrix from either **x′x** or the covariance matrix.

B. Limitations. The maximum size for **S** is 10×10.

C. Use. Assuming that **S** is in storage along with M, the number of variables, the statement

CALL CORR (M, S, R)

will cause entry into the subroutine. **R** is the correlation matrix upon exit from the subroutine. **S** and M are not altered by the subroutine.

```
      SUBROUTINE CORR(M,S,R)
      DIMENSION S(10,10),R(10,10)
      R(1,1)=1.0
      DO 10 J=2,M
      R(J,J)=1.0
      J1=J-1
      DO 10 I=1,J1
      R(I,J)= S(I,J)/SQRT(S(I,I)*S(J,J))
10    R(J,I)=R(I,J)
      RETURN
      END
```

A2.4 main for MEAN, COVAR, and CORR

This main program reads the numbers M (the number of variables) and N (the number of observations) from the first card. M is placed as far to the right as possible in the first three columns and N is placed as far to the right as possible in the second three columns. Both are punched without decimal points. The matrix **X** is read columnwise, and each **X** element is punched with a decimal point. The first variable is punched across the card in eight fields of width 10, continuing for as many cards as necessary until all elements are punched. The next variable is begun on a fresh card and so forth. The results for the variables

X_1	X_2
1.0	10.0
5.0	11.0
−2.0	−5.0
15.0	14.0
6.0	17.0
8.0	20.0

are depicted below.

```
MEANS
     .5500000E+01( 1)          .1116667E+02( 2)
COVARIANCE MATRIX
     .3470000E+02( 1 1)        .3569998E+02( 1 2)
     .3569998E+02( 2 1)        .7656641E+02( 2 2)
CORRELATION MATRIX
     .1000000E+01( 1 1)        .6926026E+00( 1 2)
     .6926026E+00( 2 1)        .1000000E+01( 2 2)
```

The main program is shown on p. 71.

```
C     MEAN,COVAR, AND CORR NEEDED
      DIMENSION X(10,100),S(10,10),R(10,10),XBAR(10)
      READ(5,5) M,N
 5    FORMAT(2I3)
      DO 10 I=1,M
 10   READ(5,15)(X(I,J),J=1,N)
 15   FORMAT(8F10.0)
      CALL MEAN(M,N,X,XBAR)
      CALL COVAR(M,N,X,S,XBAR)
      CALL CORR(M,S,R)
      DO 20 I=1,M
      DO 20 J=I,M
      S(I,J)=S(I,J)/FLOAT(N-1)
 20   S(J,I)=S(I,J)
      WRITE(6,25)(XBAR(I),I,I=1,M)
 25   FORMAT(' MEANS',/,(4(E17.7,'(',I2,')')))
      WRITE(6,30)
 30   FORMAT(' COVARIANCE MATRIX')
      DO 45 I=1,M
 45   WRITE(6,50)(S(I,J),I,J,J=1,M)
 50   FORMAT(4(E17.7,'(',2I2,')'))
      WRITE(6,60)
 60   FORMAT(' CORRELATION MATRIX')
      DO 65 I=1,M
 65   WRITE(6,50)(R(I,J),I,J,J=1,M)
      CALL EXIT
      END
```

chapter 3

Tests Concerning Means

In practical work we are often concerned with the analysis of sample means. In this context a useful distribution is one first formalized by W. S. Gosset, who wrote under the pseudonym Student (1908). In this chapter we shall discuss this distribution, which is variously known as Student's *t*, Student's distribution, and the *t* distribution, and we shall discuss Harold Hotelling's (1931) multivariate extension of it. In addition, we shall discuss the *F* distribution (or variance ratio) and the χ^2 (chi square) distribution as they relate to the problems posed in this chapter.

3.1 test that a single mean is a given constant

In the last chapter we noted that if the population variance is known or specified in advance, the hypothesis H_0: $\mu = \mu_0$ can be tested against a one- or a two-sided alternative by use of the normal distribution. The statistic is

$$z = \frac{\bar{X} - \mu_0}{\sigma_{\bar{X}}} \tag{3-1}$$

where $\sigma_{\bar{X}} = \sigma/\sqrt{n}$ for large samples. However, if σ^2 is not known or specified in advance, then it is usually estimated from the sample. A natural estimate of σ^2 is s^2, and if we replace σ with s in the definition of the standard error of the mean, we have the *estimated standard error of the mean*

$$s_{\bar{X}} = \frac{s}{\sqrt{n}} \tag{3-2}$$

Replacing $\sigma_{\bar{X}}$ with its estimate $s_{\bar{X}}$ in Eq. (3-1), we have

$$t = \frac{\bar{X} - \mu_0}{s_{\bar{X}}} \tag{3-3}$$

The statistic t, of Eq. (3-3), follows the t distribution with $n - 1$ degrees of freedom. Thus for n observations $X_1, X_2, \ldots, X_n$, with a given sample mean $\bar{X}$, only $n - 1$ of the observations are independent. That is, if the mean of a given sequence of numbers is specified, only $n - 1$ numbers can be given at random. Once these $n - 1$ numbers are known, the nth number must be such that the mean of the entire sequence is as specified. We shall use the symbol ν (nu) to denote degrees of freedom.

For example, suppose that a sample of 36 shear pins is taken at random from a large lot. It is desirable to have pins with a shear strength of 25 units. Pins with less shear strength break too often, and those with greater strength may allow damage to the equipment upon which they are used.

The implied family of hypotheses in this example is

$$H_0: \quad \mu = 25$$
$$H_1: \quad \mu \neq 25$$

and we shall assume that the test is to be conducted using $\alpha = 0.05$. From the sample it was found that $\bar{X} = 20.0$ and $s = 2.0$. Then $s_{\bar{X}} = 2/\sqrt{36} = 1/3$. The t statistic is

$$t = \frac{\bar{X} - \mu_0}{s_{\bar{X}}} = (20 - 25)3 = -15$$

and the number of degrees of freedom for the test is $\nu = n - 1 = 35$.

Table 2 in the Appendix Tables gives the right-tail areas of several t distributions. There is one t distribution associated with each number of degrees of freedom, ν. In the table at $\alpha/2 = 0.025$ and $\nu = 35$, we find the critical values for t to be ± 2.03, approximately[1]. Since the sample value of t, -15.0, is smaller than -2.03, we reject the null hypothesis and reject the lot as

[1]There is no value of t given in the table for $\nu = 35$. The value used is interpolated between the value for $\nu = 30$ and $\nu = 40$. Recall that, like the normal distribution, the t distribution is symmetric.

unsuitable. If a one-sided test had been desired, it would have been conducted in the same way, except that α rather than $\alpha/2$ would have been used in the determination of the critical value of t, and, of course, there would have been only one critical value for t.

Confidence Limits. $100(1 - \alpha)$ percent confidence limits for the mean may be set by calculating

$$\bar{X} \pm s_{\bar{X}}(t_{\alpha/2;\nu}) \tag{3-4}$$

which is an inversion of Eq. (3-3) with the critical value $t_{\alpha/2;\nu}$ replacing t. The notation $t_{\alpha/2;\nu}$ denotes that value for t associated with $\alpha/2$ and ν degrees of freedom. If we wish 95 percent confidence limits for μ in our example, then $100(1 - \alpha)$ percent $= 95$ percent and $\alpha = 0.05$. Thus $\alpha/2 = 0.025$, $\nu = 35$, and $t_{0.025;35} = 2.03$, approximately. The limits are

$$20 \pm 2.03(\tfrac{1}{3}) = 20 \pm 0.68$$

Thus the lower limit is 19.32 and the upper limit is 20.68, and we can state that upon repeated sampling, confidence limits constructed in this manner will be expected to cover the population value of the mean 95 times in 100. Notice also that this interval does not cover the value $\mu_0 = 25$. Thus, as in the last chapter, the central confidence interval is equivalent to a two-sided hypothesis test for the same value of alpha.

Additional Remarks. The t distribution converges to the standardized normal distribution as the number of degrees of freedom increases. Notice in Table 2 in the Appendix Tables that when ν approaches infinity, the area in the right tail of the associated t distribution approaches the area in the right tail of the standardized normal distribution. Thus

$$\lim_{\nu \to \infty} t_{\alpha;\nu} = z_\alpha \tag{3-5}$$

When the number of degrees of freedom is as large as or larger than 60, the normal distribution can generally be used as an approximation to the t distribution with little loss of accuracy. For this reason the t test is sometimes referred to as a small-sample technique. However, the distinction between the use of z and the use of t lies in whether or not σ^2 is known or is estimated from the sample. If σ^2 is known or specified, and if the population is $N_1(\mu, \sigma^2)$ and/or the sample size is large, then by the central limit theorem the distribution of sample means will be approximately normal with mean μ and standard error $\sigma_{\bar{X}}$. Therefore z is the appropriate test statistic. If σ^2 is estimated from the sample, and it usually is, and if the population is $N_1(\mu, \sigma^2)$, then the t statistic is appropriate. Notice that for z, normality of the population is not necessary, whereas for t it is.

Table 3 in the Appendix Tables gives selected right-tail areas of the F distribution. This distribution is based upon a pair of values for degrees of freedom, ν_1 and ν_2, and there is one F table for every level of α. When $\nu_1 = 1$, the F distribution and the t distribution are related as[2]

$$(t_{\alpha/2;\nu_2})^2 = F_{\alpha;\,\nu_1,\nu_2} \tag{3-6}$$

One final distribution will be used in this chapter and it is χ^2 (chi square). Like the t distribution, χ^2 is based upon one value for ν. Right-tail areas for the χ^2 distribution are given in Table 4 in the Appendix Tables.

Chi square is related to the normal distribution and the F distribution as

$$\chi^2_{\alpha;\,\nu} = (z_{\alpha/2})^2, \qquad \text{where } \nu = 1$$

$$\chi^2_{\alpha;\,\nu_1} = v_1 F_{\alpha;\nu_1,\nu_2}, \qquad \text{where } \nu_2 = \infty$$

It follows, therefore, that (see footnote 4)

$$(z_{\alpha/2})^2 = F_{\alpha;\,\nu_1,\nu_2}, \qquad \text{where } \nu_1 = 1 \text{ and } \nu_2 = \infty$$

3.2 test that a vector of means is a given constant vector

Suppose now that we have p sample means based upon n observations each and drawn from a multivariate normal density, $N_p(\boldsymbol{\mu}, \boldsymbol{\Sigma})$. Denote these sample means as

$$\bar{\mathbf{X}}' = [\bar{X}_1 \quad \bar{X}_2 \quad \cdots \quad \bar{X}_p]$$

This vector of means is assumed to be an estimate of $\boldsymbol{\mu}$, where

$$\boldsymbol{\mu}' = [\mu_1 \quad \mu_2 \quad \cdots \quad \mu_p]$$

We wish to test the null hypothesis that the vector of population means does not differ from the vector of constants:

$$\boldsymbol{\mu}_0' = [\mu_{01} \quad \mu_{02} \quad \cdots \quad \mu_{0p}]$$

That is, we wish to test

$$H_0: \quad \boldsymbol{\mu} = \boldsymbol{\mu}_0$$
$$H_1: \quad \boldsymbol{\mu} \neq \boldsymbol{\mu}_0$$

[2]The relationship holds under the null hypothesis. We shall not consider noncentral distributions in this text. Notice from Tables 2 and 3 in the Appendix Tables that in order to pass from values of t to values of F, given α, one must use $t_{\alpha/2;\,\nu_2}$ to correspond to $F_{\alpha;\,\nu_1,\nu_2}$ with $\nu_1 = 1$. This correspondence does not contradict Eq. (3-6) since if Prob$\{t > t_{\alpha/2;\,\nu}$ or $t < -t_{\alpha/2;\,\nu}\} = \alpha$, then Prob$\{t^2 > (t_{\alpha/2;\,\nu})^2\} = \alpha$.

By the definition of equality of vectors we are testing the hypothesis that μ_i are equal to all μ_{0i}, jointly. In other words, we are testing p null hypotheses simultaneously, i.e., $\mu_1 = \mu_{01}$ *and* $\mu_2 = \mu_{02}$ *and* $\cdots$ *and* $\mu_p = \mu_{0p}$. One method of attack for this problem would be to use the simple univariate t test of the last section repeatedly. That is, one might test H_0: $\mu_1 = \mu_{01}$, H_0: $\mu_2 = \mu_{02}$, and so forth. One problem with this approach is that the desired *simultaneous level of significance* cannot be determined exactly by the repeated use of the univariate intervals. For example, suppose that we wish to test two population means and that we set confidence intervals for μ_1 and μ_2 by the univariate method of Eq. (3-4). Denote the interval for μ_1 as I_1 and the interval for μ_2 as I_2. Suppose further that each of these intervals holds with 95 percent confidence. Then, upon repeated sampling, the probability that I_1 covers μ_1 is 0.95 and the probability that I_2 covers μ_2 is 0.95. Denote these probabilities as

$$\text{Prob}(E_1) = 0.95, \qquad \text{Prob}(E_2) = 0.95$$

where E_1 is the event that I_1 covers μ_1 and E_2 is the event that I_2 covers μ_2. The intersection of the two confidence intervals forms a rectangular confidence region, as shown by the shaded area in Figure 3.1.

Denote the joint occurrence of the events E_1 and E_2 as $E_1 \cap E_2$ (read E_1 and E_2). Then, *if* E_1 and E_2 are independent, $\text{Prob}(E_1 \cap E_2) = \text{Prob}(E_1)\,\text{Prob}(E_2)$. In our particular example the joint confidence region holds with a probability of $(0.95)^2$ *if* E_1 and E_2 are independent.

However, E_1 and E_2 are rarely independent when they apply to the kind of data encountered in business and economic research. When E_1 and E_2 are *not* independent, it can be shown that

$$\text{Prob}(E_1 \cap E_2) \geq 1 - (1 - 0.95) - (1 - 0.95) = 0.90$$

for two 95 percent confidence intervals, and we cannot determine exactly the simultaneous level of significance. This conclusion can be generalized to an arbitrary number of means.

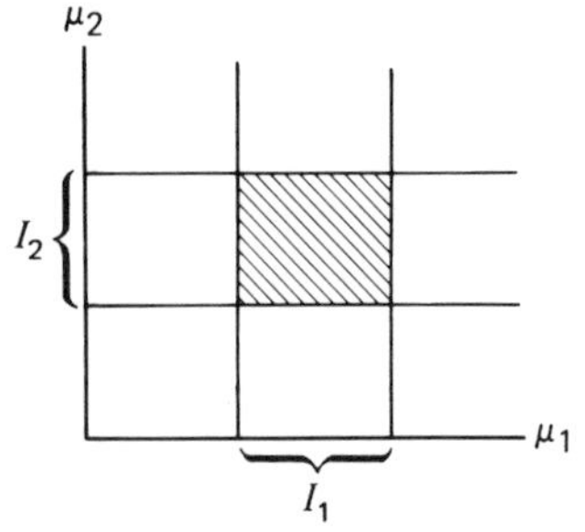

Figure 3.1. Confidence region.

In intermediate and advanced statistics we are often faced with the problem of testing several means (or setting confidence regions) simultaneously. Thus, to ensure the simultaneous level of significance which is desired, we need a test which examines all the means simultaneously.

We proceed by analogy to the case presented in the last section. Write Eq. (3-3) as

$$t = \frac{\sqrt{n}}{s}(\bar{X} - \mu_0) \tag{3-7}$$

Squaring both sides, we have for a single mean

$$t^2 = n(\bar{X} - \mu_0)(s^2)^{-1}(\bar{X} - \mu_0) \tag{3-8}$$

and by analogy Hotelling's T^2 is

$$T^2 = n(\bar{\mathbf{X}} - \boldsymbol{\mu}_0)'\mathbf{S}^{-1}(\bar{\mathbf{X}} - \boldsymbol{\mu}_0) \tag{3-9}$$

for vectors of means.[3] In Eq. (3-9) T^2 and n are scalars and the matrix S, which reduces to s^2 when there is a single mean, is defined as usual to be

$$\mathbf{S} = \frac{1}{n-1}\mathbf{x}'\mathbf{x} \tag{3-10}$$

T^2 is related to F in a manner analogous to that given by Eq. (3-6). Under the null hypothesis

$$T^2_{\alpha;\, p,\, n-p} = \frac{p(n-1)}{n-p}F_{\alpha;\, p,\, n-p} \tag{3-11}$$

Therefore the critical value for T^2 can be calculated by use of the F distribution.[4] Notice that even though we are conducting a two-sided test there is but a single critical value for T^2.

We turn to an example of the use of T^2 which involves Graduate Record Examination (GRE) scores. Both the verbal and quantitative GRE scores for a sample of 10 students who had gained admission to a graduate program in economics are shown in Table 3.1. These students were drawn from a group who were *unsuccessful* in their graduate careers. On the basis of past experience it is felt that a verbal GRE score of 590 and a quantitative GRE score of 690 are satisfactory. We wish to test whether the mean scores for this group of students differ from these criteria.

---

[3]A multiple of this statistic is sometimes used to define T^2. See, for example, Eaton and Efron (1970). Chase and Bulgren (1971) discuss the robustness of T^2.

[4]In the special case where $p = 1$, Eq. (3-11) becomes $(t_{\alpha/2;\, n-1})^2 = F_{\alpha;\, 1, n-1}$. Also, in this case $t_{\alpha/2;\, n-1} = T_{\alpha;\, 1, n-1}$.

Table 3.1 Verbal and Quantitative GRE Scores for Ten Students

	Verbal	*Quantitative*
	740	680
	670	600
	560	550
	540	520
	590	540
	590	700
	470	600
	560	540
	540	630
	500	600
Mean	576	596

The implied hypotheses are

$$H_0: \quad [\mu_1 \quad \mu_2] = [590 \quad 690]$$
$$H_1: \quad [\mu_1 \quad \mu_2] \neq [590 \quad 690]$$

which we shall test at $\alpha = 0.05$. To find the statistic T^2 of Eq. (3-9), we shall need the following vectors and matrices:

$$[\bar{\mathbf{X}} - \boldsymbol{\mu}_0]' = [576 \quad 596] - [590 \quad 690] = [-14 \quad -94]$$

$$\mathbf{x}'\mathbf{x} = \begin{bmatrix} 164 & 94 & \cdots & -76 \\ 84 & 4 & \cdots & 4 \end{bmatrix} \begin{bmatrix} 164 & 84 \\ 94 & 4 \\ \cdot & \cdot \\ \cdot & \cdot \\ \cdot & \cdot \\ -76 & 4 \end{bmatrix} = \begin{bmatrix} 56{,}240 & 17{,}240 \\ 17{,}240 & 33{,}240 \end{bmatrix}$$

$$\mathbf{S} = \frac{1}{n-1}\mathbf{x}'\mathbf{x} = \frac{1}{9}\begin{bmatrix} 56{,}240 & 17{,}240 \\ 17{,}240 & 33{,}240 \end{bmatrix} = \begin{bmatrix} 6248.9 & 1915.6 \\ 1915.6 & 3693.3 \end{bmatrix}$$

$$\mathbf{S}^{-1} = \begin{bmatrix} 0.00019028 & -0.00009869 \\ -0.00009869 & 0.00032194 \end{bmatrix}$$

Then T^2 is calculated from Eq. (3-9) as

$$T^2 = 10[-14 \quad -94]\begin{bmatrix} 0.00019028 & -0.00009869 \\ -0.00009869 & 0.00032194 \end{bmatrix}\begin{bmatrix} -14 \\ -94 \end{bmatrix} = 26.22$$

The critical value of T^2, given by Eq. (3-11), is

$$T^2_{0.05;\,2,8} = \frac{2(9)}{8}(4.46) = 10.035$$

since $F_{0.05;\,2,8} = 4.46$. The sample T^2 exceeds the critical T^2. We reject the null hypothesis, and we conclude that the population mean vector, $\boldsymbol{\mu}$, is significantly different from $\boldsymbol{\mu}_0 = [590 \quad 690]$.

Confidence Region. When p means are considered simultaneously, the confidence region (interval in one-dimensional space) is given by the boundary and interior of an ellipsoid in p-dimensional space. If all p means are considered simultaneously, then the $100(1-\alpha)$ percent confidence region will be bounded by

$$(\bar{\mathbf{X}} - \boldsymbol{\mu})'\mathbf{S}^{-1}(\bar{\mathbf{X}} - \boldsymbol{\mu}) = \frac{p(n-1)}{n(n-p)} F_{\alpha;\,p,\,n-p} \tag{3-12}$$

which follows from Eqs. (3-9) and (3-11). Clearly all terms to the right of the equality are scalars, and the terms to the left define a quadratic form in $\boldsymbol{\mu}$. Thus Eq. (3-12) defines an ellipsoid whose center is $\bar{\mathbf{X}}$. If there is a single mean, Eq. (3-12) reduces to

$$\frac{(\bar{X} - \mu)^2}{s^2} = \frac{1(n-1)}{n(n-1)} F_{\alpha;\,1,n-1}$$

Taking the square root of both sides, we have

$$\frac{\bar{X} - \mu}{s} = \pm \frac{1}{\sqrt{n}} t_{\alpha/2;\,n-1}$$

Thus μ is interior to

$$\bar{X} \pm s_{\bar{X}}(t_{\alpha/2;\,n-1})$$

which is the same result as given by Eq. (3-4).

Let us now calculate the 95 percent confidence region for $\boldsymbol{\mu}$. Following the discussion in Chaps. 1 and 2, we shall need the characteristic roots and vectors of $\mathbf{S}^{-1}$. However, since the elements of $\mathbf{S}^{-1}$ are quite small, it is advisable to calculate the roots and vectors from $\mathbf{S}$. Using SUBROUTINE CHAR, or hand calculation, we find that the roots and normalized vectors for $\mathbf{S}$ are approximately

$$\lambda_1 = 7273.7 \qquad \mathbf{P}_1' = [0.882 \quad 0.471]$$
$$\lambda_2 = 2668.5 \qquad \mathbf{P}_2' = [0.471 \quad -0.882]$$

As a check on our calculation we note that $\lambda_1 + \lambda_2 = 9942.2$, which is tr $\mathbf{S}$. Furthermore, $\mathbf{P}_1'\mathbf{P}_1$ and $\mathbf{P}_2'\mathbf{P}_2$ are both approximately unity, and $\mathbf{P}_1'\mathbf{P}_2$ is zero. Finally, $\mathbf{P}'\mathbf{S}\mathbf{P}$ gives a reasonably accurate diagonal matrix with the characteristic roots on the main diagonal. Each of these check results could be improved by carrying out the calculation of the roots and vectors to a greater number of decimal places.

We recall that if λ_1 and λ_2 are the characteristic roots of **S**, then $1/\lambda_1$ and $1/\lambda_2$ are the characteristic roots of $\mathbf{S}^{-1}$ with no change in the characteristic vectors. That is,

$$\mathbf{P'SP} = \mathbf{D} = \begin{bmatrix} \lambda_1 & 0 \\ 0 & \lambda_1 \end{bmatrix}$$

and

$$\mathbf{P'S^{-1}P} = \mathbf{D}^{-1} = \begin{bmatrix} \frac{1}{\lambda_1} & 0 \\ 0 & \frac{1}{\lambda_2} \end{bmatrix}$$

Let us set

$$\bar{\mathbf{X}} - \boldsymbol{\mu} = \mathbf{PY}$$

which is an orthogonal transformation from $\bar{\mathbf{X}} - \boldsymbol{\mu}$ to **Y**. Thus we may write the quadratic form of Eq. (3-12) as

$$\mathbf{Y'P'S^{-1}PY} = \mathbf{Y'D^{-1}Y} = \frac{1}{\lambda_1}Y_1^2 + \frac{1}{\lambda_2}Y_2^2 = 1.0035 \qquad \text{(3-13)}$$

since $[p(n-1)/n(n-p)]F_{\alpha;\, p, n-p} = 1.0035$ for a 95 percent region with $p = 2$ and $n = 10$.

Substituting the calculated characteristic roots into Eq. (3-13),

$$\frac{1}{7273.7}Y_1^2 + \frac{1}{2668.5}Y_2^2 = 1.0035$$

and solving for Y_1 in terms of Y_2,

$$Y_1 = \sqrt{7299.15795 - 2.72576Y_2^2} \qquad \text{(3-13a)}$$

Equation (3-13a) describes an ellipse with a major axis of semilength $\sqrt{(1.0035)(7273.7)} = 85.4$ and with a minor axis of semilength $\sqrt{(1.0035)(2688.5)} = 51.7$. A few values for Y_1 and Y_2 are shown below, and these and other values are plotted on the left-hand side of Figure 3.2.

Y_1	Y_2
83.82	10
78.80	20
69.61	30
54.20	40
22.02	50

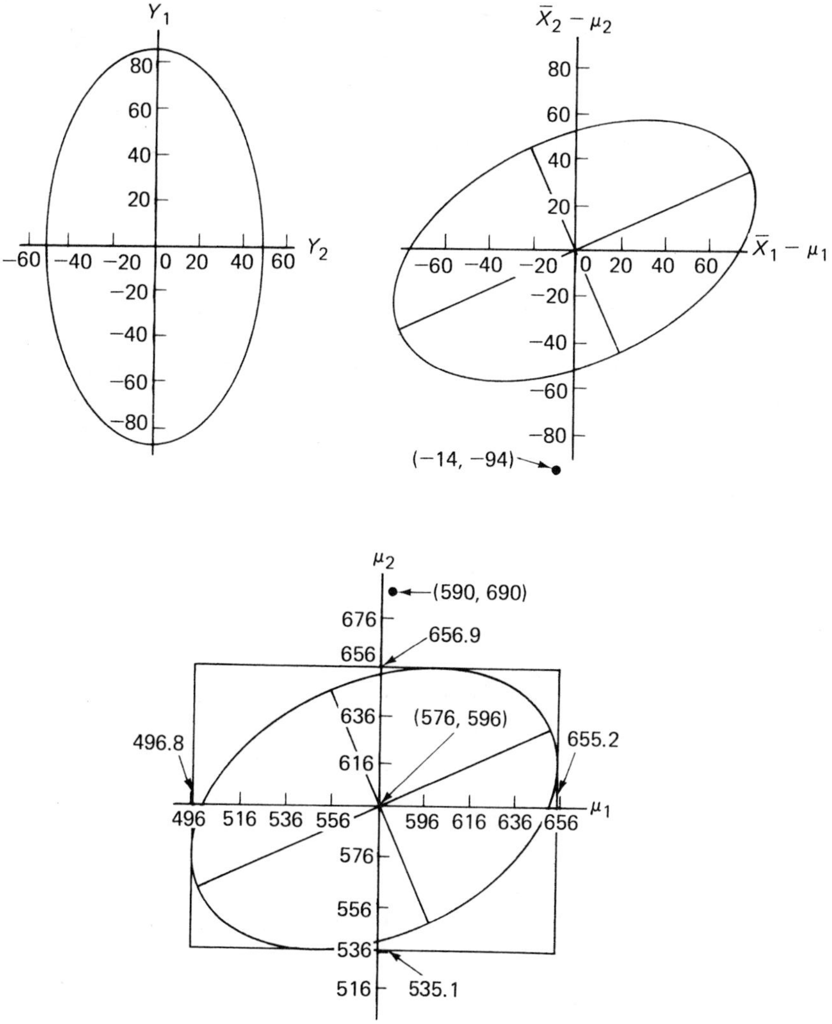

Figure 3.2. Ninety-five percent joint confidence region.

The next step in computing the ellipse consists of transforming the **Y** coordinate system into the $\bar{\mathbf{X}} - \boldsymbol{\mu}$ coordinate system. Thus for each set of **Y** coordinates we form

$$\bar{\mathbf{X}} - \boldsymbol{\mu} = \mathbf{P}\mathbf{Y} \tag{3-14}$$

which for our example is

$$\bar{\mathbf{X}} - \boldsymbol{\mu} = \begin{bmatrix} 0.882 & 0.471 \\ 0.471 & -0.882 \end{bmatrix} \begin{bmatrix} Y_1 \\ Y_2 \end{bmatrix} \quad \text{(3-14a)}$$

We illustrate these calculations with the single value $Y_2 = 10$ and $Y_1 = 83.82$, as calculated from Eq. (3-13a). These two positive values of Y actually provide four points on the ellipse drawn with respect to the Y axis. The points are, of course, as follows:

$$\begin{bmatrix} 83.82 \\ 10 \end{bmatrix} \quad \begin{bmatrix} 83.82 \\ -10 \end{bmatrix} \quad \begin{bmatrix} -83.82 \\ 10 \end{bmatrix} \quad \begin{bmatrix} -83.82 \\ -10 \end{bmatrix}$$

Premultiplying these points by $\mathbf{P}$, we convert them into coordinates for the $\bar{\mathbf{X}} - \boldsymbol{\mu}$ axes:

$$\begin{bmatrix} 78.6 \\ 30.7 \end{bmatrix} \quad \begin{bmatrix} 69.2 \\ 48.3 \end{bmatrix} \quad \begin{bmatrix} -69.2 \\ -48.3 \end{bmatrix} \quad \begin{bmatrix} 78.6 \\ -30.7 \end{bmatrix}$$

These and other values are plotted in the right diagram in Figure 3.2. As a final step we may change the units of measurement of each of the two axes so as to express the ellipse in terms of μ_1 and μ_2. That is, the center of the ellipse will be 576 and 596 rather than (0, 0). This final ellipse is shown in the lower diagram in Figure 3.2.

Also plotted in the lower diagram is the point (590, 690), or (−14, −94) in terms of $\bar{\mathbf{X}} - \boldsymbol{\mu}$. This point represents the null hypothesis and it is *not* included in the confidence region. Again, as in the univariate case, the joint confidence region is the equivalent of the hypothesis test. Had the point been included in the confidence region, the null hypothesis would have been accepted.

Linear Combinations. If X is distributed as $N_p(\boldsymbol{\mu}, \boldsymbol{\Sigma})$ and if $\mathbf{c}$ is a $p \times 1$ column vector whose elements are not all zero, then $\mathbf{c}'\mathbf{X}$ is distributed as $N_1(\mathbf{c}'\boldsymbol{\mu}, \mathbf{c}'\boldsymbol{\Sigma}\mathbf{c})$. Similarly, if $\mathbf{C}$ is a $p \times q$ matrix ($q \leq p$) with rank q, then $\mathbf{C}'\mathbf{X}$ has the distribution $N_q(\mathbf{C}'\boldsymbol{\mu}, \mathbf{C}'\boldsymbol{\Sigma}\mathbf{C})$. The ability to form linear combinations of the elements of $\boldsymbol{\mu}$ allows us to perform a wide variety of tests with T^2. We may, for example, test subsets of $\boldsymbol{\mu}$ by setting some of the columns of $\mathbf{C}$ to zero. The examples below will illustrate some of the available options, and in future sections we shall use the fact that linear combinations of normally distributed variables are themselves normally distributed.

To illustrate the use of linear combinations, consider the case where we are interested in a simultaneous hypothesis concerning the first $q \leq p$ population means. For example, we may wish to test

$$H_0\colon\ [\mu_1 \quad \mu_2 \quad \cdots \quad \mu_q] = [\mu_{01} \quad \mu_{02} \quad \cdots \mu_{0q}]$$

The new hypothesis is equivalent to

$$H_0: \quad \mathbf{C}'\boldsymbol{\mu} = \mathbf{C}'\boldsymbol{\mu}_0$$

where $\mathbf{C}'$ is a $q \times p$ matrix of rank of q:

$$\mathbf{C}' = \left.\begin{bmatrix} 1 & 0 & 0 & \cdots & 0 & 0 & \cdots & 0 \\ 0 & 1 & 0 & \cdots & 0 & 0 & \cdots & 0 \\ \cdot & & & & \cdot & \cdot & & \cdot \\ \cdot & & & & \cdot & \cdot & & \cdot \\ \cdot & & & & \cdot & \cdot & & \cdot \\ 0 & 0 & 0 & \cdots & 1 & 0 & \cdots & 0 \end{bmatrix}\right\} q$$

$$\underbrace{\qquad\qquad\qquad}_{q} \qquad \underbrace{\qquad\qquad}_{p-q}$$

In this case the T^2 statistic is [review Eq. (3-9)]

$$T^2 = n(\mathbf{C}'\bar{\mathbf{X}} - \mathbf{C}'\boldsymbol{\mu}_0)'(\mathbf{C}'\mathbf{S}\mathbf{C})^{-1}(\mathbf{C}'\bar{\mathbf{X}} - \mathbf{C}'\boldsymbol{\mu}_0) \tag{3-15}$$

with critical value

$$T^2_{\alpha;\,q,n-q} = \frac{q(n-1)}{n-q} F_{\alpha;\,q,n-q}$$

Notice that since the hypothesis concerns q population means, the degrees of freedom for T^2 are now q and $n - q$ rather than p and $n - p$. In the special case where $\mathbf{C}$ is an identity matrix of order p, the linear combination hypothesis (H_0: $\mathbf{C}'\boldsymbol{\mu} = \mathbf{C}'\boldsymbol{\mu}_0$) is the same as the full simultaneous hypothesis test (H_0: $\boldsymbol{\mu} = \boldsymbol{\mu}_0$). In this case Eq. (3-15) reduces to Eq. (3-9) and the degrees of freedom for T^2 become p and $n - p$, as before.

For our purposes, the most useful attribute of linear combination hypothesis tests is the ability to "look into" the simultaneous test for all p means in order to determine which mean, or group of means, is causing the rejection of the simultaneous null hypothesis. For the simultaneous test we know that $\mathbf{C}$ is implicitly a $p \times p$ identity matrix. Let $\mathbf{c}_i$ be the ith column vector of the $\mathbf{C}$ matrix. Then T^2 of Eq. (3-15) becomes

$$T^2 = \frac{n[\mathbf{c}_i'(\bar{\mathbf{X}} - \boldsymbol{\mu}_0)]^2}{\mathbf{c}_i'\mathbf{S}\mathbf{c}_i} \tag{3-15a}$$

If this value exceeds the critical value $T^2_{\alpha;\,p,n-p}$, we shall conclude that the simultaneous hypothesis is being rejected because of the ith mean. Notice that the degrees of freedom for T^2 are p and $n - p$ since $q = p$ when we wish to use *all* the $\mathbf{c}_i$ vectors contained in $\mathbf{C}$ for hypothesis tests. In general when $\mathbf{C}$ is an identity matrix and we wish to use *all* its $\mathbf{c}_i$ vectors, $p = q$. Of course, if $\mathbf{C}$ is of rank $q < p$, then the critical value for T^2 is $T^2_{\alpha;\,q,n-q}$.

For our example concerning the GRE scores the null hypothesis may be written as

$$H_0:\ \begin{bmatrix}1 & 0\\0 & 1\end{bmatrix}\begin{bmatrix}\mu_1\\\mu_2\end{bmatrix} = \begin{bmatrix}1 & 0\\0 & 1\end{bmatrix}\begin{bmatrix}\mu_{01}\\\mu_{02}\end{bmatrix}$$

Let us now inquire as to whether only one of the two means is causing the rejection of the null hypothesis. We test the verbal mean first. That is, let us test at $\alpha = 0.05$, with $\mathbf{c}_1' = [1\ \ 0]$:

$$\begin{aligned} H_0&:\ \mathbf{c}_1'\boldsymbol{\mu} = \mathbf{c}_1'\boldsymbol{\mu}_0 \\ H_1&:\ \mathbf{c}_1'\boldsymbol{\mu} \neq \mathbf{c}_1'\boldsymbol{\mu}_0 \end{aligned} \quad \text{or} \quad \begin{aligned} H_0&:\ \mu_1 = 590 \\ H_1&:\ \mu_1 \neq 590 \end{aligned}$$

The T^2 statistic of Eq. (3-15a) is

$$T^2 = \frac{10\{[1\ \ 0][-14\ \ -94]'\}^2}{[1\ \ 0]\begin{bmatrix}6248.9 & 1915.6\\1915.6 & 3693.3\end{bmatrix}\begin{bmatrix}1\\0\end{bmatrix}}$$

$$= \frac{10(-14)^2}{6248.9} = \frac{10(196)}{6248.9} = 0.314$$

Since the calculated value of T^2 does not exceed the critical value, 10.035, we conclude that the verbal test is not causing the rejection of the null hypothesis. Thus the quantitative test must be leading to the rejection of the null hypothesis. To verify this point, we test at $\alpha = 0.05$, with $\mathbf{c}_2' = [0\ \ 1]$:

$$\begin{aligned} H_0&:\ \mathbf{c}_2'\boldsymbol{\mu} = \mathbf{c}_2'\boldsymbol{\mu}_0 \\ H_1&:\ \mathbf{c}_2'\boldsymbol{\mu} \neq \mathbf{c}_2'\boldsymbol{\mu}_0 \end{aligned} \quad \text{or} \quad \begin{aligned} H_0&:\ \mu_2 = 690 \\ H_1&:\ \mu_2 \neq 690 \end{aligned}$$

The T^2 statistic is

$$T^2 = \frac{10\{[0\ \ 1][-14\ \ -94]'\}^2}{[0\ \ 1]\begin{bmatrix}6248.9 & 1915.6\\1915.6 & 3693.3\end{bmatrix}\begin{bmatrix}0\\1\end{bmatrix}}$$

$$= \frac{10(-94)^2}{3693.3} = \frac{10(8836)}{3693.3} = 23.9$$

Clearly, this calculated T^2 exceeds the critical value, 10.035, and we verify that the joint hypothesis $\boldsymbol{\mu} = \boldsymbol{\mu}_0$ was rejected because of the quantitative GRE scores.

The application of linear combinations is not limited to the case where all the elements of the $\mathbf{C}$ matrix are zeros or 1. For example, suppose that we wish to test the simultaneous hypothesis

$$H_0: \begin{cases} 3\mu_1 + \mu_3 + \mu_4 = K_1 \\ \mu_2 + \mu_3 = K_2 \\ \mu_1 + \mu_3 + 2\mu_4 = K_3 \end{cases}$$

Here, $q = 3$ (since there are three simultaneous null hypotheses) and the Ks are appropriate constants. The hypotheses may be written as

$$H_0: \ \mathbf{C}'\boldsymbol{\mu} = \mathbf{K}_0$$

which can be tested by use of

$$\mathbf{C}' = \begin{bmatrix} 3 & 0 & 1 & 1 & 0 & \cdots & 0 \\ 0 & 1 & 1 & 0 & 0 & \cdots & 0 \\ 1 & 0 & 1 & 2 & 0 & \cdots & 0 \end{bmatrix}$$

where $\mathbf{C}'$ is $3 \times p$. Then replace $\mathbf{C}'\boldsymbol{\mu}_0$ in Eq. (3-15) with $\mathbf{K}_0$ and use Eq. (3-15) to complete the hypothesis test.

Confidence Limits with Linear Combinations. Simultaneous confidence limits for $\mathbf{C}'\boldsymbol{\mu}$ are found by use of Eq. (3-15) and the critical value of T^2. Thus $\mathbf{C}'\boldsymbol{\mu}$ will be bounded by

$$(\mathbf{C}'\bar{\mathbf{X}} - \mathbf{C}'\boldsymbol{\mu})'(\mathbf{C}'\mathbf{S}\mathbf{C})^{-1}(\mathbf{C}'\bar{\mathbf{X}} - \mathbf{C}'\boldsymbol{\mu}) = \frac{q(n-1)}{n(n-q)}F_{\alpha;\, q,\, n-q} \tag{3-16}$$

with $100(1 - \alpha)$ percent confidence. If $\mathbf{C}$ is an identity matrix, then $q = p$ and Eq. (3-16) is formally equivalent to Eq. (3-12). If we wish $100(1 - \alpha)$ percent confidence limits for the linear combination $\mathbf{c}_i'\boldsymbol{\mu}$ (where $\mathbf{c}_i$ is a column of $\mathbf{C}$), we may write

$$\mathbf{c}_i'\bar{\mathbf{X}} \pm \sqrt{\frac{q(n-1)}{n(n-q)}\mathbf{c}_i'\mathbf{S}\mathbf{c}_i F_{\alpha;\, q,\, n-q}} \tag{3-16a}$$

For example, using the verbal and quantitative test scores, we evaluate at $\alpha = 0.05$

$$\frac{q(n-1)}{n(n-q)}F_{\alpha;\, q,\, n-q} = \frac{2(8)(4.46)}{10(8)} = 1.0035$$

since $p = q = 2$. For the verbal scores we have $\mathbf{c}_1' = [1 \ \ 0]$ so that

$$\mathbf{c}_1'\bar{\mathbf{X}} = 576$$

and

$$\mathbf{c}_1'\mathbf{S}\mathbf{c}_1 = 6248.9$$

Therefore the 95 percent confidence limits are

$$576 \pm \sqrt{(1.0035)(6248.9)} = 576 \pm 79.2$$

or

$$496.8 \leq \mu_1 \leq 655.2$$

For the quantitative score, we have $\mathbf{c}_2' = [0\ \ 1]$ so that

$$\mathbf{c}_2'\bar{\mathbf{X}} = 596$$

and

$$\mathbf{c}_2'\mathbf{S}\mathbf{c}_2 = 3693.3$$

Therefore the 95 percent confidence limits are

$$596 \pm \sqrt{(1.0035)(3693.3)} = 596 \pm 60.9$$

or

$$535.1 \leq \mu_2 \leq 656.9$$

Notice that the confidence limits for μ_1 cover 590, whereas those for μ_2 do not cover 690. Thus the confidence limits are equivalent to the hypothesis test using linear combinations.

Relationships between Confidence Intervals. We mentioned that Eq. (3-15a) or Eq. (3-16a) could be used to identify a particular mean as a cause of rejecting the simultaneous hypothesis. This procedure will be successful in many cases, but in some cases it will not. We shall explain using the example of GRE scores where there are only two means and $q = p = 2$. In the case where $p > 2$, it is difficult or impossible to visualize the joint confidence ellipsoid since it will be a figure in $p > 2$ space. Generally when $p > 2$ we shall have to project a relevant feature of the ellipsoid onto one-dimensional space. Let us now discuss such a projection.

The confidence ellipse at the bottom of Figure 3.2 has been enclosed in a box. That is, vertical and horizontal lines have been drawn tangent to the outer edges of the ellipse. These lines intersect the μ_1 axis at 496.8 and 655.2 and they intersect the μ_2 axis at 535.1 and 656.9. Thus these lines represent the dimensions of a box which will just enclose the ellipse. In this case, the dimensions of the box are given by the linear combination confidence intervals with $\mathbf{c}_1' = [1\ \ 0]$ and $\mathbf{c}_2' = [0\ \ 1]$ which we have just computed.

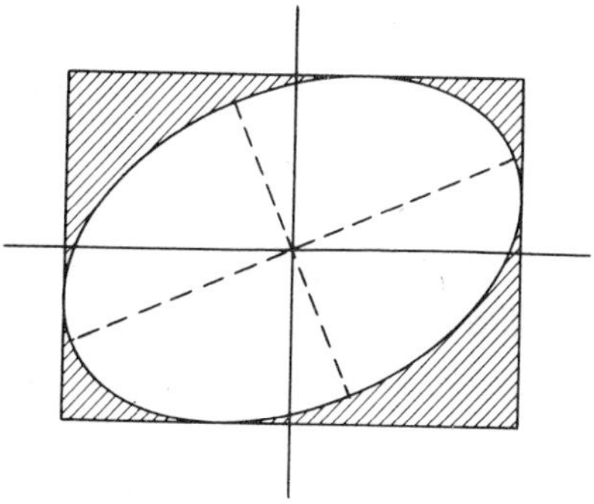

Figure 3.3. Linear combination confidence intervals.

Consider the ellipse enclosed by such a box in Figure 3.3. Any null hypothesis which lies outside the box will be rejected by use of either the ellipse or the linear combination confidence intervals. In other words, the verbal and/or the quantitative score can be identified as responsible for rejecting the simultaneous null hypothesis. Any null hypothesis which lies inside the ellipse will be accepted by both the ellipse and the linear combination confidence intervals. However, any null hypothesis which lies in the shaded area between the ellipse and the box will be rejected by the joint confidence ellipse but accepted by the linear combination confidence intervals. Thus, if the joint hypothesis has been rejected, it is not always possible to use the kind of linear combination tests described in this chapter to determine which one (ones) of the means is causing rejection.

If the cause of rejecting the simultaneous hypothesis cannot be attributed solely to either mean, then what is the cause of rejecting the simultaneous hypothesis? It must be a combination of the two means. Thus a $\mathbf{c}_i'$ vector such as $[\frac{1}{2} \quad \frac{1}{2}]$ will describe a pair of lines tangent to the joint confidence ellipse, and any $\mathbf{c}_i$ (except the zero vector) used in conjunction with Eq. (3-16a) will describe a pair of lines tangent to the joint confidence ellipse. If $p > 2$, the vector $\mathbf{c}_i$ will describe hyperplanes tangent to the joint confidence ellipsoid. Infinitely many of these tangent lines (or hyperplanes) will circumscribe the ellipse (or ellipsoid) and the shaded area will be eliminated. It is in this sense that the confidence intervals using linear combinations are simultaneous.

3.3 test that two means are equal

Another useful test concerns the equality of two means, say, μ_1 and μ_2. Under the assumption that the two populations are $N_1(\mu_1, \sigma^2)$ and $N_1(\mu_2, \sigma^2)$

we test

$$H_0: \quad \mu_1 = \mu_2$$
$$H_1: \quad \mu_1 \neq \mu_2$$

by use of the t distribution. We first form the "pooled" estimate of the common population variance, σ^2,

$$s_{**} = \frac{\sum x_1^2 + \sum x_2^2}{n_1 + n_2 - 2} \tag{3-17}$$

where $x_1 = X_1 - \bar{X}_1$ and $x_2 = X_2 - \bar{X}_2$. This pooled estimate is nothing more than an average of the two sample variances, since Eq. (3-17) can be written as

$$s_{**} = \frac{(n_1 - 1)s_{11} + (n_2 - 1)s_{22}}{(n_1 - 1) + (n_2 - 1)} \tag{3-17a}$$

where s_{11} and s_{22} are the sample variances. Next, if we assume independence, the variance of the difference between the two sample means is given by

$$s^2_{\bar{X}_1 - \bar{X}_2} = \frac{s_{**}}{n_1} + \frac{s_{**}}{n_2} = \frac{n_1 + n_2}{n_1 n_2} s_{**} \tag{3-18}$$

The test statistic is

$$t = \frac{\bar{X}_1 - \bar{X}_2}{s_{\bar{X}_1 - \bar{X}_2}} = \frac{(\bar{X}_1 - \bar{X}_2)\sqrt{n_1 n_2}}{s_* \sqrt{n_1 + n_2}} \tag{3-19}$$

with $n_1 + n_2 - 2$ degrees of freedom, and $s_* = \sqrt{s_{**}}$.

Suppose that a new treatment which protects a certain metal from corrosion is proposed. It is decided to test the new treatment against the current treatment as an aid in deciding whether the new method should be adopted. The level of significance is decided in advance to be 0.01, and the following information was obtained by subjecting random samples to a corrosive bath. Twenty-five test strips of metal treated by the old method and 20 test strips of metal treated by the new method were tested.

Statistic	*Old Method*	*New Method*
Mean life in bath	$\bar{X}_1 = 35$ units	$\bar{X}_2 = 40$ units
Variation	$\sum x_1^2 = 1220$	$\sum x_2^2 = 1190$
Sample size	$n_1 = 25$	$n_2 = 20$

Using this information, we compute

$$s_{**} = \frac{1220 + 1190}{25 + 20 - 2}$$

$$= \frac{2410}{43} = 56.05$$

and

$$s^2_{\bar{X}_1 - \bar{X}_2} = \frac{56.05}{25} + \frac{56.05}{20} = 5.045$$

so that

$$t = \frac{35 - 40}{\sqrt{5.045}} = \frac{-5}{2.2} = -2.3$$

The family of hypotheses implied by this example is

$$H_0\colon \quad \mu_1 = \mu_2$$

$$H_1\colon \quad \mu_1 < \mu_2$$

In Table 2 in the Appendix Tables we locate $t_{0.01;\,43} \doteq 2.4$. Since the sample value of t is less than 2.4 in absolute value, we *do not* reject the null hypothesis of no statistical difference in the mean life of the two batches under test.

Confidence Limits. Confidence limits may be set for the difference between two means. To simplify notation, let $\Delta\mu = \mu_1 - \mu_2$ and $\Delta\bar{X} = \bar{X}_1 - \bar{X}_2$. Then $100(1 - \alpha)$ percent confidence limits for $\Delta\mu$ are derived from Eq. (3-19) and are

$$\Delta\bar{X} \pm s_{\bar{X}_1 - \bar{X}_2}(t_{\alpha/2;\,\nu}) \tag{3-20}$$

where $\nu = n_1 + n_2 - 2$.

If we wish 99 percent confidence limits for our example, they are, since $t_{0.005;\,43} \doteq 2.7$,

$$-5 \pm 2.2(2.7)$$

or

$$-10.94 \leq \Delta\mu \leq 0.94$$

Since zero is included in this confidence interval, we conclude that there is no statistical difference between the treatments at the stated level of confidence.

3.4 test that two vectors of means are equal

Let two vectors of means, $\bar{\mathbf{X}}_1$ and $\bar{\mathbf{X}}_2$, be drawn from two multivariate normal populations, $N_p(\boldsymbol{\mu}_1, \boldsymbol{\Sigma})$ and $N_p(\boldsymbol{\mu}_2, \boldsymbol{\Sigma})$. Notice carefully that both populations are assumed to have the same covariance matrix, $\boldsymbol{\Sigma}$. We wish to test

$$H_0: \boldsymbol{\mu}_1 = \boldsymbol{\mu}_2 \qquad \text{or equivalently} \qquad H_0: \Delta\boldsymbol{\mu} = 0$$
$$H_1: \boldsymbol{\mu}_1 \neq \boldsymbol{\mu}_2 \qquad\qquad\qquad\qquad H_1: \Delta\boldsymbol{\mu} \neq 0$$

That is, we wish to test whether all elements of the two vectors of means are simultaneously equal.

Again, we proceed by analogy to the univariate case. First, form the pooled covariance matrix

$$\mathbf{S}_* = \frac{1}{n_1 + n_2 - 2}[\mathbf{x}_1'\mathbf{x}_1 + \mathbf{x}_2'\mathbf{x}_2] \tag{3-21}$$

where $\mathbf{x}_1$ is the $n_1 \times p$ matrix of observations (in deviation form) drawn from $N_p(\boldsymbol{\mu}_1, \boldsymbol{\Sigma})$ and $\mathbf{x}_2$ is the $n_2 \times p$ matrix of observations (in deviation form) drawn from $N_p(\boldsymbol{\mu}_2, \boldsymbol{\Sigma})$. Notice that the number of observations need not be the same for both populations, but, of course, the number of variables must be the same. When there is a single variable from each population, $\mathbf{S}_*$ is the same as s_{**}.

If we square both sides of Eq. (3-19), we have

$$t^2 = \frac{n_1 n_2}{n_1 + n_2}(\bar{X}_1 - \bar{X}_2)(s_{**})^{-1}(\bar{X}_1 - \bar{X}_2) \tag{3-22}$$

and replacing scalars with vectors and matrices, we get

$$T^2 = \frac{n_1 n_2}{n_1 + n_2}(\bar{\mathbf{X}}_1 - \bar{\mathbf{X}}_2)'\mathbf{S}_*^{-1}(\bar{\mathbf{X}}_1 - \bar{\mathbf{X}}_2) \tag{3-23}$$

T^2 in the form given by Eq. (3-23) is related to F under the null hypothesis as

$$T^2_{\alpha;\, p,\, n_1+n_2-p-1} = \frac{(n_1 + n_2 - 2)p}{n_1 + n_2 - p - 1} F_{\alpha;\, p,\, n_1+n_2-p-1} \tag{3-24}$$

We return to our example which uses GRE scores. Verbal and quantitative GRE scores for 10 unsuccessful and 13 successful graduate students in economics are shown in Table 3.2. We wish to determine if the vector of mean scores for the two groups of students is significantly different.

Table 3.2 Verbal and Quantitative GRE Scores for Two Groups of Students

	Successful (Group 1)		*Unsuccessful (Group 2)*	
	Verbal	*Quantitative*	*Verbal*	*Quantitative*
	750	590	740	680
	360	600	670	600
	720	750	560	550
	540	710	540	520
	570	700	590	540
	520	670	590	700
	590	790	470	600
	670	700	560	540
	620	730	540	630
	690	840	500	600
	610	680	—	—
	550	730	—	—
	590	750	—	—
Mean	598.46	710.77	576.0	596.0

For the second group of students we already know from Sec. 3.2 that

$$\mathbf{x}_2'\mathbf{x}_2 = \begin{bmatrix} 56{,}240 & 17{,}240 \\ 17{,}240 & 33{,}240 \end{bmatrix}$$

For the first group of students

$$\mathbf{x}_1'\mathbf{x}_1 = \begin{bmatrix} 121{,}569 & 25{,}615 \\ 25{,}615 & 56{,}492 \end{bmatrix}$$

Therefore, from Eq. (3-21) the pooled covariance matrix is

$$\mathbf{S}_* = \begin{bmatrix} 8467 & 2041 \\ 2041 & 4273 \end{bmatrix}$$

and its inverse is

$$\mathbf{S}_*^{-1} = \begin{bmatrix} 0.0001335 & -0.0000637 \\ -0.0000637 & 0.0002645 \end{bmatrix}$$

For Eq. (3-23) we also need

$$\frac{n_1 n_2}{n_1 + n_2} = \frac{130}{23} = 5.652$$

and

$$\bar{\mathbf{X}}_1 - \bar{\mathbf{X}}_2 = \begin{bmatrix} 22.461 \\ 114.769 \end{bmatrix}$$

Thus

$$T^2 = (5.652)[22.461 \quad 114.769]\begin{bmatrix} 0.0001335 & -0.0000637 \\ -0.0000637 & 0.0002645 \end{bmatrix}\begin{bmatrix} 22.461 \\ 114.769 \end{bmatrix}$$
$$= 18.2$$

If we wish to test the null hypothesis at $\alpha = 0.01$, we find from Eq. (3-24) that the critical value of T^2 is

$$T^2_{0.01;\,2,20} = \frac{(10 + 13 - 2)(2)(5.85)}{(10 + 13 - 2 - 1)} = 12.3$$

since $F_{0.01;\,2,\,20} = 5.85$. Therefore we reject the null hypothesis and conclude that the mean vectors are different at the stated level of significance.

Confidence Region. A $100(1 - \alpha)$ percent confidence region may be constructed for $\Delta\boldsymbol{\mu} = \boldsymbol{\mu}_1 - \boldsymbol{\mu}_2$ by use of $\Delta\bar{\mathbf{X}} = \bar{\mathbf{X}}_1 - \bar{\mathbf{X}}_2$. The ellipsoid follows from Eqs. (3-23) and (3-24). The ellipsoid is thus bounded by

$$(\Delta\bar{\mathbf{X}} - \Delta\boldsymbol{\mu})'\mathbf{S}_*^{-1}(\Delta\bar{\mathbf{X}} - \Delta\boldsymbol{\mu}) = \frac{(n_1 + n_2)}{n_1 n_2}\frac{(n_1 + n_2 - 2)p}{n_1 + n_2 - p - 1}F_{\alpha;\,p,n_1+n_2-p-1} \tag{3-25}$$

at the stated level of confidence. We leave the calculation of this region for the example of this section as an exercise since in principle it is done by the way illustrated in Sec. 3.2.

Linear Combinations. Again, linear combinations $C'\Delta\boldsymbol{\mu}$ may be tested with $\mathbf{C}$ given as a $p \times q$ matrix of rank q. Write the analogy to Eq. (3-23) as

$$T^2 = \frac{n_1 n_2}{n_1 + n_2}(\mathbf{C}'\,\Delta\bar{\mathbf{X}})'(\mathbf{C}'\mathbf{S}_*\mathbf{C})^{-1}(\mathbf{C}'\,\Delta\bar{\mathbf{X}}) \tag{3-26}$$

The critical value of T^2 is given by

$$T^2_{\alpha;\,q,n_1+n_2-q-1} = \frac{(n_1 + n_2 - 2)q}{n_1 + n_2 - q - 1}F_{\alpha;\,q,n_1+n_2-q-1}$$

An analogy to Eq. (3-15a) for the ith column vector of $\mathbf{C}$ is

$$T^2 = \frac{n_1 n_2(\mathbf{c}_i'\,\Delta\bar{\mathbf{X}})^2}{(n_1 + n_2)\mathbf{c}_i'\mathbf{S}_*\mathbf{c}_i} \tag{3-26a}$$

with a critical value of $T^2_{\alpha;\,q,n_1+n_2-q-1}$.

Let us now inquire into the reasons for the rejection of the null hypothesis that the vectors of mean GRE scores for the successful and unsuccessful graduate students were equal. We first compare the verbal scores by setting $\mathbf{c}_1' = [1 \quad 0]$. In this case Eq. (3-26a) becomes

$$T^2 = \frac{(10)(13)(22.46)^2}{(10 + 13)(8467)} = 0.34$$

The critical value of T^2 with $\alpha = 0.01$ and $q = p = 2$ is 12.3. We conclude that the verbal scores are not causing the rejection of the simultaneous null hypothesis. Therefore the quantitative scores must be causing its rejection. To verify this contention, we set $\mathbf{c}_2' = [0 \quad 1]$, and from Eq. (3-26a) we find that $T^2 = 17.41$, which exceeds 12.3. We conclude, therefore, that at the stated level of significance the quantitative GRE scores are different but that the verbal scores are not. If GRE scores are a predictor of success in graduate education in economics at this particular institution, it is the quantitative scores that seem to be significant.[5]

Confidence Limits with Linear Combinations. $100(1 - \alpha)$ percent confidence limits for $\mathbf{C}' \, \Delta\boldsymbol{\mu}$ may be set by appealing to Eq. (3-26) and the critical value for T^2. Thus $\mathbf{C}' \, \Delta\boldsymbol{\mu}$ will be bounded by

$$(\mathbf{C}' \, \Delta\bar{\mathbf{X}} - \mathbf{C}' \, \Delta\boldsymbol{\mu})'(\mathbf{C}'\mathbf{S}_*\mathbf{C})^{-1}(\mathbf{C}' \, \Delta\bar{\mathbf{X}} - \mathbf{C}' \, \Delta\boldsymbol{\mu}) = \frac{n_1 + n_2}{n_1 n_2} \frac{(n_1 + n_2 - 2)q}{n_1 + n_2 - q - 1} F_{\alpha;\, q,\, n_1+n_2-q-1} \tag{3-27}$$

at the stated level of confidence. An analogy to Eq. (3-26a) can be obtained for setting confidence limits for $\mathbf{c}_i' \, \Delta\boldsymbol{\mu}$ by changing $\mathbf{C}$ to $\mathbf{c}_i$ in Eq. (3-27). The calculation for this example is left as an exercise.

3.5 equality of covariance matrices

In the last two sections a crucial assumption was that the covariance matrices of the two normal distributions representing the populations were equal. This assumption can, and should, be tested.

Two Univariate Normal Densities. In Sec. 3.3 we assumed that the two populations were $N_1(\mu_1, \sigma^2)$ and $N_1(\mu_2, \sigma^2)$. If we assume that the samples

[5] The data used in the analysis of GRE scores in this chapter represent actual scores attained by students at a U.S. university. A "successful" student was taken to be one who completed his qualifying examination within 5 years after entering the program. Fred Westfield has speculated that the verbal score may help predict dissertation success. We have not yet checked this speculation.

are drawn from independent populations, then the joint likelihood function is

$$L = \prod_{j=1}^{n_1} f(X_{1j}) \prod_{j=1}^{n_2} f(X_{2j}) \tag{3-28}$$

where the operator $\prod$ is defined as $\prod_{j=1}^{n} f(X_{ij}) = f(X_{i1})f(X_{i2}) \cdots f(X_{in})$ and

$$f(X_{ij}) = \frac{1}{\sqrt{2\pi\sigma_{ii}}}\exp\left[-\frac{1}{2}\frac{(X_{ij}-\mu_i)^2}{\sigma_{ii}}\right], \qquad i = 1, 2$$

is the normal density function. The variance of the ith population is σ_{ii}, and n_i is the size of the sample drawn from the ith population. In Chap. 2 we showed that the maximum of the likelihood function as given by Eq. (2-26) occurs when $\bar{X} = \sum X/n$ and $SD^2 = \sum (X - \bar{X})^2/n$ are substituted for μ and σ^2, respectively. Thus in Eq. (3-28) insert $\bar{X}_1$ and $\bar{X}_2$ for μ_1 and μ_2 and SD_{11} and SD_{22} for σ_{11} and σ_{22} [see Eq. (2-26) and replace $\sum (X - \mu)^2$ with nSD] to obtain

$$\max L = \frac{1}{(2\pi)^{(n_1+n_2)/2} SD_{11}^{n_1/2} SD_{22}^{n_2/2}} \exp\left[-\frac{1}{2}(n_1 + n_2)\right]$$

Moreover, if the null hypothesis H_0: $\sigma_{11} = \sigma_{22}$ is true, then the maximum likelihood estimator of μ_i is $\bar{X}_i$ and the maximum likelihood estimator of σ_{ii} is SD_{**}. Thus under the null hypothesis

$$\max L_0 = \frac{1}{(2\pi)^{(n_1+n_2)/2} SD_{**}^{n_1/2} SD_{**}^{n_2/2}} \exp\left[-\frac{1}{2}(n_1 + n_2)\right]$$

where

$$SD_{**} = [1/(n_1 + n_2)](\sum x_1^2 + \sum x_2^2).$$

The ratio

$$L^* = \frac{\max L_0}{\max L} = \frac{SD_{11}^{n_1/2} SD_{22}^{n_2/2}}{SD_{**}^{(n_1+n_2)/2}} \tag{3-29}$$

is called the *likelihood ratio*. The ratio tends to be smaller when the null hypothesis is false than when it is true. Thus it would seem natural to reject the null hypothesis when

$$L^* \leq L^*_\alpha \tag{3-30}$$

where L^*_α is defined so that Eq. (3-30) holds with probability α when the null hypothesis is true.

For large sample sizes, $-2 \ln L^*$ approaches, under very general conditions, the chi square distribution with 1 degree of freedom. Using this approximation, we shall *reject* the null hypothesis if

$$-2 \ln L^* > \chi^2_{\alpha;\,1}$$

In the two sample univariate cases an exact test of the null hypothesis can be performed by expanding the likelihood ratio

$$L^* = \frac{[(n_1 - 1)s_{11}]^{n_1/2}[(n_2 - 1)s_{22}]^{n_2/2}}{[(n_1 - 1)s_{11} + (n_2 - 1)s_{22}]^{(n_1+n_2)/2}} \frac{(n_1 + n_2)^{(n_1+n_2)/2}}{n_1{}^{n_1/2} n_2{}^{n_2/2}} \tag{3-31}$$

where $s_{ii} = [n/(n - 1)]SD_{ii}$ is the unbiased estimator of σ_{ii}. If we define $F = s_{11}/s_{22}$, then we may write Eq. (3-31) as

$$L^* = \frac{F^{n_1/2}}{[(n_1 - 1)F + (n_2 - 1)]^{(n_1+n_2)/2}} \frac{(n_1 + n_2)^{(n_1+n_2)/2}(n_1 - 1)^{n_1/2}(n_2 - 1)^{n_2/2}}{n_1^{n_1/2} n_2^{n_2/2}} \tag{3-31a}$$

Given the sample size, L^* in Eq. (3-31a) varies as F since all other elements in that equation are constants. Thus the test of the hypothesis

$$H_0\colon \ \sigma_{11} = \sigma_{22}$$

against

$$H_1\colon \ \sigma_{11} \neq \sigma_{22}$$

is conducted by use of the statistic

$$F = \frac{s_{11}}{s_{22}} = \frac{(n_2 - 1) \sum x_1^2}{(n_1 - 1) \sum x_2^2} \tag{3-32}$$

where $s_{11} \geq s_{22}$. For a given alpha, if this F statistic exceeds $F_{\alpha/2;\, n_1-1,\, n_2-1}$, the null hypothesis is rejected.

To return to the example of Sec. 3.3, we see that

$$F = \frac{(24)(1190)}{(19)(1220)} = 1.23$$

and using $\alpha = 0.02$, $F_{0.01;\,24,19} \doteq 2.92$. Thus we do not reject the null hypothesis. Notice again that the larger of the two variances is placed in the numerator. If the smaller variance is placed in the numerator, the critical value of F is $1/F_{\alpha/2;\, n_2-1,\, n_1-1}$.

Notice that Eq. (3-32) can be written as

$$F = \frac{(n_1 - 1)s_{11}/\sigma_{11}}{(n_2 - 1)s_{22}/\sigma_{22}} \frac{1/(n_1 - 1)}{1/(n_2 - 1)} \tag{3-32a}$$

under the null hypothesis ($\sigma_{11} = \sigma_{22}$). In general the quantity

$$\frac{(n_i - 1)s_{ii}}{\sigma_{ii}}, \qquad i = 1, 2$$

is distributed as χ_i^2 with $n_i - 1$ degrees of freedom. Thus the F ratio of Eq. (3-32) can be written as

$$F = \frac{\chi_1^2/(n_1 - 1)}{\chi_2^2/(n_2 - 1)}$$

and hence the F ratio is the ratio of two independent chi squares, each divided by the number of degrees of freedom upon which it is based.

Two Multivariate Normal Densities. Another way to expand Eq. (3-29) is

$$L^* = \frac{s_{11}^{n_1/2} s_{22}^{n_2/2}}{s_{**}^{(n_1+n_2)/2}} \frac{(n_1 - 1)^{n_1/2}(n_2 - 1)^{n_2/2}(n_1 + n_2)^{(n_1+n_2)/2}}{(n_1 + n_2 - 2)^{(n_1+n_2)/2} n_1^{n_1/2} n_2^{n_2/2}} \tag{3-33}$$

Given the sample sizes, all terms in Eq. (3-33) are constants except

$$V = \frac{s_{11}^{n_1/2} s_{22}^{n_2/2}}{s_{**}^{(n_1+n_2)/2}} \tag{3-34}$$

Since L^* varies as V, we can define the critical region as

$$V \leq V_\alpha$$

Bartlett (1937) has suggested that we replace n_1 and n_2 in Eq. (3-34) with degrees of freedom ν_1 and ν_2. Thus write

$$V_1 = \frac{s_{11}^{\nu_1/2} s_{22}^{\nu_2/2}}{s_{**}^{(\nu_1+\nu_2)/2}} \tag{3-35}$$

where $\nu_1 = n_1 - 1$ and $\nu_2 = n_2 - 1$. Now

$$\begin{aligned} -2 \ln V_1 &= (\nu_1 + \nu_2) \ln s_{**} - \nu_1 \ln s_{11} - \nu_2 \ln s_{22} \\ &= \left(\sum_{j=1}^{2} \nu_j\right) \ln s_{**} - \sum_{j=1}^{2} (\nu_j \ln s_{jj}) \end{aligned} \tag{3-36}$$

If we replace scalars with matrices in Eq. (3-36), we have for p variables in each population

$$-2 \ln V_1 = \left(\sum_{j=1}^{2} \nu_j\right) \ln |\mathbf{S}_*| - \sum_{j=1}^{2} (\nu_j \ln |\mathbf{S}_j|) \tag{3-37}$$

where $\mathbf{S}_*$ is the estimated pooled covariance matrix and the $\mathbf{S}_j s$ are the estimated covariance matrices for the two populations.

Box (1949) has shown that if we use the factor

$$b = 1 - \left(\sum_{j=1}^{2} \frac{1}{v_j} - \frac{1}{\sum_{j=1}^{2} v_j}\right)\left(\frac{2p^2 + 3p - 1}{6(p+1)}\right) \tag{3-38}$$

the statistic

$$W = b(-2 \ln V_1) \tag{3-39}$$

has an approximate chi square distribution with $p(p+1)/2$ degrees of freedom for large samples. Anderson[6] explains that the χ^2 approximation is useful if

$$\omega_2 = \frac{p(p+1)\left\{(p-1)(p+2)\left[\sum_{j=1}^{2} 1/v_j^2 - 1/\sum_{j=1}^{2} v_j\right] - 6(1-b)^2\right\}}{48b^2}$$

is small.

Let us now test the hypotheses that the two $\mathbf{\Sigma}$ matrices for the GRE score example of Sec. 3.4 are equal. For Eq. (3-37) we shall need the following determinants:

$$|\mathbf{S}_*| = (8467)(4273) - (2041)^2 = 32{,}013{,}810$$

$$\begin{aligned} |\mathbf{S}_1| &= \tfrac{1}{144}|\mathbf{x}_1'\mathbf{x}_1| = \tfrac{1}{144}[(121{,}569)(56{,}492) - (25{,}615)^2] \\ &= 43{,}135{,}748 \end{aligned}$$

$$\begin{aligned} |\mathbf{S}_2| &= \tfrac{1}{81}|\mathbf{x}_2'\mathbf{x}_2| = \tfrac{1}{81}[(56{,}240)(33{,}240) - (17{,}240)^2] \\ &= 19{,}409{,}877 \end{aligned}$$

Thus Eq. (3-37) becomes

$$\begin{aligned} -2 \ln V_1 &= 21[\ln (32{,}013{,}810)] - 12[\ln (43{,}135{,}748)] \\ &\quad - 9[\ln (19{,}409{,}877)] \\ &= 21(17.2817) - 12(17.5799) - 9(16.7813) = 0.93 \end{aligned}$$

Equation (3-38) is

$$\begin{aligned} b &= 1 - (\tfrac{1}{9} + \tfrac{1}{12} - \tfrac{1}{21})\left(\frac{2(4) + 3(2) - 1}{6(3)}\right) \\ &= 1 - (0.14684)(0.72222) \\ &= 0.8939 \end{aligned}$$

Thus from Eq. (3-39)

$$W = (0.8939)(0.93) = 0.8313$$

[6]Anderson (1958), p. 255. Much of our argument is based upon his Chap. 10. For a test with $p = 2$ which uses characteristic roots, see Pillai and Al-Ani (1970).

and in the chi square distribution at $\alpha = 0.01$ and $\nu = \frac{1}{2}(2)(3) = 3$, we find that $\chi^2_{0.01;\,3} = 11.345$. Since $W < \chi^2_{\alpha;\,\nu}$, we do not reject the null hypothesis and we consider that the covariance matrices are equal at the stated level of significance. Furthermore, in this example $\omega_2 = 0.0006$, which is small, and we can be confident of the χ^2 approximation.

If the null hypothesis is rejected, another test for the equality of the mean vectors may be used. This test is described in Anderson.[7]

Questions and Problems

Sec. 3.1

1. Cowden (1957), p. 9, presents data for the warp-breaking strength (pounds) of eight specimens of cotton cloth. Test (use $\alpha = 0.01$) the null hypothesis that the population mean does not differ from 65 pounds. Use a two-sided alternative. The eight observations are 57, 65, 65, 66, 67, 67, 68, and 71.
2. Verify, by inspection of the Appendix Tables, the relationships between t and z, t and F, χ^2 and z, and z and F which are given in Sec. 3.1.

Sec. 3.2

1. Extend the argument given in the text for μ_1 and μ_2 (which states that simultaneous significance is not necessarily the same as individual significance) to the case of μ_1, μ_2, and μ_3. When will the two kinds of significance be the same? Illustrate your remarks with a diagram.
2. If a constant is added to all the GRE scores of Table 3.1, will T^2 of Eq. (3-9) be changed?
3. A certain rod is to have a tensile strength, X_1, of 10 (thousand pounds per

Sample	X_1	X_2	X_3
1	8.8	150.1	300.1
2	10.2	150.2	300.7
3	9.7	149.8	300.3
4	10.0	149.6	297.9
5	8.5	151.1	299.8
6	6.4	150.8	299.9
7	7.5	150.2	293.1
8	9.0	149.9	294.5
9	9.5	149.2	300.2
10	9.0	150.0	300.0

[7]Anderson (1958), pp. 118ff.

square inch); a diameter, X_2, of 150 millimeters; and a length, X_3, of 300 millimeters. A sample of 10 such rods is given to you by a prospective vendor. Determine whether the vendor's product meets your specifications. Use $\alpha = 0.01$. Give a complete analysis which includes, perhaps, the use of linear combinations. Be sure to state your assumptions.

Sec. 3.3

1. Ohlin (1967), p. 40, presents figures for various countries regarding yields (100 kilograms per hectare) of maize and rice. Use $\alpha = 0.05$ to set for equality of average yield for maize and rice.

	Yield	
Country	*Maize*	*Rice*
Colombia	11.2	19.5
Israel	40.4	9.7
Argentina	17.7	33.6
Chile	20.7	26.9
Mexico	9.4	22.5
Taiwan	17.5	32.1
U.A.R.	24.0	52.3
Japan	25.9	50.5

Sec. 3.4

1. The following data on income of college teachers comes from the *AAUP Bulletin*, June 1971, p. 242. The figures give average salaries for four ranks in two types of institutions. Category I institutions are those which regularly offer the doctorate degree. Category IIA institutions offer degrees above the baccalaureate but are not included in category I. Use $\alpha = 0.05$, and test the

					Category I				
Professor	\$20,080	\$17,260	\$18,290	\$19,740	\$21,000	\$21,470	\$18,110	\$17,310	\$19,610
Associate	14,090	13,450	13,960	14,520	15,040	14,970	14,040	13,680	14,550
Assistant	11,520	11,380	11,610	11,820	12,030	11,900	11,600	13,360	11,960
Instructor	8,760	8,760	9,030	9,190	9,270	9,390	8,490	8,510	8,990
					Category IIA				
Professor	\$17,470	\$15,930	\$15,630	\$16,930	\$18,750	\$17,720	\$14,760	\$14,980	\$16,300
Associate	13,340	12,980	12,570	13,660	14,660	13,830	12,370	12,630	13,270
Assistant	10,950	10,740	10,710	11,360	12,120	11,460	10,550	10,530	11,010
Instructor	9,040	8,700	8,820	9,160	9,700	9,710	8,580	8,570	8,620

hypothesis that the mean salaries are equal between the two categories. Use linear combinations (if necessary) to complete the analysis. State your assumptions.

Sec. 3.5

1. Test the assumption of equality of covariance matrices for Problem 1 of Sec. 3.3 and Problem 1 of Sec. 3.4. Use $\alpha = 0.05$. Do the results of these tests modify your previous conclusions? Comment on the use of $\alpha = 0.01$ for your tests.

APPENDIX

A3.1 main for test that a vector of means is a given constant vector

A. Description. This main program calculates the actual T^2 of Eq. (3-9) and the critical T^2 of Eq. (3-11). It also prints the covariance matrix and its inverse.

B. Limitations. The maximum number of variables is 10 and the maximum number of observations on each variable is 100.

C. Use. On the first data card punch M, the number of variables, in columns 1–3 and punch N, the number of observations on each variable, in columns 4–6. Place these numbers as far to the right in their fields as possible and do not punch the decimal point. In columns 7–16 punch the critical value for F with the decimal point. For the example of Sec. 3.2 the first card would look as follows:

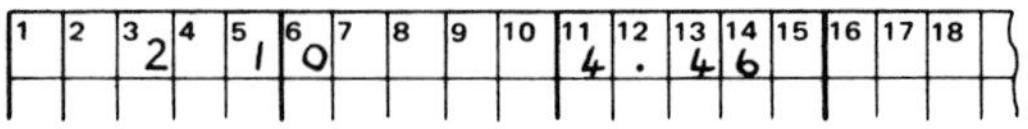

On the second card punch the hypothetical means with the decimal point in columns 1–10, 11–20, and so forth. Continue to a third card if $M > 8$. For the example of Sec. 3.2 the second card looks as follows:

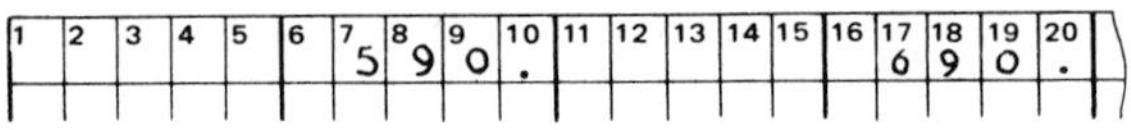

On the remaining data cards punch the sample observations. The format, as usual, is to start with the first variable; punch it with decimal point across the card in eight fields of width 10, continuing for as many cards as necessary. Start the second variable on a fresh card and repeat the procedure. The first variable in the deck should correspond to the first hypothetical mean on the second card, the second variable should correspond to the second hypothetical mean, and so forth. The main program and the output for the example of Sec. 3.2 are shown below.

```
ACTUAL MEANS
     .5760000E+03( 1)        .5960000E+03( 2)
HYPOTHETICAL MEANS
     .5900000E+03( 1)        .6900000E+03( 2)
COVARIANCE MATRIX
     .6248887E+04( 1 1)        .1915555E+04( 1 2)
     .1915555E+04( 2 1)        .3693333E+04( 2 2)
INVERSE OF COVARIANCE MATRIX
     .1902812E-03( 1 1)       -.9868966E-04( 1 2)
    -.9868966E-04( 2 1)        .3219435E-03( 2 2)
T-SQR.ACTUAL=    .2622235E+02     T-SQR.CRITICAL=    .1003500E+02
```

```
C       MEAN,COVAR, AND INVS NEEDED
        DIMENSION X(10,100),S(10,10),XBAR(10),XBD(10)
        READ(5,5) M,N,F
 5      FORMAT(2I3,F10.0)
        READ(5,15)(XBD(I),I=1,M)
        DO 10 I=1,M
10      READ(5,15)(X(I,J),J=1,N)
15      FORMAT(8F10.0)
        CALL MEAN(M,N,X,XBAR)
        WRITE(6,20)(XBAR(I),I,I=1,M)
20      FORMAT(' ACTUAL MEANS',/,(4(E17.7,'(',I2,')')))
        WRITE(6,30)(XBD(I),I,I=1,M)
30      FORMAT(' HYPOTHETICAL MEANS',/,(4(E17.7,'(',I2,')')))
        CALL COVAR(M,N,X,S,XBAR)
        DO 35 I=1,M
        DO 35 J=1,M
35      S(I,J)=S(I,J)/FLOAT(N-1)
        WRITE(6,40)
40      FORMAT(' COVARIANCE MATRIX')
        DO 45 I=1,M
45      WRITE(6,50)(S(I,J),I,J,J=1,M)
50      FORMAT(4(E17.7,'(',2I2,')'))
        CALL INVS(S,M)
        WRITE(6,55)
55      FORMAT(' INVERSE OF COVARIANCE MATRIX')
        DO 60 I=1,M
        XBD(I)=XBAR(I)-XBD(I)
60      WRITE(6,50)(S(I,J),I,J,J=1,M)
        DO 80 I=1,M
        XBAR(I)=0.0
        DO 80 J=1,M
80      XBAR(I)=XBAR(I)+XBD(J)*S(J,I)
        T=0.0
        DO 90 I=1,M
90      T=T+XBD(I)*XBAR(I)
        T=T*FLOAT(N)
        TC=F*FLOAT(M)*FLOAT(N-1)/FLOAT(N-M)
        WRITE(6,100) T,TC
100     FORMAT(' T-SQR.ACTUAL=',E15.7,4X,'T-SQR.CRITICAL=',E15.7)
        CALL EXIT
        END
```

A3.2 main for test that two vectors of means are equal

A. Description. This main program calculates the actual T^2 of Eq. (3-23) and the critical T^2 of Eq. (3-24). The means for both sets of variables and the pooled covariance matrix and its inverse are also printed.

B. Limitations. Each of the two sets of variables may have a maximum of 10 variables with a maximum of 100 observations on each variable.

C. Use. On card 1 punch the following without the decimal point and as far to the right in the field as possible:

Column	
1–3	M (number of variables in each set)
4–6	N_1 (number of observations on the variables in the first set)
7–9	N_2 (number of observations on the variables in the second set)

Finally, in columns 10–19 punch the critical value of F with the decimal point.

The remaining data cards contain the sample values. The first k cards of this group contain the observations of the first variable of the first set, punched with the decimal point across the card in eight fields of width 10. The next k cards contain the observations on the second variable of the first set punched in the same way as the first variable. After all variables in the first set are assembled, the observations on the variables in the second set are assembled in the same way. The entire data deck for the data of Sec. 3.4 appears as follows. Notice that the data for the second group were entered first. Thus the order of entry of the two groups does not matter.

1	2	3	4	5	6	7	8	9	10	11	12	13	14	15	16	17	18	19	20	21	22	23	24	25	26	27	28	29	30
		2		1	0		1	3			5	.	8	5															
						7	4	0	.							6	7	0	.							5	6	0	.
						5	4	0	.							5	0	0	.										
						6	8	0	.							6	0	0	.							5	5	0	.
						6	3	0	.							6	0	0	.										
						7	5	0	.							3	6	0	.							7	2	0	.
						6	2	0	.							6	9	0	.							6	1	0	.
						5	9	0	.							6	0	0	.							7	5	0	.
						7	3	0	.							8	4	0	.							6	8	0	.

The main program together with the output for the illustration of Sec. 3.4 are shown below.

```
MEANS FOR FIRST SET
     .5760000E+03( 1)        .5960000E+03( 2)
MEANS FOR SECOND SET
     .5984614E+03( 1)        .7107690E+03( 2)
POOLED COVARIANCE MATRIX
     .8467086E+04( 1 1)        .2040732E+04( 1 2)
     .2040732E+04( 2 1)        .4272961E+04( 2 2)
INVERSE OF POOLED COVARIANCE MATRIX
     .1334677E-03( 1 1)       -.6374305E-04( 1 2)
    -.6374305E-04( 2 1)        .2644728E-03( 2 2)
T-SQR.ACTUAL=    .1821304E+02     T-SQR.CRITICAL=    .1228500E+02
```

```
C       MEAN,COVAR, AND INVS NEEDED
        DIMENSION X(10,100),X1(10,100),A(10,10),S(10,10),XBAR(10),
       1XBD(10)
        READ(5,5) M,N1,N2,F
 5      FORMAT(3I3,F10.0)
        DO 10 I=1,M
 10     READ(5,15)(X(I,J),J=1,N1)
 15     FORMAT(8F10.0)
        DO 20 I=1,M
 20     READ(5,15)(X1(I,J),J=1,N2)
        CALL MEAN(M,N1,X,XBAR)
        WRITE(6,25)(XBAR(I),I,I=1,M)
 25     FORMAT(' MEANS FOR FIRST SET',/,(4(E17.7,'(',I2,')')))
        CALL COVAR(M,N1,X,S,XBAR)
        DO 40 I=1,M
        XBD(I)=XBAR(I)
        DO 40 J=1,M
 40     A(I,J)=S(I,J)
        CALL MEAN(M,N2,X1,XBAR)
        WRITE(6,45)(XBAR(I),I,I=1,M)
 45     FORMAT(' MEANS FOR SECOND SET',/,(4(E17.7,'(',I2,')')))
        CALL COVAR(M,N2,X1,S,XBAR)
        DO 50 I=1,M
        XBD(I)=XBD(I)-XBAR(I)
        DO 50 J=1,M
```

```
50    A(I,J)=(A(I,J)+S(I,J))/FLOAT(N1+N2-2)
      WRITE (6,55)
55    FORMAT(' POOLED COVARIANCE MATRIX')
      DO 60 I=1,M
60    WRITE(6,65)(A(I,J),I,J,J=1,M)
65    FORMAT(4(E17.7,'(',2I2,')'))
      CALL INVS(A,M)
      WRITE(6,70)
70    FORMAT(' INVERSE OF POOLED COVARIANCE MATRIX')
      DO 75 I=1,M
75    WRITE(6,65)(A(I,J),I,J,J=1,M)
      DO 80 I=1,M
      XBAR(I)=0.0
      DO 80 J=1,M
80    XBAR(I)=XBAR(I)+XBD(J)*A(J,I)
      T=0.0
      DO 90 I=1,M
90    T=T+ XBD(I)*XBAR(I)
      T=T*FLOAT(N1*N2)/FLOAT(N1+N2)
      TC=F*FLOAT(N1+N2-2)*FLOAT(M)/FLOAT(N1+N2-M-1)
100   FORMAT(' T-SQR,ACTUAL=',E15.7,4X,'T-SQR,CRITICAL=',E15.7)
      WRITE(6,100)T,TC
      CALL EXIT
      END
```

chapter 4

Linear Regression

By use of the methods of the previous chapter we were able to reject the null hypothesis that two means are equal. In so doing we were able to establish an interesting statistical finding which arose upon classifying the observations, but we were not able to establish an explicit relationship of the means to some other variable. Economists are usually interested in explicit relationships which explain why means are statistically different, and as a result they often build models to explain such differences. A very useful technique in this regard is that of linear regression.[1]

4.1 a consumption pattern

Consider a hypothetical consumption pattern for U.S. households. In Table 4.1, m sample observations on per capita consumption, Y, drawn from each of n per capita income brackets are represented. Also shown are the conditional sample means, $\bar{Y}_i$, and the conditional population means, μ_i (the means are conditional upon a given value of X_i). Economic theory and

[1]The term regression was coined by Sir Francis Galton to explain a biological phenomenon. Current practice is to use the term to refer to the fitting of curves to sample observations.

intuition tell us that per capita consumption should depend positively on per capita income. Therefore as income (measured by the midvalues of the income class intervals, X_i) rises, so should the conditional mean values for consumption, μ_i. Thus we reason that μ_i depends in a positive way on X_i.

Table 4.1 Hypothetical Per Capita Income and Consumption Figures

Per Capita Income (Class Midvalue)	*Per Capita Consumption*		
	Population Mean	*Sample Mean*	*Observations*
X_1	μ_1	$\bar{Y}_1$	$Y_{11} \cdots Y_{1j} \cdots Y_{1m}$
$\vdots$	$\vdots$	$\vdots$	$\vdots \quad \vdots \quad \vdots$
X_i	μ_i	$\bar{Y}_i$	$Y_{i1} \cdots Y_{ij} \cdots Y_{im}$
$\vdots$	$\vdots$	$\vdots$	$\vdots \quad \vdots \quad \vdots$
X_n	μ_n	$\bar{Y}_n$	$Y_{n1} \cdots Y_{nj} \cdots Y_{nm}$

The hypothetical observations of Table 4.1 are plotted in Figure 4.1. Each solid point represents one household. The *independent* variable, income, is plotted on the horizontal axis, and the *dependent* variable, consumption, is plotted on the vertical axis. A plot of the sample observations is often called a *scatter diagram*. Finally, the circles in Figure 4.1 represent the conditional sample means, $\bar{Y}_i$, and the crosses represent the conditional population means, μ_i.

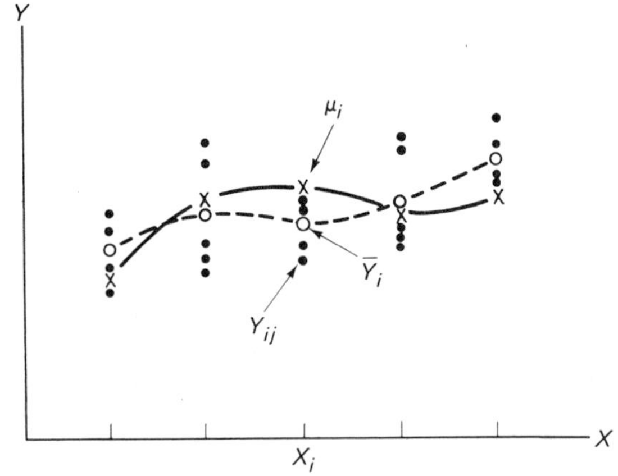

Figure 4.1. Hypothetical sample observations, means, and regression relationships.

Suppose that the specification that μ_i depends in some functional way on X_i, i.e.,

$$\mu_i = f(X_i), \qquad i = 1, 2, \ldots, n \tag{4-1}$$

is given by the solid curve in Figure 4.1. We call this curve a *population regression function.* Since each conditional sample mean, $\bar{Y}_i$, is an unbiased estimate of its corresponding conditional population mean, μ_i, the dashed curve in Figure 4.1, which passes through each $\bar{Y}_i$, is *one* reasonable estimate of the population regression function. Call this curve a *sample* or *estimated regression function.*

This illustration indicates the general nature of the regression problem, which is to establish a functional relationship between a single dependent variable and one or more independent variables. While the general nature of the analysis is illustrated, the problem as stated is too vague to be of much practical significance. The sample regression function may be satisfactory to some persons but unsatisfactory to others. Also, the specific nature of the relationship is difficult to establish from a graphic analysis of the scatter diagram. Therefore, to make the problem concrete, we need to make some simplifying assumptions.

4.2 a linear regression model

Suppose that Eq. (4-1) is linear. That is, let it have the form

$$\mu_i = \beta_0 + \beta_1 X_i, \qquad i = 1, 2, \ldots, n \tag{4-2}$$

where β_0 and β_1 are unknown constants. This special case of Eq. (4-1) is one of the most simple functional relationships in mathematics. The general nature of the regression problem is not changed by this assumption, but we have assumed that the μ_is lie along a straight line such as that shown in Figure 4.2.

As before, for any given per capita income class midvalue, X_i, there is a distribution of per capita consumption, Y_{ij}. As shown in Chap. 2, if the samples are random,

$$E(Y_{ij}) = \mu_i \tag{4-3}$$

for $i = 1, 2, \ldots, n$. Assume that for any given X_i the distribution of Y_{ij} has the constant variance

$$\text{var}(Y_{ij}) = \sigma_i^2 = \sigma^2 \tag{4-4}$$

Then, for any X_i

$$\text{var}(\bar{Y}_i) = \frac{\sigma^2}{m} \tag{4-5}$$

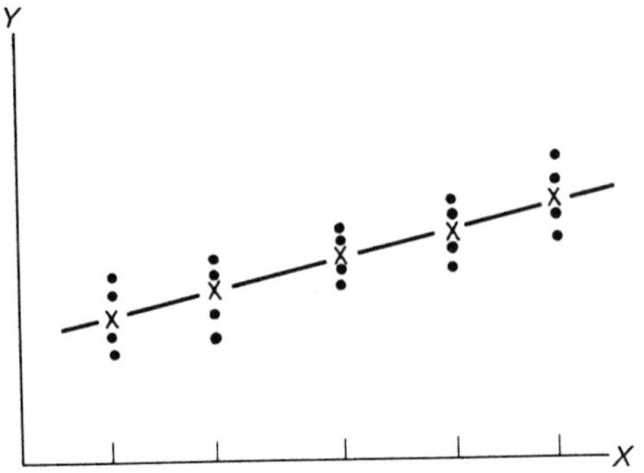

Figure 4.2. Linear regression relationships.

This assumption that the variance of each Y_{ij} distribution is constant for all X_is is known as the assumption of *homoscedasticity*. We have already made this assumption in various places. For example, in Sec. 2.5 we assumed that there were several populations with equal variances when we discussed maximum likelihood estimation. Also, in Sec. 3.3 we assumed two populations with equal variances when we tested the equality of two means.

Since $\bar{Y}_i$ is an unbiased estimate of μ_i, then the *error*

$$\epsilon_i = \bar{Y}_i - \mu_i \tag{4-6}$$

is a random variable with expected value

$$\begin{aligned} E(\epsilon_i) &= E(\bar{Y}_i) - E(\mu_i) \\ &= \mu_i - \mu_i \\ &= 0 \end{aligned} \tag{4-7}$$

Furthermore, the variance of ϵ_i is

$$\begin{aligned} \text{var}(\epsilon_i) &= \text{var}(\bar{Y}_i) - \text{var}(\mu_i) \\ &= \frac{\sigma^2}{m} \end{aligned} \tag{4-8}$$

which follows from Eq. (4-5) and the fact that μ_i is a constant for any given value of i and thus has zero variance.

Sometimes we refer to ϵ_i as the *disturbance* of the regression equation. Thus from Eqs. (4-2) and (4-6) we may write

$$\bar{Y}_i = \beta_0 + \beta_1 X_i + \epsilon_i \tag{4-9}$$

We also notice that since ϵ_i and $\bar{Y}_i$ differ only by a constant, μ_i, their distributions will be identical except for their means. In addition, we assume that the X_i are statistically constant (nonstochastic) and thus ϵ_i and X_i are statistically independent (have zero covariance) for all i:

$$\text{cov}(\epsilon_i, X_i) = E(\epsilon_i X_i) = 0$$

In most practical work in economics and business, either because the data are not available or because they are too expensive to collect, we generally have only one observation for the dependent variable corresponding to each value of the independent variable. That is, $m = 1$, and Y_{i1}, or simply Y_i, is the only observation corresponding to X_i. Nevertheless, if we assume that this Y_i value is drawn at random from the population Y_{ij}, then Y_i is still an unbiased estimator of μ_i. This fact follows because we may think of the single Y_i as $\bar{Y}_i$ with $m = 1$. In this case Eq. (4-9) becomes

$$Y_i = \beta_0 + \beta_1 X_i + \epsilon_i \tag{4-9a}$$

and it is still true under our existing assumptions that for all i

$$E(Y_i) = \mu_i = \beta_0 + \beta_1 X_i$$
$$E(\epsilon_i) = 0$$
$$\text{var}(Y_i) = \text{var}(\epsilon_i) = \sigma^2$$

A further assumption that is often made is that the Y_i, $i = 1, 2, \ldots, n$, are independently distributed. That is,

$$\text{cov}(Y_i, Y_{i'}) = E(\epsilon_i \epsilon_{i'}) = 0 \qquad \text{for } i \neq i'$$

In words, this assumption states that if, say, $i = 1$ and Y_1 is drawn from the population corresponding to $i = 1$, then the fact that this specific observation was drawn will have no effect on the value of $Y_{i'}$ drawn from the population associated with, say, the value $i' = 2$.

The linear regression model of Eq. (4-9a) can be extended to include K independent variables:

$$Y_i = \beta_0 + \beta_1 X_{1i} + \beta_2 X_{2i} + \cdots + \beta_K X_{Ki} + \epsilon_i \tag{4-10}$$

If $K = 1$, we have Eq. (4-9a), and the regression relation is a straight line in two-dimensional space. If $K > 1$, then the regression function is a hyperplane in $(K + 1)$-dimensional space. Extending our basic assumptions to include K independent variables, we have

$$E(Y_i) = \mu_i = \beta_0 + \beta_1 X_{1i} + \cdots + \beta_K X_{Ki} \tag{4-11}$$

which defines the population mean of each Y_i. Furthermore, the variance of

Y_i is assumed constant for all i and the Y_is are assumed to be distributed independently. Both of these assumptions may be expressed compactly as

$$\begin{aligned} \operatorname{cov}(Y_i, Y_{i'}) &= 0 \qquad \text{for } i \neq i' \\ &= \sigma^2 \qquad \text{for } i = i' \end{aligned} \tag{4-12}$$

Alternatively, we may assume that the error, ϵ_i, has mean

$$E(\epsilon_i) = 0 \tag{4-13}$$

has constant variance for all i, and is distributed independently for $i \neq i'$:

$$\begin{aligned} E(\epsilon_i \epsilon_{i'}) &= 0 \qquad \text{for } i \neq i' \\ &= \sigma^2 \qquad \text{for } i = i' \end{aligned} \tag{4-14}$$

Last, the distribution of ϵ_i is independent of X_{ki}:

$$E(X_{ki}\epsilon_i) = 0 \qquad \text{for all } i = 1, 2, \ldots, n \quad \text{and } k = 1, 2, \ldots, K$$

4.3 estimation methods

In this section we shall discuss two methods of estimating the regression parameters β_k of Eq. (4-10) as well as the common variance, σ^2. Following the discussion of Chap. 2, we shall treat the method of least-squares and the method of maximum likelihood.

Least-Squares Estimation. In Chap. 2 we used the method of least-squares to estimate a single constant population mean, μ. For simplicity let us now relate this estimation to regression where there is but a single independent variable, X_i. Later in the section we shall extend the discussion to K independent variables.

For a given value of X_i, the population mean of Y_i depends on the parameters β_0 and β_1 [see Eq. (4.2)]. Thus we need to estimate these two parameters in order to estimate μ_i. The estimate will then be

$$\hat{\mu}_i = \hat{\beta}_0 + \hat{\beta}_1 X_i \tag{4-15}$$

where the caret over a quantity indicates that it is an estimated quantity.

The method of least-squares gives estimates of β_0 and β_1 such that the sum of the squared sample errors (or *residuals*), e_i, where $e_i = Y_i - \hat{\mu}_i$, is a minimum. That is, we seek to minimize

$$\sum_{i=1}^{n} e_i^2 = \sum_{i=1}^{n} (Y_i - \hat{\mu}_i)^2 \tag{4-16}$$

In Figure 4.3 a vertical line is drawn for a particular value of X_i. Also shown is a particular value of the dependent variable, Y_i. The solid line is the population regression line and the dashed line is the sample regression line. The vertical distance between Y_i and $\hat{\mu}_i$ is the error e_i since the sample regression line is being used to estimate $\hat{\mu}_i$. Thus we often write

$$\hat{Y}_i = \hat{\mu}_i = \hat{\beta}_0 + \hat{\beta}_1 X_i$$

while keeping in mind that $\hat{\mu}_i$ is not to be confused with $\bar{Y}_i$, the conditional sample mean, or $\bar{Y}$, the *unconditional* sample mean.

To minimize the sum of the squared sample residuals, we note that

$$\sum_{i=1}^{n} e_i^2 = \sum_{i=1}^{n} (Y_i - \hat{\beta}_0 - \hat{\beta}_1 X_i)^2$$

and we set the first partial derivatives with respect to $\hat{\beta}_0$ and $\hat{\beta}_1$ equal to zero,

$$\frac{\partial \sum e^2}{\partial \hat{\beta}_0} = -2 \sum (Y - \hat{\beta}_0 - \hat{\beta}_1 X) = 0$$

$$\frac{\partial \sum e^2}{\partial \hat{\beta}_1} = -2 \sum X(Y - \hat{\beta}_0 - \hat{\beta}_1 X) = 0$$

where the summation extends from 1 to n. Upon simplification we have

$$\begin{aligned} \sum Y &= n\hat{\beta}_0 + \hat{\beta}_1 \sum X \\ \sum XY &= \hat{\beta}_0 \sum X + \hat{\beta}_1 \sum X^2 \end{aligned} \tag{4-17}$$

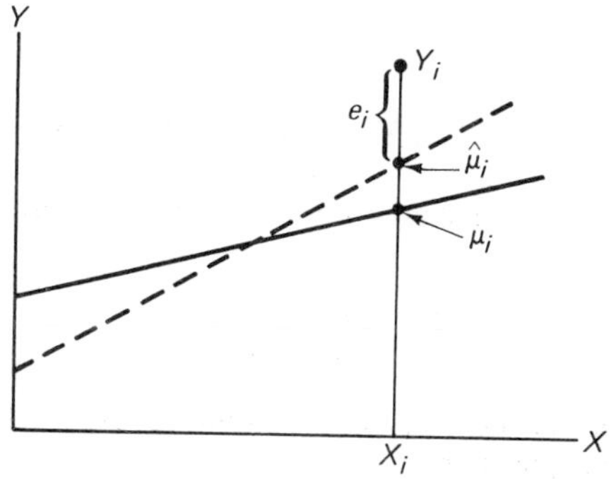

Figure 4.3. Population and sample regression lines.

When solved simultaneously for $\hat{\beta}_0$ and $\hat{\beta}_1$ these two equations (often called *normal equations*)[2] will give the estimates that we seek.

Write the two equations in matrix notation:

$$\begin{bmatrix} n & \sum X \\ \sum X & \sum X^2 \end{bmatrix} \begin{bmatrix} \hat{\beta}_0 \\ \hat{\beta}_1 \end{bmatrix} = \begin{bmatrix} \sum Y \\ \sum XY \end{bmatrix} \tag{4-17a}$$

Then

$$\begin{bmatrix} \hat{\beta}_0 \\ \hat{\beta}_1 \end{bmatrix} = \begin{bmatrix} n & \sum X \\ \sum X & \sum X^2 \end{bmatrix}^{-1} \begin{bmatrix} \sum Y \\ \sum XY \end{bmatrix} \tag{4-18}$$

is one possible computing method. However, computing can be simplified in the case of one independent variable by manipulation of Eq. (4-18). Clearly (see footnote 11 in Chap. 1),

$$\begin{bmatrix} n & \sum X \\ \sum X & \sum X^2 \end{bmatrix}^{-1} = \frac{1}{n \sum X^2 - (\sum X)^2} \begin{bmatrix} \sum X^2 & -\sum X \\ -\sum X & n \end{bmatrix}$$

so that from Eq. (4-18)

$$\begin{aligned} \hat{\beta}_0 &= \frac{\sum Y \sum X^2 - \sum X \sum XY}{n \sum X^2 - (\sum X)^2} \\ \hat{\beta}_1 &= \frac{n \sum XY - \sum X \sum Y}{n \sum X^2 - (\sum X)^2} \end{aligned} \tag{4-19}$$

Recall further that[3]

$$\begin{aligned} \sum x^2 &= \sum (X - \bar{X})^2 \\ &= \sum X^2 - \frac{(\sum X)^2}{n} \end{aligned} \tag{4-20}$$

and that[4]

$$\sum xy = \sum (X - \bar{X})(Y - \bar{Y}) = \sum XY - \sum X \sum Y/n \tag{4-21}$$

Then it follows by direct substitution into Eq. (4-19) that

$$\hat{\beta}_1 = \frac{\sum xy}{\sum x^2} \tag{4-22}$$

---

[2]The term normal does not imply any connection with the normal distribution but rather refers to its geometric interpretation.

[3]$\sum (X - \bar{X})^2 = \sum X^2 - 2\bar{X} \sum X + n(\bar{X})^2 = \sum X^2 - (\sum X)^2/n$.

[4]$\sum (X - \bar{X})(Y - \bar{Y}) = \sum XY - \bar{Y} \sum X - \bar{X} \sum Y + n\bar{X}\bar{Y} = \sum XY - \sum X \sum Y/n$.

and

$$\hat{\beta}_0 = \bar{Y} - \hat{\beta}_1 \bar{X} \tag{4-23}$$

where $\bar{Y} = \sum Y/n$ and $\bar{X} = \sum X/n$ are the unconditional sample means.

In Eq. (4-22) it is clear that $\hat{\beta}_1$ takes the same sign as $\sum xy$, the sample covariation between X and Y. Furthermore, as we shall stress again, in Eq. (4-23) if $\hat{\beta}_1$ is zero, then $\hat{\beta}_0 = \bar{Y}$.

By definition the quantities $\sum x^2$ and $\sum xy$ may be computed *directly* by use of the first form of Eqs. (4-20) and (4-21) or *indirectly* by use of the second form of these equations. For computation on a desk calculator the indirect method is easier, but for calculation on a digital computer the direct method is more accurate and is preferred.

Table 4.2 shows 10 observations on per capita consumption and per capita income. Since these observations run across time, the series are referred to as *time series*. Thus the data are of a different type than the hypothetical *static* or *cross-section* data of Table 4.1. The main problem with economic time series is that the Y_i observations are usually not independent. That is, if we sample a value of Y_i which is above $\bar{Y}$, the odds are great that the next Y_i value sampled will be above $\bar{Y}$. The Y_is are said to be *autocorrelated* (correlated with themselves). We shall have more to say about problems in regression created by temporal data, but for the moment let us use these observations to illustrate calculation.

Table 4.2 Per Capita Disposable Income and Per Capita Consumption, 1958 Prices

Year	*Consumption, Y*	*Income, X*
1958	1666	1831
1959	1735	1881
1960	1749	1883
1961	1755	1909
1962	1813	1968
1963	1865	2013
1964	1945	2123
1965	2044	2235
1966	2122	2332
1967	2162	2401
	$\sum Y = 18{,}856$	$\sum X = 20{,}576$ $\sum X^2 = 42{,}711{,}264$
	$\sum XY = 39{,}116{,}913$	

Source: *Economic Report of the President* (Washington, D.C.: Government Printing Office), Jan. 1969, p. 245.

Using the calculations reported in Table 4.2, we find that

$$\sum x^2 = \sum X^2 - \frac{(\sum X)^2}{n}$$

$$= 42{,}711{,}264 - \frac{(20{,}576)^2}{10} = 374{,}086.4$$

$$\sum xy = \sum XY - \frac{(\sum X)(\sum Y)}{n}$$

$$= 39{,}116{,}913 - \frac{(20{,}576)(18{,}856)}{10} = 318{,}807.4$$

Thus

$$\hat{\beta}_1 = \frac{\sum xy}{\sum x^2} = \frac{318{,}807.4}{374{,}086.4} = 0.8522$$

and

$$\hat{\beta}_0 = \bar{Y} - \hat{\beta}_1\bar{X} = 1885.6 - (0.8522)(2057.6) = 132.1$$

so that the estimated regression line is

$$\hat{Y}_i = 132.1 + 0.8522X_i$$

The scatter of points and the sample regression line are shown in Figure 4.4.

The fact that linear regression is closely related to the estimation of means should be clear by now. Two additional observations should serve to clarify further this close relationship. First, the sample linear regression line must pass through the unconditional sample means, $\bar{X}$ and $\bar{Y}$, as indicated in Figure 4.4 by the horizontal line at $\bar{Y}$ and the vertical line at $\bar{X}$. Thus when $X_i = \bar{X}$,

$$\hat{Y}_i = \hat{\beta}_0 + \hat{\beta}_1\bar{X}$$

but by Eq. (4-23)

$$\hat{Y}_i = \bar{Y} - \hat{\beta}_1\bar{X} + \hat{\beta}_1\bar{X} = \bar{Y}$$

Second, if $\hat{\beta}_1 = 0$ (i.e., the sample regression line is flat), then for all X_is

$$\hat{Y}_i = \hat{\beta}_0 = \bar{Y}$$

Therefore in the special case where the slope of the sample regression line is zero, the sample regression line is the unconditional sample mean of Y. This is precisely the estimate of the population mean that the method of least-squares gave us in Chap. 2. In general, we can say that in regression we attempt to use information given by one or more independent variables to aid us in the estimation of the conditional means of the dependent variable. If none of the independent variables are of use to us in this attempt, then the slopes associated with these useless independent variables will be zero, and

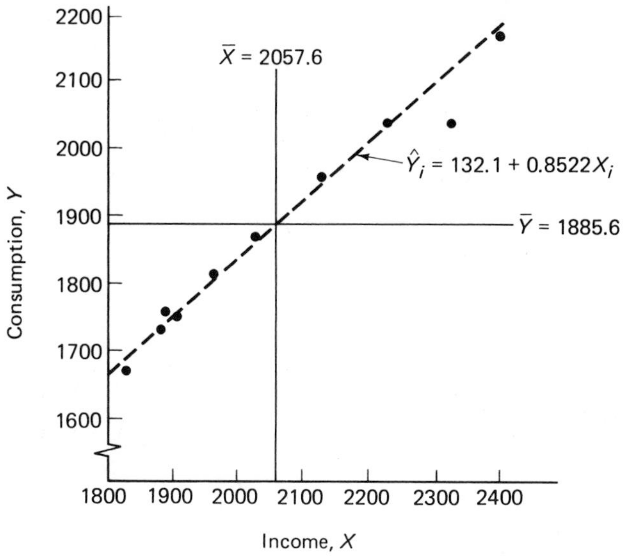

Figure 4.4. Scatter diagram of data in Table 4.2; $\hat{Y}_i = 132.1 + 0.8522X_i$.

the sample regression line takes on the value of the unconditional sample mean value of the dependent variable.

Let us now write the linear regression model with one independent variable in a more general way which uses matrix algebra. Define the following matrices:

$$\mathbf{Y}' = [Y_1 \quad Y_2 \quad \cdots \quad Y_n]$$

$$\mathbf{X} = \begin{bmatrix} 1 & X_1 \\ 1 & X_2 \\ \cdot & \cdot \\ \cdot & \cdot \\ \cdot & \cdot \\ 1 & X_n \end{bmatrix}$$

For the data of Table 4.2 these two matrices become

$$\mathbf{Y}' = [1666 \quad 1735 \quad \cdots \quad 2162]$$

$$\mathbf{X} = \begin{bmatrix} 1 & 1831 \\ 1 & 1881 \\ \cdot & \cdot \\ \cdot & \cdot \\ \cdot & \cdot \\ 1 & 2401 \end{bmatrix}$$

Clearly,

$$\mathbf{X'Y} = \begin{bmatrix} 1 & 1 & \cdots & 1 \\ X_1 & X_2 & \cdots & X_n \end{bmatrix} \begin{bmatrix} Y_1 \\ Y_2 \\ \cdot \\ \cdot \\ \cdot \\ Y_n \end{bmatrix} = \begin{bmatrix} \sum Y \\ \sum XY \end{bmatrix} \tag{4-24}$$

and

$$\mathbf{X'X} = \begin{bmatrix} 1 & 1 & \cdots & 1 \\ X_1 & X_2 & \cdots & X_n \end{bmatrix} \begin{bmatrix} 1 & X_1 \\ 1 & X_2 \\ \cdot & \cdot \\ \cdot & \cdot \\ \cdot & \cdot \\ 1 & X_n \end{bmatrix} = \begin{bmatrix} n & \sum X \\ \sum X & \sum X^2 \end{bmatrix} \tag{4-25}$$

If we define the vector

$$\hat{\boldsymbol{\beta}}' = [\hat{\beta}_0 \quad \hat{\beta}_1]$$

then the normal equations of Eq. (4-17) can be expressed as

$$(\mathbf{X'X})\hat{\boldsymbol{\beta}} = \mathbf{X'Y} \tag{4-26}$$

and their solution is

$$\hat{\boldsymbol{\beta}} = (\mathbf{X'X})^{-1}\mathbf{X'Y} \tag{4-27}$$

Furthermore, the sample regression equation can be written as

$$\mathbf{Y} = \mathbf{X}\hat{\boldsymbol{\beta}} + \mathbf{e} \tag{4-28}$$

where

$$\mathbf{e}' = [e_1 \quad e_2 \quad \cdots \quad e_n] = (\mathbf{Y} - \mathbf{X}\hat{\boldsymbol{\beta}})'$$

or, alternatively, as

$$\hat{\mathbf{Y}} = \mathbf{X}\hat{\boldsymbol{\beta}} \tag{4-29}$$

where

$$\hat{\mathbf{Y}}' = [\hat{Y}_1 \quad \hat{Y}_2 \quad \cdots \quad \hat{Y}_n] \tag{4-30}$$

In the case where $K > 1$ independent variables are involved, the definitions of $\mathbf{Y}$, $\hat{\mathbf{Y}}$, and $\mathbf{e}$ remain unchanged, but the definitions of $\hat{\boldsymbol{\beta}}$ and $\mathbf{X}$ become

$$\hat{\boldsymbol{\beta}}' = [\hat{\beta}_0 \quad \hat{\beta}_1 \quad \cdots \quad \hat{\beta}_K]$$

$$\mathbf{X} = \begin{bmatrix} 1 & X_{11} & \cdots & X_{K1} \\ 1 & X_{12} & \cdots & X_{K2} \\ \cdot & & & \cdot \\ \cdot & & & \cdot \\ \cdot & & & \cdot \\ 1 & X_{1n} & \cdots & X_{Kn} \end{bmatrix} \tag{4-31}$$

The method of least-squares still minimizes the sum of the squared sample errors. In general we minimize

$$\begin{aligned} \sum e^2 &= \mathbf{e}'\mathbf{e} \\ &= (\mathbf{Y} - \mathbf{X}\hat{\boldsymbol{\beta}})'(\mathbf{Y} - \mathbf{X}\hat{\boldsymbol{\beta}}) \end{aligned} \tag{4-32}$$

Equation (4-32) is a quadratic form. To minimize $\sum e^2$, we shall need to equate the first partial derivatives with respect to $\hat{\boldsymbol{\beta}}$ to zero. Expanding Eq. (4-32), we have

$$\sum e^2 = \mathbf{Y}'\mathbf{Y} - 2\hat{\boldsymbol{\beta}}'\mathbf{X}'\mathbf{Y} + \hat{\boldsymbol{\beta}}'\mathbf{X}'\mathbf{X}\hat{\boldsymbol{\beta}}$$

Therefore, using the results of Sec. 1.5,

$$\frac{\partial \sum e^2}{\partial \hat{\boldsymbol{\beta}}} = -2\mathbf{X}'\mathbf{Y} + 2\mathbf{X}'\mathbf{X}\hat{\boldsymbol{\beta}} \tag{4-33}$$

Setting Eq. (4-33) equal to zero and arranging terms, we have the general normal equations

$$(\mathbf{X}'\mathbf{X})\hat{\boldsymbol{\beta}} = \mathbf{X}'\mathbf{Y}$$

which are the same as Eq. (4-26), and their solution is obviously Eq. (4-27).

The scores for 27 riveting trainees are shown in Table 4.3. The dependent variable is their ability to set rivets. The two independent variables are the trainees' scores on a manual-dexterity test, X_1, and the trainees' scores on a finger-dexterity test, X_2. The object of the analysis is to study the two tests' ability to predict rivet-setting capability. If the tests are useful predictors, they may then be used to evaluate prospective employees.

The $\mathbf{X}'\mathbf{X}$ matrix is

$$\begin{bmatrix} 1 & 1 & \cdots & 1 \\ X_{11} & X_{12} & \cdots & X_{1n} \\ X_{21} & X_{22} & \cdots & X_{2n} \end{bmatrix} \begin{bmatrix} 1 & X_{11} & X_{21} \\ 1 & X_{12} & X_{22} \\ \cdot & \cdot & \cdot \\ \cdot & \cdot & \cdot \\ \cdot & \cdot & \cdot \\ 1 & X_{1n} & X_{2n} \end{bmatrix} = \begin{bmatrix} n & \sum X_1 & \sum X_2 \\ \sum X_1 & \sum X_1^2 & \sum X_1 X_2 \\ \sum X_2 & \sum X_1 X_2 & \sum X_2^2 \end{bmatrix}$$

Table 4.3 Ability to Set Rivets and Two Test Scores

	Number of Rivets Set per Man, Y	*Manual-Dexterity Score, X_1*	*Finger-Dexterity Score, X_2*
	230	135	107
	81	93	67
	100	108	81
	212	138	93
	216	123	81
	156	116	86
	201	119	86
	194	112	96
	164	128	80
	166	116	86
	146	125	78
	196	114	89
	202	128	84
	203	129	80
	201	125	99
	195	120	86
	180	126	92
	174	136	95
	120	104	82
	198	116	76
	189	112	80
	184	109	85
	174	113	75
	168	113	87
	143	104	69
	131	103	65
	130	125	84
Sum	4654	3190	2269
Mean	172.370	118.148	84.037

Source: Croxton, Cowden, and Bolch (1969), p. 230.

and is symmetric. For the data of this example

$$\mathbf{X}'\mathbf{X} = \begin{bmatrix} 1 & 1 & \cdots & 1 \\ 135 & 93 & \cdots & 125 \\ 107 & 67 & \cdots & 84 \end{bmatrix} \begin{bmatrix} 1 & 135 & 107 \\ 1 & 93 & 67 \\ \cdot & \cdot & \cdot \\ \cdot & \cdot & \cdot \\ \cdot & \cdot & \cdot \\ 1 & 125 & 84 \end{bmatrix} = \begin{bmatrix} 27 & 3190 & 2269 \\ 3190 & 380{,}040 & 269{,}820 \\ 2269 & 269{,}820 & 193{,}021 \end{bmatrix}$$

The $\mathbf{X'Y}$ matrix is

$$\begin{bmatrix} 1 & 1 & \cdots & 1 \\ X_{11} & X_{12} & \cdots & X_{1n} \\ X_{21} & X_{22} & \cdots & X_{2n} \end{bmatrix} \begin{bmatrix} Y_1 \\ Y_2 \\ \cdot \\ \cdot \\ \cdot \\ Y_n \end{bmatrix} = \begin{bmatrix} \sum Y \\ \sum X_1 Y \\ \sum X_2 Y \end{bmatrix}$$

which for our example comes to

$$\begin{bmatrix} 1 & 1 & \cdots & 1 \\ 135 & 93 & \cdots & 125 \\ 107 & 67 & \cdots & 84 \end{bmatrix} \begin{bmatrix} 230 \\ 81 \\ \cdot \\ \cdot \\ \cdot \\ 130 \end{bmatrix} = \begin{bmatrix} 4654 \\ 556{,}637 \\ 396{,}486 \end{bmatrix}$$

The sample regression equation will be of the form

$$\hat{Y}_i = \hat{\beta}_0 + \hat{\beta}_1 X_{1i} + \hat{\beta}_2 X_{2i}$$

To find these estimated parameters, we shall need the inverse of $\mathbf{X'X}$, which is approximately

$$(\mathbf{X'X})^{-1} = \begin{bmatrix} 4.724735663 & -0.030042200 & -0.013544844 \\ -0.030042200 & 0.000540150 & -0.000401912 \\ -0.013544844 & -0.000401912 & 0.000726227 \end{bmatrix}$$

Then

$$\begin{bmatrix} \hat{\beta}_0 \\ \hat{\beta}_1 \\ \hat{\beta}_2 \end{bmatrix} = \begin{bmatrix} 4.724735663 & -0.030042200 & -0.013544844 \\ -0.030042200 & 0.000540150 & -0.000401912 \\ -0.013544844 & -0.000401912 & 0.000726227 \end{bmatrix} \begin{bmatrix} 4654 \\ 556{,}637 \\ 396{,}486 \end{bmatrix}$$

$$= \begin{bmatrix} -104.020 \\ 1.498 \\ 1.182 \end{bmatrix}$$

by Eq. (4-27). Therefore the estimated equation is

$$\hat{Y}_i = -104.02 + 1.498 X_{1i} + 1.182 X_{2i}$$

Just as we may estimate the parameters of a regression equation with one independent variable by use of deviations about the means of the variables, so we can (and should) estimate the regression parameters by this method in the case of many independent variables. Except for the intercept, $\hat{\beta}_0$, the estimates are

$$\hat{\boldsymbol{\beta}}^* = \begin{bmatrix} \hat{\beta}_1 \\ \cdot \\ \cdot \\ \cdot \\ \hat{\beta}_K \end{bmatrix} = (\mathbf{x}'\mathbf{x})^{-1}\mathbf{x}'\mathbf{y} \tag{4-34}$$

where

$$\mathbf{y} = \begin{bmatrix} y_1 \\ y_2 \\ \cdot \\ \cdot \\ \cdot \\ y_n \end{bmatrix}, \qquad \mathbf{x} = \begin{bmatrix} x_{11} & x_{21} & \cdots & x_{K1} \\ x_{12} & x_{22} & \cdots & x_{K2} \\ \cdot & & & \cdot \\ \cdot & & & \cdot \\ \cdot & & & \cdot \\ x_{1n} & x_{2n} & \cdots & x_{Kn} \end{bmatrix} \tag{4-35}$$

As usual, $y_i = Y_i - \bar{Y}$ and $x_{ik} = X_{ik} - \bar{X}_k$ for $k = 1, 2, \ldots, K$. The intercept is found by

$$\hat{\beta}_0 = \bar{Y} - \bar{\mathbf{X}}'\hat{\boldsymbol{\beta}}^* \tag{4-36}$$

where $\bar{\mathbf{X}}' = [\bar{X}_1 \quad \bar{X}_2 \quad \cdots \quad \bar{X}_K]$.

Using our present example, the reader may verify that

$$\mathbf{x}'\mathbf{x} = \begin{bmatrix} \sum x_1^2 & \sum x_1x_2 \\ \sum x_1x_2 & \sum x_2^2 \end{bmatrix} = \begin{bmatrix} 3147.41 & 1741.85 \\ 1741.85 & 2340.96 \end{bmatrix}$$

and that

$$\mathbf{x}'\mathbf{y} = \begin{bmatrix} \sum x_1 y \\ \sum x_2 y \end{bmatrix} = \begin{bmatrix} 6775.52 \\ 5377.62 \end{bmatrix},$$

$$(\mathbf{x}'\mathbf{x})^{-1} = \begin{bmatrix} 0.000540150 & -0.000401912 \\ -0.000401912 & 0.000726227 \end{bmatrix}$$

Notice that $(\mathbf{x}'\mathbf{x})^{-1}$ is a submatrix of $(\mathbf{X}'\mathbf{X})^{-1}$. By Eq. (4-34) we have

$$\hat{\boldsymbol{\beta}}^* = \begin{bmatrix} 0.000540150 & -0.000401912 \\ -0.000401912 & 0.000726227 \end{bmatrix} \begin{bmatrix} 6775.52 \\ 5377.62 \end{bmatrix}$$

$$= \begin{bmatrix} 1.498 \\ 1.182 \end{bmatrix}$$

The intercept is found by use of Eq. (4-36), $\bar{Y} = 172.370$ and $\bar{\mathbf{X}}' =$

[118.148 84.037]:

$$\hat{\beta}_0 = 172.370 - [118.148 \quad 84.037]\begin{bmatrix}1.498\\1.182\end{bmatrix}$$
$$= 172.370 - 276.317 = -104.0$$

Some accuracy was lost in the calculation of $\hat{\beta}_0$ because we rounded the slopes before calculating $\hat{\beta}_0$. For greater accuracy the calculated slopes should be rounded *after* calculating $\hat{\beta}_0$.

The vector of sample residuals, **e**, is given by least-squares as

$$\mathbf{e} = \mathbf{Y} - \hat{\mathbf{Y}} = \mathbf{Y} - \mathbf{X}\hat{\boldsymbol{\beta}}$$

Now $\hat{\boldsymbol{\beta}} = (\mathbf{X}'\mathbf{X})^{-1}\mathbf{X}'\mathbf{Y}$, and in the population $\mathbf{Y} = \mathbf{X}\boldsymbol{\beta} + \boldsymbol{\epsilon}$. By direct substitution

$$\begin{aligned}\mathbf{e} &= \mathbf{X}\boldsymbol{\beta} + \boldsymbol{\epsilon} - \mathbf{X}[(\mathbf{X}'\mathbf{X})^{-1}\mathbf{X}'(\mathbf{X}\boldsymbol{\beta} + \boldsymbol{\epsilon})]\\ &= \mathbf{X}\boldsymbol{\beta} + \boldsymbol{\epsilon} - \mathbf{X}(\mathbf{X}'\mathbf{X})^{-1}\mathbf{X}'\mathbf{X}\boldsymbol{\beta} - \mathbf{X}(\mathbf{X}'\mathbf{X})^{-1}\mathbf{X}'\boldsymbol{\epsilon}\\ &= \mathbf{X}\boldsymbol{\beta} + \boldsymbol{\epsilon} - \mathbf{X}\boldsymbol{\beta} - \mathbf{X}(\mathbf{X}'\mathbf{X})^{-1}\mathbf{X}'\boldsymbol{\epsilon}\\ &= [\mathbf{I}_n - \mathbf{X}(\mathbf{X}'\mathbf{X})^{-1}\mathbf{X}']\boldsymbol{\epsilon} \qquad (4\text{-}37)\end{aligned}$$

where $\mathbf{I}_n$ is an $n \times n$ identity matrix.

Equation (4-37) shows that the sample vector of residuals is a linear function of the population vector of errors. Write Eq. (4-37) as

$$\mathbf{e} = \mathbf{M}\boldsymbol{\epsilon} \qquad (4\text{-}37a)$$

where

$$\mathbf{M} = \mathbf{I}_n - \mathbf{X}(\mathbf{X}'\mathbf{X})^{-1}\mathbf{X}'$$

is a symmetric, *idempotent* matrix. An idempotent matrix is one where $\mathbf{M}'\mathbf{M} = \mathbf{M}$. One symmetric idempotent matrix that we have already encountered is the identity matrix.[5]

The sum of the squared sample residuals is

$$\begin{aligned}\mathbf{e}'\mathbf{e} &= \boldsymbol{\epsilon}'\mathbf{M}'\mathbf{M}\boldsymbol{\epsilon} = \boldsymbol{\epsilon}'\mathbf{M}\boldsymbol{\epsilon}\\ &= \sum_{i=1}^{n} M_{ii}\epsilon_i^2 + \sum_{i \neq i'} M_{ii'}\epsilon_i\epsilon_{i'} \qquad (4\text{-}38)\end{aligned}$$

The assumption of independence [i.e., the assumption that $E(\epsilon_i\epsilon_{i'}) = 0$ for $i \neq i'$] allows us to drop the second term in Eq. (4-38) upon taking the

[5] To verify that **M** is idempotent, write it as $\mathbf{M} = \mathbf{I}_n - \mathbf{A}$, where $\mathbf{A} = \mathbf{X}(\mathbf{X}'\mathbf{X})^{-1}\mathbf{X}'$. Then expand $\mathbf{M}'\mathbf{M}$.

expected value of $\mathbf{e}'\mathbf{e}$:

$$E(\mathbf{e}'\mathbf{e}) = \sum_{i=1}^{n} M_{ii}E(\epsilon_i^2) = \sigma^2 \sum_{i=1}^{n} M_{ii} = \sigma^2(n - K - 1)$$

since $\sum_{i=1}^{n} M_{ii} = \text{tr}\,\mathbf{M} = \text{tr}\,\mathbf{I}_n - \text{tr}(\mathbf{X}'\mathbf{X}(\mathbf{X}'\mathbf{X})^{-1}) = n - (K + 1)$. Therefore an unbiased estimator of σ^2 is

$$\hat{\sigma}^2 = \frac{\mathbf{e}'\mathbf{e}}{n - K - 1} \tag{4-39}$$

The statistic $\hat{\sigma}$ is called the *standard error of estimate*. Among other things it is an index of the closeness of fit of the sample regression relation to the sample Y values, since $\mathbf{e}'\mathbf{e}$ will approach zero as the sample regression function approaches passing through each of the sample Y values. For our present example (see the Appendix to this chapter)

$$\mathbf{e}'\mathbf{e} = [5.23071 \quad \cdots \quad -52.59375]\begin{bmatrix} 5.23071 \\ \cdot \\ \cdot \\ \cdot \\ -52.59375 \end{bmatrix} = 18{,}202$$

so that

$$\hat{\sigma}^2 = \frac{18{,}202}{27 - 2 - 1} = 758.4$$

and the standard error of estimate is $\hat{\sigma} = \sqrt{758.4} = 27.54$.

Maximum Likelihood Estimation. If we make all the assumptions necessary for least-squares estimation and in addition assume that the Y_is (or ϵ_is) follow some specific distribution in the population, we may then use maximum likelihood estimation. Suppose that we assume that the Y_is are normally distributed:

$$f(Y_i) = \frac{1}{\sqrt{2\pi\sigma^2}} \exp\left[-\frac{1}{2\sigma^2}(Y_i - \mu_i)^2\right]$$

Then, the likelihood function is given by $\prod_{i=1}^{n} f(Y_i)$. Following the approach of Sec. 2.5, if we replace $(Y_i - \mu_i)^2$ above with $(\mathbf{Y} - \mathbf{X}\boldsymbol{\beta})'(\mathbf{Y} - \mathbf{X}\boldsymbol{\beta})$, take the natural logarithm of the likelihood function, and set the first partial derivatives with respect to $\boldsymbol{\beta}$ and σ^2 equal to zero, then the estimates which emerge are

$$\hat{\boldsymbol{\beta}} = (\mathbf{X}'\mathbf{X})^{-1}\mathbf{X}'\mathbf{Y}$$

and

$$\hat{\sigma}^2 = \frac{\mathbf{e}'\mathbf{e}}{n}$$

Thus under the assumption of normality the maximum likelihood and least-squares estimates of $\boldsymbol{\beta}$ are identical. However, the maximum likelihood estimate of σ^2 is biased.

4.4 some hypothesis tests and confidence intervals

In the previous section we saw that while the assumption that the Y_is, or ϵ_is, are normally distributed is not necessary for least-squares estimation, it is an assumption that causes least-squares and maximum likelihood estimation of the parameters of the regression equation to be identical. In this section we shall make the assumption that the Y_is, or ϵ_is, are normally distributed. We remark that this assumption, in addition to those previously made, is sufficient but not necessary for hypothesis testing. Other assumptions concerning the population distribution are possible; however, the assumption of normality is the one most commonly made in practical work.

Let us first show that $\hat{\boldsymbol{\beta}}$ is an unbiased estimator of $\boldsymbol{\beta}$. By Eq. (4-27) and the population regression $\mathbf{Y} = \mathbf{X}\boldsymbol{\beta} + \boldsymbol{\epsilon}$, we have

$$\begin{aligned} \hat{\boldsymbol{\beta}} &= (\mathbf{X}'\mathbf{X})^{-1}\mathbf{X}'\mathbf{Y} \\ &= (\mathbf{X}'\mathbf{X})^{-1}\mathbf{X}'(\mathbf{X}\boldsymbol{\beta} + \boldsymbol{\epsilon}) \end{aligned} \tag{4-40}$$

Under the assumptions that the matrix $\mathbf{X}$ is a statistical constant (*non-stochastic*) and that $E(\boldsymbol{\epsilon}) = 0$, we have

$$\begin{aligned} E(\hat{\boldsymbol{\beta}}) &= (\mathbf{X}'\mathbf{X})^{-1}(\mathbf{X}'\mathbf{X})E(\boldsymbol{\beta}) + (\mathbf{X}'\mathbf{X})^{-1}(\mathbf{X}')E(\boldsymbol{\epsilon}) \\ &= \mathbf{I}\boldsymbol{\beta} = \boldsymbol{\beta} \end{aligned} \tag{4-41}$$

where $\mathbf{I}$ is the identity matrix resulting from the multiplication $(\mathbf{X}'\mathbf{X})^{-1}(\mathbf{X}'\mathbf{X})$. Note that the assumption that $\mathbf{X}$ is a statistical constant (that it does not vary upon repeated sampling) is explicit in our previous discussions since this is a condition that will cause $E(\epsilon_i X_i)$ to be zero.

Having determined the expected value of $\hat{\boldsymbol{\beta}}$, we now turn to the estimation of its covariance matrix. In the previous section we assumed that

$$\begin{aligned} E(\epsilon_i \epsilon_{i'}) &= 0, \qquad i \neq i' \\ &= \sigma^2, \qquad i = i' \end{aligned}$$

These two assumptions can be expressed by writing

$$E(\boldsymbol{\epsilon}\boldsymbol{\epsilon}') = \boldsymbol{\Sigma} = \sigma^2\mathbf{I} \tag{4-42}$$

That is, we assume that the population covariance matrix of the errors, $\boldsymbol{\Sigma}$, is an $n \times n$ diagonal matrix with σ^2 on its main diagonal. Since the covariance matrix contains the variances on its main diagonal and covariances elsewhere, writing the population covariance matrix in this way is the same as stating that the errors are distributed independently with constant variance. That is, we may write[6]

$$\boldsymbol{\Sigma} = \sigma^2 \begin{bmatrix} 1 & 0 & \cdots & 0 \\ 0 & 1 & \cdots & 0 \\ \vdots & & & \vdots \\ 0 & 0 & \cdots & 1 \end{bmatrix} = \begin{bmatrix} \sigma^2 & 0 & \cdots & 0 \\ 0 & \sigma^2 & \cdots & 0 \\ \vdots & & & \vdots \\ 0 & 0 & \cdots & \sigma^2 \end{bmatrix} \tag{4-42a}$$

The covariance matrix of $\hat{\boldsymbol{\beta}}$ is found by evaluating

$$E\{[\hat{\boldsymbol{\beta}} - E(\hat{\boldsymbol{\beta}})][\hat{\boldsymbol{\beta}} - E(\hat{\boldsymbol{\beta}})]'\} = E[(\hat{\boldsymbol{\beta}} - \boldsymbol{\beta})(\hat{\boldsymbol{\beta}} - \boldsymbol{\beta})']$$

since $E(\hat{\boldsymbol{\beta}}) = \boldsymbol{\beta}$. Furthermore, from Eq. (4-40) we see that

$$\hat{\boldsymbol{\beta}} - \boldsymbol{\beta} = (\mathbf{X}'\mathbf{X})^{-1}\mathbf{X}'\boldsymbol{\epsilon} \tag{4-40a}$$

so that

$$E[(\hat{\boldsymbol{\beta}} - \boldsymbol{\beta})(\hat{\boldsymbol{\beta}} - \boldsymbol{\beta})'] = E\{[(\mathbf{X}'\mathbf{X})^{-1}\mathbf{X}'\boldsymbol{\epsilon}][\boldsymbol{\epsilon}'\mathbf{X}(\mathbf{X}'\mathbf{X})^{-1}]\}$$

since $[(\mathbf{X}'\mathbf{X})^{-1}\mathbf{X}'\boldsymbol{\epsilon}]' = [\boldsymbol{\epsilon}'\mathbf{X}(\mathbf{X}'\mathbf{X})^{-1}]$ by virtue of the fact that $(\mathbf{X}'\mathbf{X})^{-1}$ is symmetric and therefore unchanged by transposition. Upon further simplification we find that

$$\begin{aligned} E[(\hat{\boldsymbol{\beta}} - \boldsymbol{\beta})(\hat{\boldsymbol{\beta}} - \boldsymbol{\beta})'] &= \sigma^2\mathbf{I}(\mathbf{X}'\mathbf{X})^{-1} \\ &= \sigma^2(\mathbf{X}'\mathbf{X})^{-1} \end{aligned} \tag{4-43}$$

The variance σ^2 is usually unknown and is estimated from the sample. An unbiased estimate is

$$\hat{\sigma}^2 = \frac{\mathbf{e}'\mathbf{e}}{n - K - 1}$$

[6]There is a slight modification of our notation here. We write σ^2 for the elements on the main diagonal of the matrix rather than σ_{ii}, to stress the fact that only one variance is involved.

as discussed in the last section. Therefore the estimated covariance matrix is

$$\mathbf{S}_{\beta} = \hat{\sigma}^2(\mathbf{X}'\mathbf{X})^{-1} \tag{4-44}$$

Using the data of the last section concerning the ability to set rivets, we have

$$\mathbf{S}_{\beta} = \begin{bmatrix} s_{\beta_{00}} & s_{\beta_{01}} & s_{\beta_{02}} \\ s_{\beta_{10}} & s_{\beta_{11}} & s_{\beta_{12}} \\ s_{\beta_{20}} & s_{\beta_{21}} & s_{\beta_{22}} \end{bmatrix} \tag{4-44a}$$

For our example

$$\mathbf{S}_{\beta} = 758.4 \begin{bmatrix} 4.724735663 & -0.030042200 & -0.013544844 \\ -0.030042200 & 0.000540150 & -0.000401912 \\ -0.013544844 & -0.000401912 & 0.000726227 \end{bmatrix}$$

$$= \begin{bmatrix} 3583.24 & -22.78 & -10.272 \\ -22.78 & 0.410 & -0.305 \\ -10.272 & -0.305 & 0.551 \end{bmatrix}$$

since $\hat{\sigma}^2 = 758.4$.

The square root of the elements on the main diagonal of $\mathbf{S}_{\beta}$ are usually called the estimated *standard errors* of the coefficients.

There is one point concerning calculation that is worthy of mention. If calculation is done using $(\mathbf{x}'\mathbf{x})^{-1}$ rather than $(\mathbf{X}'\mathbf{X})^{-1}$, then $s_{\beta_{00}}, s_{\beta_{01}}, \ldots, s_{\beta_{0K}}$ cannot be calculated directly from $(\mathbf{x}'\mathbf{x})^{-1}$. Actually, since we are generally less interested in the variance and covariances of the intercept, $\hat{\beta}_0$, this row of $\mathbf{S}_{\beta}$ is often never calculated. For example, the program in the Appendix to this chapter does not calculate the standard error of the intercept. However, if the full $\mathbf{S}_{\beta}$ matrix is desired, it is easily obtained. Recall first that $(\mathbf{x}'\mathbf{x})^{-1}$ is a submatrix of $(\mathbf{X}'\mathbf{X})^{-1}$. That is, we may write

$$(\mathbf{X}'\mathbf{X})^{-1} = \left[\begin{array}{c|c} \mathbf{A} & \mathbf{C}' \\ \hline \mathbf{B} & (\mathbf{x}'\mathbf{x})^{-1} \end{array}\right]$$

where $\mathbf{A}$ is 1×1, $\mathbf{B}$ is $K \times 1$, and $\mathbf{C}'$ is $1 \times K$. Using the method of partitioned inversion (see Sec. 1.4), we see that is easy to show that

$$\mathbf{A} = \frac{1}{n} - \mathbf{C}'\bar{\mathbf{X}}$$

where

$$\bar{\mathbf{X}}' = [\bar{X}_1 \quad \bar{X}_2 \quad \cdots \quad \bar{X}_K],$$

$$\mathbf{C}' = (-1)\bar{\mathbf{X}}'(\mathbf{x}'\mathbf{x})^{-1}$$

and, of course, $\mathbf{B} = \mathbf{C}$. Using our example,

$$\mathbf{C}' = (-1)\bar{\mathbf{X}}'(\mathbf{x}'\mathbf{x})^{-1}$$
$$= (-1)[118.148 \quad 84.037]\begin{bmatrix} 0.000540150 & -0.000401912 \\ -0.000401912 & 0.000726227 \end{bmatrix}$$
$$= [-0.0300422 \quad -0.01354484]$$

which agrees with our previous results. Furthermore,

$$\mathbf{A} = \frac{1}{27} - [-0.0300422 \quad -0.01354484]\begin{bmatrix} 118.148 \\ 84.037 \end{bmatrix}$$
$$= \frac{1}{27} + 4.687694 = 4.72473$$

which also agrees with our previous results when rounding error is taken into account.

Having discussed the estimation of the covariance matrix of $\hat{\boldsymbol{\beta}}$ and having obtained the expected value of $\hat{\boldsymbol{\beta}}$, we now turn to the sampling distribution of $\hat{\boldsymbol{\beta}}$. From Eq. (4-27)

$$\hat{\boldsymbol{\beta}} = (\mathbf{X}'\mathbf{X})^{-1}\mathbf{X}'\mathbf{Y}$$

Now, since $(\mathbf{X}'\mathbf{X})^{-1}\mathbf{X}'$ is assumed constant, $\hat{\boldsymbol{\beta}}$ is a linear function of $\mathbf{Y}$. Since we have assumed that $\mathbf{Y}$ is normally distributed, it follows that $\hat{\boldsymbol{\beta}}$ is normally distributed. This result follows because linear functions of variables that are normally distributed are also normally distributed (as we mentioned in Sec. 3.2 when dealing with linear combinations). However, since the covariance matrix of $\hat{\boldsymbol{\beta}}$ is usually estimated, $\hat{\boldsymbol{\beta}}$ will follow Student's distribution which will be used for hypothesis tests.

Test that a Single Coefficient Is a Given Constant. The test of the hypothesis

$$H_0\colon \quad \beta_k = \beta_{0k}$$

where β_k is a single population regression coefficient and β_{0k} is a hypothetical constant value for this coefficient, can be conducted by use of

$$t = \frac{\hat{\beta}_k - \beta_{0k}}{\sqrt{s_{\beta_{kk}}}} \tag{4-45}$$

The variance $s_{\beta_{kk}}$ is the kth diagonal element of $\mathbf{S}_\beta$ and its square root, s_{β_k}, is the *standard error* of the coefficient $\hat{\beta}_k$. The null hypothesis may be tested against a one- or a two-sided alternative, and the t statistic of Eq. (4-45) follows the t distribution with $n - K - 1$ degrees of freedom.

Of considerable importance is the test of the hypothesis that a β_k is zero.

As noted previously, if a slope is determined not to be significantly different from zero, then the variable associated with that slope is making no statistical contribution to the regression equation. Less frequently, but nevertheless occasionally, we are interested in testing the hypothesis that a slope is a constant other than zero. In addition, we are sometimes interested in testing whether or not the intercept, β_0, is statistically equal to some constant such as zero.

Still using our example concerning the ability to set rivets, let us test the hypothesis that the finger-dexterity test is making no statistical contribution to the regression equation. This test will have a one-sided alternative since we reason, a priori, that an increase in measured finger dexterity should lead to an improved ability to set rivets. Therefore we test

$$H_0: \quad \beta_2 = 0$$
$$H_1: \quad \beta_2 > 0$$

Let us conduct the test at $\alpha = 0.05$.

From previous calculation, we find that the t statistic of Eq. (4-45) becomes

$$t = \frac{1.1822 - 0}{\sqrt{0.551}} = \frac{1.1822}{0.742} = 1.59$$

In the t distribution at $\alpha = 0.05$ and $\nu = n - K - 1 = 24$, we find $t_{0.05;\,24} = 1.711$. Since the actual value of t does not exceed the critical value, we do not reject the null hypothesis at the stated level of significance.

Had a two-sided alternative hypothesis been specified, we would have conducted the test in the same way except that $t_{\alpha/2;\,\nu}$, rather than $t_{\alpha;\,\nu}$, would have been chosen for the critical value of t. In applications of regression analysis to problems in economics, we seldom use a two-sided alternative hypothesis in univariate tests involving slopes. This statement follows because economic theory generally specifies in advance the sign of the slope. Thus we would not expect, a priori, that the slope of a consumption function would be negative.

As a practical matter, the test just described can be conducted very quickly by observing the following rule of thumb. If the hypothetical value of the slope is zero (and it usually is), the t statistic of Eq. (4-45) can be written as the ratio of the estimated slope to its estimated standard error. Turning to Table 3 in the Appendix Tables, we note that the critical value of t is less than 2.0 at the 0.05 level when ν exceeds 5. Thus, since we nearly always have $\nu > 5$ (that is $n - K - 1$ greater than 5) in economic applications of regression analysis, we use the rule of thumb of rejecting the null hypothesis whenever the absolute value of the sample slope is greater than two times its standard error.

Confidence intervals for a single coefficient may be set in the usual way by an inversion of Eq. (4-45). Thus for a $100(1-\alpha)$ percent interval for β_k we evaluate

$$\hat{\beta}_k \pm (t_{\alpha/2;\,\nu})\sqrt{s_{\beta_{kk}}} \tag{4-46}$$

Joint Hypothesis Tests. Having tested the hypothesis that the slope β_2 is zero, we are now tempted to test the hypothesis that the slope β_1 is zero. If we use the ordinary univariate t statistic, we might test

$$H_0:\ \beta_1 = 0$$
$$H_1:\ \beta_1 > 0$$

by use of

$$t = \frac{1.498}{\sqrt{0.410}} = \frac{1.498}{0.640} = 2.34$$

Using the rule of thumb just discussed, we reject the null hypothesis and conclude that the manual-dexterity test is making a significant statistical contribution (at $\alpha = 0.05$) to the regression.

Each of the two tests just performed is univariate. The tests are strictly analogous to the univariate tests which were pointed out to be deficient in Sec. 3.2. Thus the simultaneous level of significance for these two tests will not necessarily be 0.05 because of lack of independence.

As the student may have anticipated, the simultaneous test can be developed by use of Hotelling's T^2. Reasoning by analogy to Eq. (4-45), we find that

$$T^2 = (\hat{\boldsymbol{\beta}} - \boldsymbol{\beta}_0)'\mathbf{S}_{\hat{\beta}}^{-1}(\hat{\boldsymbol{\beta}} - \boldsymbol{\beta}_0) \tag{4-47}$$

where $\hat{\boldsymbol{\beta}}$ is the vector of estimated parameters and $\boldsymbol{\beta}_0$ is the vector of hypothetical constants. $\mathbf{S}_{\beta}$ is defined by Eq. (4-44). Under the null hypothesis that $\boldsymbol{\beta} = \boldsymbol{\beta}_0$, the critical value of T^2 will be given by

$$T^2_{\alpha;\,K+1,n-K-1} = (K+1)F_{\alpha;\,K+1,n-K-1} \tag{4-48}$$

There is a strict analogy between Eqs. (4-47) and (4-48) and Eqs. (3-9) and (3-11). If the reader notices that it is *not* necessary to multiply all elements of $\mathbf{S}_{\beta}$ by $1/n$ in order to obtain the estimated covariance matrix (as it was for $\mathbf{S}$) and if he notices that p, of Chap. 3, is equivalent to $K+1$ in this chapter, the analogy will become transparent.

Let us now test the joint hypothesis that both slopes, in the rivet-setting example, are zero and that the intercept is -100.0. The null hypothesis is

$$H_0:\ [\beta_0 \quad \beta_1 \quad \beta_2] = [-100.0 \quad 0 \quad 0]$$

which we shall test at $\alpha = 0.05$. Notice that we need not invert $\mathbf{S}_\beta$ to find $\mathbf{S}_{\hat{\beta}}^{-1}$. Indeed, such a computing method will sometimes lead to a serious rounding error. Instead we need only note that $\mathbf{S}_{\hat{\beta}}^{-1} = (\mathbf{X}'\mathbf{X})(1/\hat{\sigma}^2)$. Thus the T^2 statistic of Eq. (4-47) becomes

$$T^2 = [-4.02 \quad 1.498 \quad 1.182] \begin{bmatrix} 0.03560 & 4.20622 & 2.99182 \\ 4.20622 & 501.108 & 355.775 \\ 2.99182 & 355.775 & 254.511 \end{bmatrix} \begin{bmatrix} -4.020 \\ 1.498 \\ 1.182 \end{bmatrix}$$
$$= 2660.83$$

where the 3×3 matrix is $\mathbf{S}_{\hat{\beta}}^{-1}$ (approximately). From Eq. (4-48) the critical value of T^2 is

$$\begin{aligned} T^2_{0.05;\,3,24} &= (3)F_{0.05;\,3,24} \\ &= (3)(3.01) \\ &= 9.03 \end{aligned}$$

Since the calculated value of T^2 exceeds the critical value, we reject the null hypothesis.

A $100(1-\alpha)$ percent joint confidence ellipsoid (region) may be calculated by use of Eqs. (4-47) and (4-48). The region will be bounded by

$$(\hat{\boldsymbol{\beta}} - \boldsymbol{\beta})'\mathbf{S}_{\hat{\beta}}^{-1}(\hat{\boldsymbol{\beta}} - \boldsymbol{\beta}) = (K+1)F_{\alpha;\,K+1,n-K-1} \tag{4-49}$$

Again, the resulting ellipsoid will be in $(K+1)$-dimensional space so that, as in the previous chapter, the ellipsoid cannot be visualized unless $K+1 \leq 3$.

Linear Combinations. Just as we could test hypotheses and set confidence limits using linear combinations when we were dealing with means, so we may use linear combinations when dealing with regression parameters. Let $\mathbf{C}$ be a $(K+1) \times q$ matrix of rank $q \leq K+1$. Then, the linear combination $\mathbf{C}'\hat{\boldsymbol{\beta}}$ has a q-dimensional normal distribution under our previous assumptions. By analogy to Eq. (3-15) of Chap. 3, the hypothesis

$$H_0\colon\ \mathbf{C}'\boldsymbol{\beta} = \mathbf{C}'\boldsymbol{\beta}_0$$

may be tested by use of

$$T^2 = (\mathbf{C}'\hat{\boldsymbol{\beta}} - \mathbf{C}'\boldsymbol{\beta}_0)'(\mathbf{C}'\mathbf{S}_\beta\mathbf{C})^{-1}(\mathbf{C}'\hat{\boldsymbol{\beta}} - \mathbf{C}'\boldsymbol{\beta}_0) \tag{4-50}$$

with critical value

$$T^2_{\alpha;\,q,n-K-1} = qF_{\alpha;\,q,n-K-1}$$

Notice that if $\mathbf{C}$ is an identity matrix of order $K+1$, then Eq. (4-50) reduces to Eq. (4-49) and that the second number of degrees of freedom for T^2 and F is $n - K - 1$ rather than $n - q$. This degree of freedom holds because the variance $\mathbf{C}'\mathbf{S}_\beta\mathbf{C}$ is still obtained from the estimate $\hat{\sigma}^2$, which has $n - K - 1$ degrees of freedom.

Joint confidence regions may be set with linear combinations. A $100(1 - \alpha)$ percent confidence region for a linear combination of the regression parameters will be bounded by

$$(\mathbf{C}'\hat{\boldsymbol{\beta}} - \mathbf{C}'\boldsymbol{\beta})'(\mathbf{C}'\mathbf{S}_\beta\mathbf{C})^{-1}(\mathbf{C}'\hat{\boldsymbol{\beta}} - \mathbf{C}'\boldsymbol{\beta}) = qF_{\alpha;\, q,\, n-K-1} \tag{4-51}$$

Let us consider three possible applications of linear combination hypothesis tests using our rivet-setting example. First, we might wish to test the following joint hypothesis

$$H_0:\quad \begin{cases} \beta_0 + \beta_1 + \beta_2 = -100.0 \\ \qquad \beta_1 - \beta_2 = 0.6 \end{cases}$$

This simultaneous hypothesis has little intuitive meaning within the context of the present example, but such tests sometimes arise in econometric applications where production functions or difference equations are involved. The hypothesis is equivalent to

$$H_0:\quad \mathbf{C}'\boldsymbol{\beta} = \boldsymbol{\gamma}$$

with the alternative

$$H_1:\quad \mathbf{C}'\boldsymbol{\beta} \neq \boldsymbol{\gamma}$$

where

$$\mathbf{C}' = \begin{bmatrix} 1 & 1 & 1 \\ 0 & 1 & -1 \end{bmatrix} \quad \text{and} \quad \boldsymbol{\gamma}' = [-100.0 \quad 0.6]$$

The test is carried out by use of the T^2 statistic of Eq. (4-50) with $\boldsymbol{\gamma}$ replacing $\mathbf{C}'\boldsymbol{\beta}_0$ and $q = 2$.

A second application concerns the hypothesis

$$H_0:\quad c_0\beta_0 + c_1\beta_1 + \cdots + c_K\beta_K = \gamma$$

This single hypothesis is equivalent to

$$H_0:\quad \mathbf{c}'\boldsymbol{\beta} = \gamma$$

where $\mathbf{c}'$ is a row vector $\mathbf{c}' = [c_0 \quad c_1 \quad \cdots \quad c_K]$. In the rivet-setting example we may wish to test

$$H_0:\quad \beta_1 - \beta_2 = 0.6$$

which is one of the two joint hypotheses given in our first illustration directly above. The $\mathbf{c}'$ vector is $\mathbf{c}' = [0 \quad 1 \quad -1]$ and Eq. (4-50) becomes

$$T^2 = \frac{(\mathbf{c}'\hat{\boldsymbol{\beta}} - \gamma)^2}{\mathbf{c}'\mathbf{S}_\beta \mathbf{c}}$$

with γ replacing $\mathbf{c}'\boldsymbol{\beta}_0$ and $q = 1$. The critical value is $T^2_{\alpha;\,1,n-K-1} = F_{\alpha;\,1,n-K-1}$. Clearly, for this single hypothesis, the critical value is nothing more than the square of Student's t since $T^2_{\alpha;\,1,n-K-1} = t^2_{\alpha/2;\,n-K-1}$.

A third application of linear combinations concerns a simultaneous test that a subset of the full $\boldsymbol{\beta}$ vector is equal to some constant vector $\boldsymbol{\beta}_0$. Very often we shall wish to ignore the intercept, β_0, in such a test. After all, it is the slopes, not the intercept, which give us an indication of the importance of changes in the independent variables. In this case the hypothesis is

$$H_0\colon\ [\beta_1 \quad \beta_2 \quad \cdots \quad \beta_K] = [\beta_{01} \quad \beta_{02} \quad \cdots \quad \beta_{0K}]$$

which is equivalent to

$$H_0\colon\ \mathbf{C}'\boldsymbol{\beta} = \mathbf{C}'\boldsymbol{\beta}_0$$

with the $\mathbf{C}$ matrix of rank $q = K$ given by

$$\mathbf{C}' = \underbrace{\begin{bmatrix} 0 & 1 & 0 & \cdots & 0 \\ 0 & 0 & 1 & \cdots & 0 \\ \cdot & & & & \cdot \\ \cdot & & & & \cdot \\ \cdot & & & & \cdot \\ 0 & 0 & 0 & \cdots & 1 \end{bmatrix}}_{q+1=K+1} \left.\vphantom{\begin{bmatrix}0\\0\\0\\0\\0\\0\end{bmatrix}}\right\} q = K$$

The first column of $\mathbf{C}'$ (which corresponds to β_0, which we wish to ignore) is zero. If we wished to ignore only β_1, the second column of $\mathbf{C}'$ would contain only zeros, while the first column would have a 1 in its first position and zeros elsewhere.[7]

Continuing to suppose that we wish to ignore the intercept, we may write the null hypothesis as

$$H_0\colon\ \boldsymbol{\beta}^* = \boldsymbol{\beta}_0^*$$

[7]In general the rank of $\mathbf{C}$ will be equal to $K + 1$ minus the number of coefficients ignored. If we ignore β_0 and β_1, then $\mathbf{C}'$ will contain zeros in its first two columns and 1s on the main diagonal of the remaining $K + 1 - 2$ columns. In the extreme case where all coefficients except one are ignored, the rank of $\mathbf{C}$ is 1 and Eq. (4-50) reduces to the univariate statistic reported earlier as Eq. (4-45). Once again we find that the univariate statistics are simply special cases of the multivariate ones.

since, by previous notation, we have defined $\boldsymbol{\beta}^*$ to contain only the slopes. We may also define, for convenience,

$$\mathbf{S}_{\beta^*} = \mathbf{C}'\mathbf{S}_{\beta}\mathbf{C}$$

which, as the reader may verify, is the $\mathbf{S}_{\beta}$ matrix with the first row and column deleted. The T^2 statistic then becomes

$$T^2 = (\hat{\boldsymbol{\beta}}^* - \boldsymbol{\beta}_0^*)'\mathbf{S}_{\beta^*}^{-1}(\hat{\boldsymbol{\beta}}^* - \boldsymbol{\beta}_0^*) \tag{4-47a}$$

with critical value

$$T^2_{\alpha;\,K,n-K-1} = KF_{\alpha;\,K,n-K-1} \tag{4-48a}$$

A very common usage of Eq. (4-47a) is to test the hypothesis that all the slopes (but not the intercept) are simultaneously equal to zero. This test is, therefore, an *overall* test of the regression equation, since if we fail to reject the null hypothesis, then we conclude that none of the independent variables is making a statistical contribution to the explanation of the dependent variable. For our present example let us test

$$H_0\colon\ \boldsymbol{\beta}^* = \boldsymbol{\beta}_0^*$$
$$H_1\colon\ \boldsymbol{\beta}^* \neq \boldsymbol{\beta}_0^*$$

where $(\hat{\boldsymbol{\beta}}^*)' = [1.498 \quad 1.182]$ and $(\boldsymbol{\beta}_0^*)' = [0 \quad 0]$. Thus

$$\mathbf{S}_{\beta^*} = \begin{bmatrix} 0.410 & -0.305 \\ -0.305 & 0.551 \end{bmatrix}$$

and

$$\mathbf{S}_{\beta^*}^{-1} = \begin{bmatrix} 4.1501 & 2.2967 \\ 2.2967 & 3.0867 \end{bmatrix}$$

so that

$$\begin{aligned} T^2 &= [1.498 \quad 1.182]\begin{bmatrix} 4.1501 & 2.2967 \\ 2.2967 & 3.0867 \end{bmatrix}\begin{bmatrix} 1.498 \\ 1.182 \end{bmatrix} \\ &= 21.76 \end{aligned}$$

The critical value of T^2 by Eq. (4-48a) and with $\alpha = 0.05$ is

$$\begin{aligned} T^2_{0.05;\,2,24} &= (2)F_{0.05;\,2,24} \\ &= (2)(3.40) = 6.80 \end{aligned}$$

Thus we reject the null hypothesis that all slopes are zero.

The test of the null hypothesis that all the slopes are zero against the alternative that *not all* slopes are zero is generally accomplished by use of *analysis of variance* and the F ratio. Since analysis of variance will occupy most of the next chapter, we shall give it only brief attention here. Write the regression equation in deviation form[8]:

$$\mathbf{y} = \mathbf{x}\hat{\boldsymbol{\beta}}^* + \mathbf{e} \tag{4-52}$$

Now

$$\begin{aligned} \mathbf{e}'\mathbf{e} &= (\mathbf{y} - \mathbf{x}\hat{\boldsymbol{\beta}}^*)'(\mathbf{y} - \mathbf{x}\hat{\boldsymbol{\beta}}^*) \\ &= \mathbf{y}'\mathbf{y} - (\hat{\boldsymbol{\beta}}^*)'(\mathbf{x}'\mathbf{x})\hat{\boldsymbol{\beta}}^* \end{aligned} \tag{4-53}$$

since from the normal equations

$$(\mathbf{x}'\mathbf{x})\hat{\boldsymbol{\beta}}^* = \mathbf{x}'\mathbf{y}$$

Thus Eq. (4-53) becomes

$$\mathbf{e}'\mathbf{e} = \mathbf{y}'\mathbf{y} - (\hat{\boldsymbol{\beta}}^*)'(\mathbf{x}'\mathbf{y})$$

or

$$\mathbf{y}'\mathbf{y} = (\hat{\boldsymbol{\beta}}^*)'(\mathbf{x}'\mathbf{y}) + \mathbf{e}'\mathbf{e} \tag{4-54}$$

Equation (4-54) is a fundamental one since it decomposes (or makes ready for analysis) the variation in $\mathbf{y}$ into two component parts. Since $\mathbf{y}'\mathbf{y}$ measures the variation of the dependent variable, the left-hand side of Eq. (4-54) is often referred to as *total variation*. It is the variation in the dependent variable to be "explained" by use of the regression equation. Also, $\mathbf{e}$ measures the sample error, or residual, and so $\mathbf{e}'\mathbf{e}$ is often referred to as *unexplained variation*. The remaining term, $(\hat{\boldsymbol{\beta}}^*)'(\mathbf{x}'\mathbf{y})$, is called the *explained variation*. Thus we often write Eq. (4-54) verbally as

Total variation = explained variation + unexplained variation

If explained variation is "large" relative to unexplained variation, we reason that some or all of the slopes are significantly different from zero. Hence a test of the hypothesis that all slopes are jointly equal to zero against the alternative that not all slopes are zero is often conducted by use of the ratio

$$F = \frac{(\hat{\boldsymbol{\beta}}^*)'(\mathbf{x}'\mathbf{y})/K}{\mathbf{e}'\mathbf{e}/(n - K - 1)} \tag{4-55}$$

---

[8]Write $\mathbf{Y} = \hat{\beta}_0 + \mathbf{X}^*\hat{\boldsymbol{\beta}}^* + \mathbf{e}$ and $\bar{\mathbf{Y}} = \hat{\beta}_0 + (\bar{\mathbf{X}}^*)\hat{\boldsymbol{\beta}}^*$, where $\bar{\mathbf{Y}}$ is $n \times 1$ with all elements equal to $\bar{Y}$, $\mathbf{X}^*$ is $n \times K$ and the same as $\mathbf{X}$ except that the first column of $\mathbf{X}$ has been deleted, and $\bar{\mathbf{X}}^*$ is $n \times K$ with all rows equal to $\bar{\mathbf{X}}$. Upon subtracting the second equation from the first, we arrive at Eq. (4-52).

This ratio follows the F distribution with K and $n - K - 1$ degrees of freedom. Thus, if F exceeds $F_{\alpha; K, n-K-1}$, the null hypothesis is rejected.

Let us now show the relationship between Eqs. (4-47a) and (4-55). From Eq. (4-47a) and the null hypothesis (i.e., $\boldsymbol{\beta}_0^* = 0$)

$$T^2 = (\hat{\boldsymbol{\beta}}^*)'[\hat{\sigma}^2(\mathbf{x}'\mathbf{x})^{-1}]^{-1}(\hat{\boldsymbol{\beta}}^*)$$

We are allowed to substitute $(\mathbf{x}'\mathbf{x})^{-1}$ for $(\mathbf{X}'\mathbf{X})^{-1}$ since $(\mathbf{x}'\mathbf{x})^{-1}$ is a submatrix of $(\mathbf{X}'\mathbf{X})^{-1}$, and since we only use this portion of $(\mathbf{X}'\mathbf{X})^{-1}$ in the test. Furthermore, from Eq. (4-39), $\hat{\sigma}^2 = \mathbf{e}'\mathbf{e}/(n - K - 1)$ so that

$$T^2 = \frac{(\hat{\boldsymbol{\beta}}^*)'(\mathbf{x}'\mathbf{x})(\hat{\boldsymbol{\beta}}^*)}{\mathbf{e}'\mathbf{e}/(n - K - 1)}$$

Furthermore, since $(\mathbf{x}'\mathbf{x})\hat{\boldsymbol{\beta}}^* = \mathbf{x}'\mathbf{y}$,

$$T^2 = \frac{(\hat{\boldsymbol{\beta}}^*)'(\mathbf{x}'\mathbf{y})}{\mathbf{e}'\mathbf{e}/(n - K - 1)}$$

Finally, from Eq. (4-48a), $F = T^2/K$, so that

$$F = \frac{(\hat{\boldsymbol{\beta}}^*)'(\mathbf{x}'\mathbf{y})/K}{\mathbf{e}'\mathbf{e}/(n - K - 1)}$$

For our present example

$$(\hat{\boldsymbol{\beta}}^*)'(\mathbf{x}'\mathbf{y}) = [1.498 \quad 1.182]\begin{bmatrix} 6775.52 \\ 5377.62 \end{bmatrix}$$
$$= 16{,}506.07$$

and, as already calculated,

$$\mathbf{e}'\mathbf{e} = 18{,}202$$

so that

$$F = \frac{16{,}506.07/2}{18{,}202.0/24} = \frac{8253.0}{758.42} = 10.9$$

which is the previously calculated T^2 divided by $K = 2$ (i.e., $21.74/2 = 10.9$). Furthermore, this calculated F exceeds $F_{0.05; 2, 24} = 3.40$ so that we reject the null hypothesis just as we did using T^2.

Confidence regions using Eq. (4-49) are also very useful. We leave the calculation of the confidence region for the present example as an exercise, and use instead an example reported by Griliches (1967).

Griliches fitted an equation where two of the slopes ended up taking on the values $\hat{\beta}_1 = 1.15$ and $\hat{\beta}_2 = -0.33$. According to theoretical expectations, which we omit here, the sum of these two slopes in the population should be less than 1. As Griliches points out, it is not legitimate to use univariate confidence intervals to examine this expectation. In fact, had he used the two univariate confidence intervals he would have arrived in this case at an erroneous conclusion. Figure 4.5 shows a straight line representing $\beta_1 + \beta_2 = 1$. Any point to the left of this line gives $\beta_1 + \beta_2 < 1$. The center of the box drawn in this figure represents the point $\hat{\beta}_2, \hat{\beta}_1$ and the box itself represents the boundaries of the two univariate confidence intervals set by use of Eq. (4-46), each with 95 percent confidence. Notice that part of the box extends into the region to the right of the straight line so that the contention that the sum of the coefficients is greater than 1 would be accepted by use of the two univariate intervals. The ellipse shown in Figure 4.5 shows the joint confidence region generated by Eq. (4-49). No part of the region to the right of the straight line is covered by this elliptical joint region, and so it is concluded at $\alpha = 0.05$ that the sum of the coefficients is less than unity.

We stress here that Griliches' test was not one which tested the hypothesis that $\beta_1 + \beta_2 = 1$. That test would have been carried out by use of linear combinations. Rather, by use of the joint confidence ellipse, Griliches was able to test all joint null hypotheses which would have specified $\beta_1 + \beta_2 < 1$, i.e., the region to the left of the line $\beta_1 + \beta_2 = 1$. This example should serve to point out once again the generality of the joint confidence region.

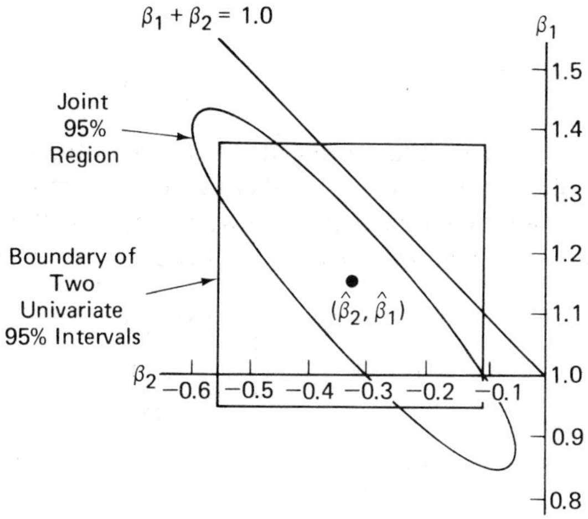

Figure 4.5. Comparison of univariate confidence limits and a joint confidence region for regression slopes.

Source: Griliches (1967), p. 31

4.5 prediction

One of the main reasons for formulating a regression equation is so that the equation can be used in prediction. Very often it is necessary to predict the population mean μ_i for a given set of values, X_i. Recall that $\hat{\mu}_i = \hat{Y}_i$

and that

$$\hat{\mathbf{Y}} = \mathbf{X}\hat{\boldsymbol{\beta}}$$

Therefore

1. If $\hat{\boldsymbol{\beta}}$ is normally distributed, then $\hat{\mathbf{Y}}$ is also normally distributed.
2. $\hat{\mathbf{Y}}$ is an unbiased estimator of $\boldsymbol{\mu}$ (where $\boldsymbol{\mu}$ is the vector of conditional means); $E(\hat{\mathbf{Y}}) = E(\mathbf{X}\hat{\boldsymbol{\beta}}) = \mathbf{X}E(\hat{\boldsymbol{\beta}}) = \mathbf{X}\boldsymbol{\beta} = \boldsymbol{\mu}$.
3. The variance of $\hat{\mathbf{Y}}$ is

$$\begin{aligned}\text{var}(\hat{\mathbf{Y}}) &= E[(\mathbf{X}\hat{\boldsymbol{\beta}} - \mathbf{X}\boldsymbol{\beta})(\mathbf{X}\hat{\boldsymbol{\beta}} - \mathbf{X}\boldsymbol{\beta})'] \\ &= E[\mathbf{X}(\hat{\boldsymbol{\beta}} - \boldsymbol{\beta})(\hat{\boldsymbol{\beta}} - \boldsymbol{\beta})'\mathbf{X}'] \\ &= \mathbf{X}E[(\hat{\boldsymbol{\beta}} - \boldsymbol{\beta})(\hat{\boldsymbol{\beta}} - \boldsymbol{\beta})']\mathbf{X}' \\ &= \sigma^2\mathbf{X}(\mathbf{X}'\mathbf{X})^{-1}\mathbf{X}'\end{aligned}$$

If we wish an unbiased prediction of the population mean, μ_i, associated with the set of values X_i, such a prediction can be obtained by use of

$$\hat{Y}_i = \hat{\mu}_i = \mathbf{X}_i\hat{\boldsymbol{\beta}} \tag{4-56}$$

where

$$\mathbf{X}_i = [1 \quad X_{1i} \quad X_{2i} \quad \cdots \quad X_{Ki}]$$

and the vector $\mathbf{X}_i$ may lie within or outside the range of the sample observations. For example, if it is desired to predict the mean rivet-setting ability of all trainees who score 100 on the manual-dexterity test and 100 on the finger-dexterity test, then the regression equation of Sec. 4.2 gives the prediction

$$\begin{aligned}\hat{Y}_i &= [1 \quad 100 \quad 100]\begin{bmatrix}-104.020 \\ 1.498 \\ 1.182\end{bmatrix} \\ &= 163.98\end{aligned}$$

The variance of the prediction will be

$$\text{var}(\hat{Y}_i) = \text{var}(\hat{\mu}_i) = \sigma^2\mathbf{X}_i(\mathbf{X}'\mathbf{X})^{-1}\mathbf{X}_i' \tag{4-57}$$

However, since σ^2 is usually unknown, it is replaced by $\hat{\sigma}^2$, giving the estimated variance

$$\begin{aligned}s^2_{\hat{Y}_i} &= \hat{\sigma}^2\mathbf{X}_i(\mathbf{X}'\mathbf{X})^{-1}\mathbf{X}_i' \\ &= \mathbf{X}_i\mathbf{S}_\beta\mathbf{X}_i'\end{aligned} \tag{4-58}$$

For our example

$$s_{\hat{Y}_i}^2 = [1 \quad 100 \quad 100] \begin{bmatrix} 3583.240 & -22.780 & -10.272 \\ -22.780 & 0.410 & -0.305 \\ -10.272 & -0.305 & 0.551 \end{bmatrix} \begin{bmatrix} 1 \\ 100 \\ 100 \end{bmatrix}$$

$$= 482.84$$

Test that a Single Prediction is a Given Constant. To test the hypothesis that a prediction is a constant, i.e.,

$$H_0: \quad \mu_i = \mu_{0i}$$

we form the statistic

$$t = \frac{\hat{\mu}_i - \mu_{0i}}{\sqrt{s_{\hat{Y}_i}^2}} \tag{4-59}$$

which follows the t distribution with $n - K - 1$ degrees of freedom.

For our particular example, let us test the hypothesis that the population mean is 160 against a two-sided alternative. The t statistic becomes

$$t = \frac{163.98 - 160}{\sqrt{482.84}} = \frac{3.98}{21.97} = 0.181$$

and at $\alpha = 0.05$ the critical value of t is $t_{0.025;\,24} = 2.064$. Since the sample t does not exceed the critical value, we do not reject the null hypothesis.

A $100(1 - \alpha)$ percent confidence (or prediction) interval for μ_i may be set by use of Eq. (4-59). Thus the interval will be bounded by

$$\hat{Y}_i \pm (t_{\alpha/2;\, n-K-1})\sqrt{s_{\hat{Y}_i}^2} \tag{4-60}$$

In the case where $K = 1$ (there is one independent variable), the estimate of the population μ_i is given by

$$\hat{Y}_i = \hat{\beta}_0 + \hat{\beta}_1 X_i$$

Now in this case [see the discussion of Eq. (4-19)]

$$\mathbf{S}_{\hat{\beta}} = \hat{\sigma}^2(\mathbf{X}'\mathbf{X})^{-1}$$

$$= \frac{\hat{\sigma}^2}{n \sum x^2} \begin{bmatrix} \sum X^2 & -\sum X \\ -\sum X & n \end{bmatrix}$$

Therefore

$$s_{\hat{Y}_i}^2 = [1 \quad X_i] \mathbf{S}_{\hat{\beta}} \begin{bmatrix} 1 \\ X_i \end{bmatrix}$$

$$= \frac{\hat{\sigma}^2}{n \sum x^2} \{\sum X^2 - 2X_i \sum X + nX_i^2\}$$

$$= \frac{\hat{\sigma}^2}{\sum x^2} \left\{\frac{1}{n} \sum X^2 - 2X_i\bar{X} + X_i^2\right\}$$

Recalling that $\sum X^2 = \sum x^2 + n(\bar{X})^2$, we have

$$s_{\hat{Y}_i}^2 = \hat{\sigma}^2\left\{\frac{1}{n} + \frac{(X_i - \bar{X})^2}{\sum x^2}\right\} \tag{4-61}$$

Notice that $s_{\hat{Y}_i}^2$ increases as X_i departs in absolute value from $\bar{X}$. Therefore the confidence interval given by Eq. (4-60) will increase as $(X_i - \bar{X})^2$ increases. This is a commonsense result since it states that the most reliable (minimum variance) prediction can be achieved when $X_i = \bar{X}$. Surely this should be the case since the sample regression line must pass through $\bar{X}$. Thus as we move away from the center of the sample scatter, the reliability of the prediction decreases.

Equation (4-60) gives the confidence interval for the population mean μ_i. Sometimes we may be more interested in setting a confidence interval for a Y_i rather than for a μ_i. For example, if a trainee makes a given score on the two tests, we may be interested in predicting his individual rivet-setting ability rather than the mean rivet-setting ability of all persons who make the same scores as this trainee.

In predicting an individual value of Y_i (rather than the mean value μ_i) associated with $\mathbf{X}_i$, it can be shown that an unbiased estimate of Y_i is still $\hat{Y}_i = \mathbf{X}_i\hat{\boldsymbol{\beta}}$. Furthermore, since $Y_i = \mathbf{X}_i\boldsymbol{\beta} + \epsilon_i$, the error of the individual prediction is equal to

$$\begin{aligned} Y_i - \hat{Y}_i &= \mathbf{X}_i\boldsymbol{\beta} + \epsilon_i - \mathbf{X}_i\hat{\boldsymbol{\beta}} \\ &= \mathbf{X}_i(\boldsymbol{\beta} - \hat{\boldsymbol{\beta}}) + \epsilon_i \end{aligned}$$

We note that $\hat{Y}_i$ is an unbiased estimate of the individual value of Y_i since the expected value of the error is zero. The variance of the individual prediction is

$$\begin{aligned} E[(Y_i - \hat{Y}_i)]^2 &= E\{[\mathbf{X}_i(\boldsymbol{\beta} - \hat{\boldsymbol{\beta}}) + \epsilon_i]^2\} \\ &= E\{[\mathbf{X}_i(\hat{\boldsymbol{\beta}} - \boldsymbol{\beta})]^2\} + E(\epsilon_i^2) - E\{2\mathbf{X}_i(\hat{\boldsymbol{\beta}} - \boldsymbol{\beta})\epsilon_i\} \end{aligned}$$

The term $E\{2\mathbf{X}_i(\hat{\boldsymbol{\beta}} - \boldsymbol{\beta})\epsilon_i\}$ expands into $2\mathbf{X}_iE\{((\mathbf{X}'\mathbf{X})^{-1}\mathbf{X}'\boldsymbol{\epsilon})\epsilon_i\}$ from Eq. (4-40a). Thus, since the Xs are assumed to be statistical constants and $\boldsymbol{\epsilon}$ is independent of ϵ_i, the term vanishes upon taking the expected value. Therefore the variance of the individual prediction is equal to

$$\sigma^2\mathbf{X}_i(\mathbf{X}'\mathbf{X})^{-1}\mathbf{X}_i' + \sigma^2$$

or

$$\sigma^2[1 + \mathbf{X}_i(\mathbf{X}'\mathbf{X})^{-1}\mathbf{X}_i'] \tag{4-62}$$

Replacing σ^2 with its estimate, we have the estimate of the variance:

$$\hat{\sigma}^2[1 + \mathbf{X}_i(\mathbf{X}'\mathbf{X})^{-1}\mathbf{X}_i'] \tag{4-63}$$

or

$$\hat{\sigma}^2 + \mathbf{X}_i\mathbf{S}_{\hat{\beta}}\mathbf{X}_i'$$

A $100(1 - \alpha)$ percent confidence (or prediction) interval for Y_i will be bounded by

$$\hat{Y}_i \pm (t_{\alpha/2;\, n-K-1})\sqrt{\hat{\sigma}^2 + \mathbf{X}_i\mathbf{S}_{\hat{\beta}}\mathbf{X}_i'} \tag{4-64}$$

In the special case where $K = 1$ the reader may verify that

$$\hat{\sigma}^2 + \mathbf{X}_i\mathbf{S}_{\hat{\beta}}\mathbf{X}_i' = \hat{\sigma}^2\left[1 + \frac{1}{n} + \frac{(X_i - \bar{X})^2}{\sum x^2}\right]$$

Thus the confidence interval of Eq. (4-64) will increase as $(X_i - \bar{X})^2$ increases. This interval will be wider than the one for μ_i and reflects greater uncertainty in predicting a specific value for Y_i than in predicting the mean value μ_i.

Joint Tests. Suppose that we have several predictions. For example, given several income figures we may wish to use the consumption function of this chapter to predict several consumption figures. If we have p such predictions, they will be given by

$$\begin{bmatrix} \hat{Y}_1 \\ \hat{Y}_2 \\ \cdot \\ \cdot \\ \cdot \\ \hat{Y}_p \end{bmatrix} = \begin{bmatrix} 1 & X_{11} & \cdots & X_{K1} \\ 1 & X_{12} & \cdots & X_{K2} \\ \cdot & & & \cdot \\ \cdot & & & \cdot \\ \cdot & & & \cdot \\ 1 & X_{1p} & \cdots & X_{Kp} \end{bmatrix} \begin{bmatrix} \hat{\beta}_0 \\ \hat{\beta}_1 \\ \cdot \\ \cdot \\ \cdot \\ \hat{\beta}_K \end{bmatrix} \tag{4-65}$$

or

$$\hat{\mathbf{Y}}_* = \mathbf{X}_*\hat{\boldsymbol{\beta}} \tag{4-65a}$$

The joint hypothesis

$$H_0\colon\ \boldsymbol{\mu}_* = \boldsymbol{\mu}_{0*}$$

where $\boldsymbol{\mu}_*$ is the vector of means corresponding to $\hat{\mathbf{Y}}_*$ and $\boldsymbol{\mu}_{0*}$ is a vector of constants, can be tested by use of Hotelling's T^2:

$$T^2 = (\hat{\mathbf{Y}}_* - \boldsymbol{\mu}_{0*})'\mathbf{S}_{\hat{Y}_*}^{-1}(\hat{\mathbf{Y}}_* - \boldsymbol{\mu}_{0*}) \tag{4-66}$$

The matrix $\mathbf{S}_{\hat{Y}_*}$ is the covariance matrix for the p estimates, $\hat{\mathbf{Y}}_*$. That is,

$$\mathbf{S}_{\hat{Y}_*} = \begin{bmatrix} s_{\hat{Y}_1\hat{Y}_1} & s_{\hat{Y}_1\hat{Y}_2} & \cdots & s_{\hat{Y}_1\hat{Y}_p} \\ \cdot & & & \cdot \\ \cdot & & & \cdot \\ \cdot & & & \cdot \\ s_{\hat{Y}_1\hat{Y}_p} & s_{\hat{Y}_2\hat{Y}_p} & \cdots & s_{\hat{Y}_p\hat{Y}_p} \end{bmatrix}$$
$$= \mathbf{X}_*\mathbf{S}_\beta\mathbf{X}'_*$$

Under the null hypothesis, the critical value of T^2 is given by

$$T^2_{\alpha;\,p,n-K-1} = pF_{\alpha;\,p,n-K-1}$$

and a $100(1-\alpha)$ percent joint confidence region is bounded by

$$(\hat{\mathbf{Y}}_* - \boldsymbol{\mu}_{0*})'\mathbf{S}^{-1}_{\hat{Y}_*}(\hat{\mathbf{Y}}_* - \boldsymbol{\mu}_{0*}) = pF_{\alpha;\,p,n-K-1}$$

If it is desired to test joint hypotheses and set joint confidence regions for specific $\mathbf{Y}_i$ values rather than $\boldsymbol{\mu}_i$ values, then the last four equations remain in force except that [analogous to Eq. (4-63)] $\mathbf{S}_{\hat{Y}_*}$ is replaced by

$$\hat{\sigma}^2\mathbf{I} + \mathbf{X}_*\mathbf{S}_\beta\mathbf{X}'_*$$

where $\mathbf{I}$ is the identity matrix of the same order as $\mathbf{X}_*\mathbf{S}_\beta\mathbf{X}'_*$.

4.6 test of the equality of coefficients for different regression equations

In Sec. 4.4 we were interested in testing the coefficients in a single regression equation. Sometimes we are interested in testing whether the population coefficients are equal in different regression equations (see especially Sec. 5.4). Table 4.4 shows the selling price, Y, and floor space, X, of several residential dwellings, marketed in the Memphis, Tennessee area in 1970. The houses have been classified by an experienced realtor according to their condition. The condition classifications are fair, good, and excellent. Suppose that three regression equations are specified, one for each condition classification:

$$\begin{aligned} Y_{1i} &= \beta_{10} + \beta_{11}X_{1i} + \epsilon_{1i}, \qquad i = 1, 2, \ldots, n_1 \\ Y_{2i} &= \beta_{20} + \beta_{21}X_{2i} + \epsilon_{2i}, \qquad i = 1, 2, \ldots, n_2 \\ Y_{3i} &= \beta_{30} + \beta_{31}X_{3i} + \epsilon_{3i}, \qquad i = 1, 2, \ldots, n_3 \end{aligned} \tag{4-67}$$

The first subscript indicates the classification (i.e., 1 = fair, 2 = good, and 3 = excellent). Figure 4.6 shows the sample Y values plotted against the

Table 4.4 Selling Price, Floor Space (square feet), and Condition of Twenty-two Residential Dwellings

	Condition					
	Fair (1)		*Good (2)*		*Excellent (3)*	
	Value, Y_{1i}	*Square feet*, X_{1i}	*Value*, Y_{2i}	*Square feet*, X_{2i}	*Value*, Y_{3i}	*Square feet*, X_{3i}
	$10,600	843	$14,650	1317	$16,650	1318
	10,750	907	12,850	1080	27,500	1507
	7,550	824	17,900	1688	16,450	1292
	8,800	672	21,977	1738	33,300	2350
	8,300	698	13,900	1040	33,000	1732
	5,298	723	13,100	1404	30,212	2010
	—	—	19,750	1558	22,900	1749
	—	—	11,600	976	20,180	1762
Mean	8549.66	777.83	15,715.88	1350.13	25,024.00	1715.00

Source: Data courtesy of Mike Harris.

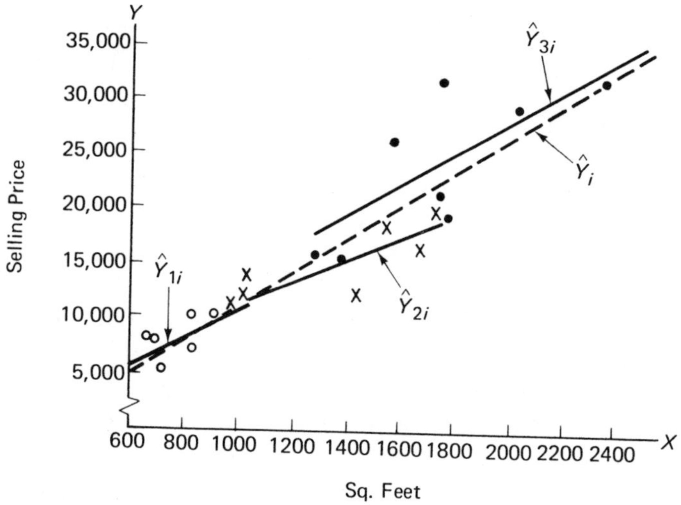

Figure 4.6. Scatter diagram of data of Table 4.4 and regression lines.

sample X values. Each pair of observations is distinguishable as belonging to one of the three classifications: circles for houses rated fair, crosses for houses rated good, and dots for houses rated excellent. The three estimated regression lines are also plotted as solid lines and are

$$\begin{aligned} Y_{1i} &= -1135.36 + 12.45X_{1i} + e_{1i} \\ Y_{2i} &= 981.17 + 10.91X_{2i} + e_{2i} \\ Y_{3i} &= -606.98 + 14.95X_{3i} + e_{3i} \end{aligned} \tag{4-68}$$

with the sum of the squared residuals for each equation being

$$\mathbf{e}_1'\mathbf{e}_1 = \sum_{i=1}^{6} e_{1i}^2 = 13{,}976{,}855.58$$

$$\mathbf{e}_2'\mathbf{e}_2 = \sum_{i=1}^{8} e_{2i}^2 = 22{,}719{,}035.91$$

$$\mathbf{e}_3'\mathbf{e}_3 = \sum_{i=1}^{8} e_{3i}^2 = 141{,}617{,}937.46$$

We want to test whether these three regression lines have the same coefficients. That is, we wish to test

H_0: $\beta_{10} = \beta_{20} = \beta_{30} = \beta_0$ and $\beta_{11} = \beta_{21} = \beta_{31} = \beta_1$
H_1: Not all corresponding coefficients are equal

For this test we reason as follows. Under the hypotheses that all slopes and all intercepts are equal, one regression line will serve in place of three. This line is

$$Y_i = \beta_0 + \beta_1 X_i + \epsilon_i, \qquad i = 1, 2, \ldots, n \tag{4-69}$$

where $n = n_1 + n_2 + n_3$. That is, under the hypothesis we can pool all the data together, and Eq. (4-69) will fit just as well as Eqs. (4-67). More specifically, the sum of the squared residuals of Eq. (4-69) should not be significantly different from the total of the sum of squared residuals of the three individual equations.

The estimate of the grand regression (shown as the dashed line in Figure 4.6) is

$$Y_i = -4078.13 + 16.00X_i + e_i \tag{4-70}$$

and was obtained by regressing all 22 X and Y values of Table 4.4. The sum of the squared residuals for this equation is

$$\mathbf{e}'\mathbf{e} = \sum_{i=1}^{22} e_i^2 = 244{,}351{,}749.56$$

The total sum of squared residuals from the three individual equations is

$$S_1 = \mathbf{e}_1'\mathbf{e}_1 + \mathbf{e}_2'\mathbf{e}_2 + \mathbf{e}_3'\mathbf{e}_3 = 178{,}313{,}828.95 \qquad (4\text{-}71)$$

with $n - 6$ [or $(n_1 - 2) + (n_2 - 2) + (n_3 - 2) = n - 6$] degrees of freedom. Furthermore, let

$$S_0 = \mathbf{e}'\mathbf{e} \qquad (4\text{-}72)$$

with $n - 2$ degrees of freedom. The discrepancy, $S_0 - S_1$, will be attributable to sampling error if the null hypothesis is true. Call this discrepancy

$$S_2 = S_0 - S_1 = 66{,}037{,}920.61 \qquad (4\text{-}73)$$

with 4 [or $(n - 2) - (n - 6) = 4$] degrees of freedom. It can be shown that the null hypothesis can be tested by use of

$$F = \frac{S_2/4}{S_1/(n-6)}$$

which follows the F distribution with 4 and $n - 6$ degrees of freedom. This F ratio compares the discrepancy with the total sum of squares from the three equations, both adjusted for their corresponding numbers of degrees of freedom. In our case

$$F = \frac{66{,}037{,}920.61/4}{178{,}313{,}828.95/16} = 1.48$$

and this F ratio does not exceed $F_{0.05;\,4,16} \doteq 3.01$. Therefore we find no difference between the coefficients at the stated level of significance.

In general, suppose that we have p equations

$$\begin{aligned} \mathbf{Y}_1 &= \mathbf{X}_1\boldsymbol{\beta}_1 + \boldsymbol{\epsilon}_1 \\ &\vdots \\ \mathbf{Y}_p &= \mathbf{X}_p\boldsymbol{\beta}_p + \boldsymbol{\epsilon}_p \end{aligned} \qquad (4\text{-}74)$$

with $n_1, n_2, \ldots, n_p$ observations in each of the equations, respectively, and $\boldsymbol{\beta}_i' = [\beta_{i0} \quad \beta_{i1} \quad \cdots \quad \beta_{iK}]$. Thus we have p regression equations with K variables in each equation. We wish to test

$$H_0\colon\ \boldsymbol{\beta}_1 = \boldsymbol{\beta}_2 = \cdots = \boldsymbol{\beta}_p = \boldsymbol{\beta}$$
$$H_1\colon\ \text{Not all } \boldsymbol{\beta}_i\text{s are equal}$$

To carry out this test, we first pool all the data and estimate the grand regression equation

$$\mathbf{Y} = \mathbf{X}\hat{\boldsymbol{\beta}} + \mathbf{e} \tag{4-75}$$

which is based upon $n = n_1 + n_2 + \cdots + n_p$ observations. Then we calculate the sum of squared residuals for each individual regression equation and total these sums, forming

$$S_1 = \sum_{i=1}^{p} \mathbf{e}_i'\mathbf{e}_i \tag{4-76}$$

This total is based upon $n - p(K + 1)$ degrees of freedom. Similarly, we calculate the sum of the squared residuals for Eq. (4-75)

$$S_0 = \mathbf{e}'\mathbf{e} \tag{4-77}$$

which is based upon $n - K - 1$ degrees of freedom. The discrepancy is

$$S_2 = S_0 - S_1$$

and is based upon $(n - K - 1) - (n - p(K + 1)) = (p - 1)(K + 1)$ degrees of freedom. The F ratio is defined as

$$F = \frac{S_2/(p - 1)(K + 1)}{S_1/(n - p(K + 1))} \tag{4-78}$$

with $(p - 1)(K + 1)$ and $n - p(K + 1)$ degrees of freedom.

Subsets of Coefficients. As we have remarked before, in economic research the intercept of a regression equation is generally of less consequence than the slope. Therefore suppose that we are interested in testing the hypothesis that only the slopes are equal among several regression equations. That is, we wish to test for our residential dwelling example the hypotheses

$$\begin{aligned} H_0&: \quad \beta_{11} = \beta_{21} = \beta_{31} = \beta \\ H_1&: \quad \text{Not all slopes are equal} \end{aligned} \tag{4-79}$$

Under the hypothesis that all slopes are equal, the effect of floor space upon selling price is the same for the three condition classifications, and the model of Eqs. (4-67) becomes

$$\begin{aligned} Y_{1i} &= \beta_{10} + \beta X_{1i} + \epsilon_{1i}, \qquad i = 1, 2, \ldots, n_1 \\ Y_{2i} &= \beta_{20} + \beta X_{2i} + \epsilon_{2i}, \qquad i = 1, 2, \ldots, n_2 \\ Y_{3i} &= \beta_{30} + \beta X_{3i} + \epsilon_{3i}, \qquad i = 1, 2, \ldots, n_3 \end{aligned} \tag{4-80}$$

In matrix form, by pooling the observations, we have

$$\begin{bmatrix} Y_{11} \\ \cdot \\ \cdot \\ \cdot \\ Y_{1n_1} \\ Y_{21} \\ \cdot \\ \cdot \\ \cdot \\ Y_{2n_2} \\ Y_{31} \\ \cdot \\ \cdot \\ \cdot \\ Y_{3n_3} \end{bmatrix} = \begin{bmatrix} 1 & 0 & 0 & X_{11} \\ \cdot & \cdot & \cdot & \cdot \\ \cdot & \cdot & \cdot & \cdot \\ \cdot & \cdot & \cdot & \cdot \\ 1 & 0 & 0 & X_{1n_1} \\ 0 & 1 & 0 & X_{21} \\ \cdot & \cdot & \cdot & \cdot \\ \cdot & \cdot & \cdot & \cdot \\ \cdot & \cdot & \cdot & \cdot \\ 0 & 1 & 0 & X_{2n_2} \\ 0 & 0 & 1 & X_{31} \\ \cdot & \cdot & \cdot & \cdot \\ \cdot & \cdot & \cdot & \cdot \\ \cdot & \cdot & \cdot & \cdot \\ 0 & 0 & 1 & X_{3n_3} \end{bmatrix} \begin{bmatrix} \beta_{10} \\ \beta_{20} \\ \beta_{30} \\ \beta \end{bmatrix} + \begin{bmatrix} \epsilon_{11} \\ \cdot \\ \cdot \\ \cdot \\ \epsilon_{1n_1} \\ \epsilon_{21} \\ \cdot \\ \cdot \\ \cdot \\ \epsilon_{2n_2} \\ \epsilon_{31} \\ \cdot \\ \cdot \\ \cdot \\ \epsilon_{3n_3} \end{bmatrix}$$

The coefficient β is called the average (or pooled) slope. It will be encountered again in Sec. 5.4.

More compactly, we may write

$$\mathbf{Y} = \mathbf{W}\boldsymbol{\delta} + \boldsymbol{\epsilon} \tag{4-81}$$

The least-squares estimate of Eq. (4-81) is

$$\boldsymbol{\delta} = (\mathbf{W}'\mathbf{W})^{-1}\mathbf{W}'\mathbf{Y} \tag{4-82}$$

with $\mathbf{e}$ as the sample residual. Under the hypothesis we have

$$S_0 = \mathbf{e}'\mathbf{e}$$

with $n - 4$ degrees of freedom. For our example

$$S_0 = 184{,}241{,}946.90$$

The discrepancy between S_0 and the total of the sum of squares of residuals for the three equations (178,313,828.95) is

$$\begin{aligned} S_2 = S_0 - S_1 &= 184{,}241{,}946.90 - 178{,}313{,}828.95 \\ &= 5{,}928{,}117.95 \end{aligned}$$

with 2 [or $(n - 4) - (n - 6)$] degrees of freedom. Thus the F ratio is

$$F = \frac{S_2/2}{S_1/16} = 0.27 \tag{4-83}$$

with 2 and 16 degrees of freedom. Of course, this value does not exceed $F_{0.05;\,2,16}$, and we do not reject the null hypothesis.

Again, in general, suppose that we have a set of p equations

$$\begin{aligned} \mathbf{Y}_1 &= \mathbf{X}_1\boldsymbol{\alpha}_1 + \mathbf{Z}_1\boldsymbol{\beta}_1 + \boldsymbol{\epsilon}_1 \\ &\vdots \\ \mathbf{Y}_p &= \mathbf{X}_p\boldsymbol{\alpha}_p + \mathbf{Z}_p\boldsymbol{\beta}_p + \boldsymbol{\epsilon}_p \end{aligned} \tag{4-84}$$

where each $\boldsymbol{\alpha}_i$ is $q \times 1$ and each $\boldsymbol{\beta}_i$ is $r \times 1$. For any equation $q + r = K + 1$, the number of independent variables plus the intercept. In Eq. (4-84) we have split the coefficients into two sets: a set represented by $\boldsymbol{\alpha}$, which is *not* the one that we are interested in testing, and a set represented by $\boldsymbol{\beta}$, which is the one that we are interested in testing for equality. Hence we wish to test

$$H_0:\ \boldsymbol{\beta}_1 = \boldsymbol{\beta}_2 = \cdots = \boldsymbol{\beta}_p = \boldsymbol{\beta}$$
$$H_1:\ \text{Not all } \boldsymbol{\beta}\text{s are equal}$$

Again, we pool the observations under the hypothesis

$$\begin{bmatrix} \mathbf{Y}_1 \\ \mathbf{Y}_2 \\ \vdots \\ \mathbf{Y}_p \end{bmatrix} = \begin{bmatrix} \mathbf{X}_1 & \mathbf{0} & \mathbf{0} & \cdots & \mathbf{0} & \mathbf{Z}_1 \\ \cdot & \mathbf{X}_2 & & & \mathbf{0} & \mathbf{Z}_2 \\ \vdots & & \ddots & & & \vdots \\ \mathbf{0} & \mathbf{0} & \mathbf{0} & \cdots & \mathbf{X}_p & \mathbf{Z}_p \end{bmatrix} \begin{bmatrix} \boldsymbol{\alpha}_1 \\ \boldsymbol{\alpha}_2 \\ \vdots \\ \boldsymbol{\alpha}_p \\ \boldsymbol{\beta} \end{bmatrix} + \begin{bmatrix} \boldsymbol{\epsilon}_1 \\ \boldsymbol{\epsilon}_2 \\ \vdots \\ \boldsymbol{\epsilon}_p \end{bmatrix} \tag{4-85}$$

In Eq. (4-85) each $\mathbf{Y}_i$ is an $n_i \times 1$ vector, each $\mathbf{X}_i$ is an $n_i \times q$ matrix, and each $\mathbf{Z}_i$ is an $n_i \times r$ matrix. The remainder of the items in the matrix to the right of the equality sign are zero. Again, we may write Eq. (4-85) compactly as $\mathbf{Y} = \mathbf{W}\boldsymbol{\delta} + \boldsymbol{\epsilon}$, whose solution is Eq. (4-82). The test procedure is the same as that given by Eqs. (4-76)–(4-78) except that the number of degrees of freedom is different, i.e.,

$$F = \frac{S_2/r(p-1)}{S_1/(n - p(K+1))} \tag{4-86}$$

with $r(p-1)$ and $n - p(K+1)$ degrees of freedom.

Questions and Problems

Sec. 4.2

1. If the assumptions of Eq. (4-14) hold, the distribution of $\boldsymbol{\epsilon}$ is said to be spherical. Thus the population covariance matrix is of the form $\boldsymbol{\Sigma} = \sigma^2\mathbf{I}$, where $\mathbf{I}$ is an $n \times n$ identity matrix. The presence of nonzero elements off the main diagonal of $\boldsymbol{\Sigma}$ is referred to as the problem of *autocorrelation.* The presence of unequal elements on the main diagonal of $\boldsymbol{\Sigma}$ is called the problem of *heteroscedasticity.* Suppose that $\boldsymbol{\Sigma}^*$ is the population covariance matrix for $\boldsymbol{\epsilon}$ with $\boldsymbol{\Sigma}^*$ demonstrating both autocorrelation and heteroscedasticity. Discuss a transformation matrix $\mathbf{T}$ such that $\mathbf{T}\boldsymbol{\Sigma}^*\mathbf{T}' = \mathbf{I}$ (an $n \times n$ identity matrix). Can $\mathbf{I}$ represent a well-behaved $\boldsymbol{\Sigma}$ matrix?

Sec. 4.3

1. *Generalized least squares* (a continuation of Problem 1 of Sec. 4.2). Let $\mathbf{Y} = \mathbf{XB} + \boldsymbol{\epsilon}$ have a covariance matrix $\boldsymbol{\Sigma}^*$ which we wish to convert to $\mathbf{I}$ by pre- and postmultiplication by $\mathbf{T}$ and $\mathbf{T}'$, respectively. Show that such conversion implies that we write the regression function as $\mathbf{TY} = \mathbf{TX}\boldsymbol{\beta} + \mathbf{T}\boldsymbol{\epsilon}$. Write the least-squares estimate of $\boldsymbol{\beta}$.

2. Table 6.2 of Chap. 6 shows data for consumption, Y; current profits, X_1; lagged profits, X_2; and current wages, X_3. Regress Y on the three X variables. Discuss any interesting features of the estimated coefficients that might impress you from an economic point of view.

Sec. 4.4

1. Using the data of Table 6.2, the calculations that you completed in Problem 2 of Sec. 4.3, and ignoring any possible problems that might arise because the data are economic time series,
   a. Test the univariate hypotheses that each of the coefficients is zero. Use $\alpha = 0.05$.
   b. Test the hypothesis that all slopes are jointly zero. Use $\alpha = 0.05$.
   c. Draw a joint confidence ellipse for β_1 and β_3. Since these coefficients both represent marginal propensities to consume, they should be greater than zero and $\beta_1 + \beta_3 < 1$. Does your ellipse tend to confirm this a priori rule? Use $\alpha = 0.05$.

Sec. 4.5

1. For the consumption function of Problem 2 of Sec. 4.3, suppose that $X_1 = 12.0$, $X_2 = 20.0$, and $X_3 = 32.0$. What is your prediction of the mean value of consumption? Test, at $\alpha = 0.05$, the hypothesis that this prediction does not differ from 50.

2. Discuss the precision of estimating the mean value of consumption with $X_3 = 200.0$ as opposed to the precision of the estimate done in Problem 1 on p. 147.

Sec. 4.6

1. Betancourt (1971) presents some figures for the average income, Y, and consumption, C, for several occupational groups in Chile. The data are differentiated between urban and rural workers. Test the hypothesis that the marginal propensities to consume are equal for rural and urban workers. Use $\alpha = 0.05$.

Occupation Group	*Urban*		*Rural*	
	C	*Y*	*C*	*Y*
0	198.46	191.42	118.18	114.13
1	483.31	507.18	291.95	287.85
2	278.07	277.60	272.76	272.41
3	234.25	230.60	170.67	173.60
4	163.05	162.45	109.88	107.16
5	112.83	113.50	127.75	130.24
6	202.05	210.75	160.10	154.35
7	148.05	143.04	138.27	136.78
8	148.59	145.46	116.66	115.07
9	131.57	127.43	119.48	118.25

APPENDIX

A4.1 subroutine REG

A. Description. The following subroutine calculates the $\hat{\boldsymbol{\beta}}$ vector for linear regression. It also calculates the standard errors of each $\hat{\beta}_k$, $k = 1, 2, \ldots, K$, and the means and standard deviation of each variable. In addition the program calculates the multiple coefficient of determination (to be discussed in the next chapter), the standard error of estimate, and the F ratio [see Eq. (5-6)]. Also, the actual and the estimated values of the dependent variable along with the sample error (or residual) vector, $\mathbf{e}$, are calculated. The matrix $(\mathbf{x}'\mathbf{x})^{-1}$ is also shown.

B. Limitations. A maximum of nine independent variables (not counting the intercept), with a maximum of 100 observations on each variable, is allowed. These maximum sizes can be increased by altering the DIMENSION statements in the subroutine and the related programs. The **x′x** matrix, of course, cannot be singular.

C. Use. Assuming that M (the total number of variables, dependent and independent), N (the number of observations on each variable), and **X** (the data matrix to be described below) are in storage, the statement

CALL REG (M, N, X, XBAR, S, A, AA, B, SEE, YHAT, YRES, R2, F, SER, SSE)

will cause entry into the subroutine. The arguments to the subroutine are as follows.

M	total number of variables
N	total number of observations on each variable
X	data matrix
XBAR	vector of means
S	covariation matrix
A	**x′x** before inversion, $(\mathbf{x'x})^{-1}$ after inversion
AA	intercept
B	vector of slopes
SEE	standard error of estimate
YHAT	vector of estimated values of dependent variable
YRES	vector of residuals, or errors
R2	multiple coefficient of determination
F	F ratio of Eq. (5-6)
SER	vector of standard errors for the slopes
SSE	sum of the squared errors

Subroutine REG also uses subroutines MEAN, COVAR, and INVS. The subroutine is shown below.

```
      SUBROUTINE REG(M,N,X,XBAR,S,A,AA,B,SEE,YHAT,YRES,R2,F,SER,SSE)
C     MEAN,COVAR, AND INVS NEEDED
      DIMENSION X(10,100),XBAR(10),S(10,10),A(10,10),
     1B(10),YHAT(100),YRES(100),SER(10)
      CALL MEAN (M,N,X,XBAR)
      M1=M-1
      CALL COVAR(M,N,X,S,XBAR)
      DO 20 I=1,M1
      DO 20 J=1,M1
   20 A(I,J)=S(I+1,J+1)
      CALL INVS(A,M1)
      DO 25 I=1,M1
      B(I)=0.0
      DO 25 J=1,M1
```

```
25    B(I)=B(I)+S(1,J+1)*A(J,I)
      AA=0.0
      DO 27 I=1,M1
27    AA=AA+XBAR(I+1)*B(I)
      AA=XBAR(1)-AA
      SSE=0.0
      DO 35 I=1,N
      YHAT(I)=0.0
      DO 30 J=2,M
30    YHAT(I)=YHAT(I)+B(J-1)*X(J,I)
      YHAT(I)=YHAT(I)+AA
      YRES(I)=X(1,I)-YHAT(I)
35    SSE=SSE+YRES(I)*YRES(I)
      R2=(S(1,1)-SSE)/S(1,1)
      F=(R2/FLOAT(M1))/((1.-R2)/FLOAT(N-M))
      SEE = SQRT(SSE/FLOAT(N-M))
      DO 52 I=1,M1
52    SER(I)=SEE*SQRT(A(I,I))
      RETURN
      END
```

A4.2 main for regression

A. Description. This main program basically has the function of reading in the data matrix and printing the results from subroutine REG. In addition, the main program calls subroutine CORR and prints out the correlation matrix.

B. Limitations. Same as for REG.

C. Use. The first data card contains M (the total number of variables, dependent and independent) in columns 1–3, right-adjusted and punched without a decimal point. In columns 4–6 on this same card, punch N (the number of observations on each variable), right-adjusted and punched without a decimal point. Then, starting on a fresh card, punch the values of the dependent variable in eight fields of width 10 with the decimal point. Continue from card to card until all observations are punched for the dependent variable. Starting on a fresh card, punch the first independent variable in the same way. Continue punching the independent variables in this way until all are punched. It is important to remember to start each new variable on a fresh card and that the *first* variable in the deck is taken to be dependent.

The main program is shown below.

```
C     REG AND CORR NEEDED
      DIMENSION X(10,100),XBAR(10), S(10,10),A(10,10),R(10,10),
     1B(10),YHAT(100),YRES(100),SER(10)
      READ(5,5) M,N
5     FORMAT(2I3)
      DO 10 I=1,M
10    READ(5,15)(X(I,J),J=1,N)
15    FORMAT(8F10.0)
```

```
C       TRANSFORMATIONS MAY BE ADDED HERE
        CALL REG(M,N,X,XBAR,S,A,AA,B,SEE,YHAT,YRES,R2,F,SER,SSE)
        CALL CORR( M,S,R)
        M1=M-1
        WRITE(6,22)
 22     FORMAT(' INVERSE OF X PRIME X IN DEVIATION FORM')
        DO 23 I=1,M1
 23     WRITE(6,24)(A(I,J),I,J,J=1,M1)
 24     FORMAT(4(E16.8,'(',2I2,')'))
        WRITE(6,45)
 45     FORMAT(' CORRELATION MATRIX( ZERO IS DEPENDENT VARIABLE)')
        DO 50 I=1,M
 50     S(I,I)=SQRT(S(I,I)/FLOAT(N-1))
        DO 51 I=1,M
        I1=I-1
 51     WRITE(6,55)(R(I,J+1),I1,J,J=0,M1)
 55     FORMAT(10(F7.4,'(',2I2,')'))
        WRITE(6,60)
 60     FORMAT(' MULTIPLE REGRESSION AND SELECTED UNIVARIATE STATISTICS')
        WRITE(6,115)
        WRITE(6,62)
 62     FORMAT(8X,'VARIABLE',19X,'MULTIPLE',23X,'UNIVARIATE',/,9X,'NUMBER'
       1,19X,'REGRESSION',22X,'STATISTICS',/,26X,'SLOPE',10X,'STD.ERROR',
       211X,'MEAN', 7X,'STD.DEVIATION')
        WRITE(6,115)
        WRITE(6,65) XBAR(1),S(1,1),B(1),SER(1),XBAR(2),S(2,2)
 65     FORMAT(' DEPENDENT ',15X,'...',14X,'...',3X,2F17.4,/,' INDEPENDENT
       1  1',4F17.4)
        IF(M1-1) 82,82,69
 69     CONTINUE
        DO 70 I=2,M1
 70     WRITE(6,80) I,B(I),SER(I),XBAR(I+1),S(I+1,I+1)
 80     FORMAT(13X,I2,4F17.4)
 82     WRITE(6,115)
        WRITE(6,85) AA,R2,SEE,F
        WRITE(6,115)
 85     FORMAT(' INTERCEPT=',F11.4,' MULT. R-SQUARED=',F6.4 ,' STD.ERROR O
       1F EST.=',F11.4,' F=',F11.4)
        WRITE(6,90)
 90     FORMAT(' ACTUAL AND ESTIMATED VALUES OF DEPENDENT VARIABLE',/,9X,'
       1ACTUAL',10X,'ESTIMATED',9X,'RESIDUAL')
        DO 100 I=1,N
 100    WRITE(6,110) X(1,I), YHAT(I),YRES(I)
 110    FORMAT(3F17.5)
 115    FORMAT(' -----------------------------------------------------------
       1--------------------------------')
        CALL EXIT
        END
```

The data preparation for the two-variable example (consumption and income) of this chapter is shown below. The result of the calculation is also shown.

1	2	3	4	5	6	7	8	9	10	11	12	13	14	15	16	17	18	19	20	21	22	23	24	25	26	27	28	29	30
		2		1	0																								
					1	6	6	6	.						1	7	3	5	.						1	7	4	9	.
					2	1	2	2	.						2	1	6	2	.										
					1	8	3	1	.						1	8	8	1	.						1	8	8	3	.
					2	3	3	2	.						2	4	0	1	.										

```
INVERSE OF X PRIME X IN DEVIATION FORM
 .26731805E-05( 1 1)
CORRELATION MATRIX( ZERO IS DEPENDENT VARIABLE)
1.0000( 0 0)   .9972( 0 1)
 .9972( 1 0) 1.0000( 1 1)
MULTIPLE REGRESSION AND SELECTED UNIVARIATE STATISTICS
-------------------------------------------------------------------------------------------
    VARIABLE                   MULTIPLE                              UNIVARIATE
     NUMBER                   REGRESSION                             STATISTICS
                     SLOPE               STD.ERROR          MEAN          STD.DEVIATION
-------------------------------------------------------------------------------------------
DEPENDENT             ...                   ...          1885.5999          174.2336
INDEPENDENT   1      .8522                 .0225         2057.5999          203.8753
-------------------------------------------------------------------------------------------
INTERCEPT=    132.0540 MULT. R-SQUARED= .9944 STD.ERROR OF EST.=     13.7813 F=  1430.4990
-------------------------------------------------------------------------------------------
ACTUAL AND ESTIMATED VALUES OF DEPENDENT VARIABLE
      ACTUAL             ESTIMATED          RESIDUAL
    1666.00000          1692.48511         -26.48511
    1735.00000          1735.09644          -.09644
    1749.00000          1736.80103         12.19897
    1755.00000          1758.95898         -3.95898
    1813.00000          1809.24048          3.75952
    1865.00000          1847.59058         17.40942
    1945.00000          1941.33594          3.66406
    2044.00000          2036.78540          7.21460
    2122.00000          2119.45166          2.54834
    2162.00000          2178.25562        -16.25562
```

Finally, we show the data preparation for the three-variable problem of this chapter which deals with the ability to set rivets. The output for this problem is also shown.

Columns 1–6	Columns 7–10	Columns 17–20	Columns 27–30
3 (col 3), 2 (col 5), 7 (col 6)			
	230.	81.	100.
	164.	166.	146.
	180.	174.	120.
	143.	131.	130.
	135.	93.	108.
	128.	116.	125.
	126.	136.	104.
	104.	103.	125.
	107.	67.	81.
	80.	86.	78.
	92.	95.	82.
	69.	65.	84.

```
INVERSE OF X PRIME X IN DEVIATION FORM
  .54014940E-03( 1 1)   -.40191156E-03( 1 2)
 -.40191156E-03( 2 1)    .72622765E-03( 2 2)
CORRELATION MATRIX( ZERO IS DEPENDENT VARIABLE)
1.0000( 0 0)  .6482( 0 1)  .5966( 0 2)
 .6482( 1 0) 1.0000( 1 1)  .6417( 1 2)
 .5966( 2 0)  .6417( 2 1) 1.0000( 2 2)
MULTIPLE REGRESSION AND SELECTED UNIVARIATE STATISTICS
---------------------------------------------------------------------------------------------
     VARIABLE                        MULTIPLE
      NUMBER                        REGRESSION                          UNIVARIATE
                         SLOPE               STD.ERROR                  STATISTICS
                                                             MEAN        STD.DEVIATION
---------------------------------------------------------------------------------------------
DEPENDENT                 ...                  ...          172.3704        36.5388
INDEPENDENT   1          1.4985               .6400         118.1481        11.0025
              2          1.1822               .7421          84.0370         9.4888
---------------------------------------------------------------------------------------------
INTERCEPT=  -104.0190 MULT. R-SQUARED= .4756 STD.ERROR OF EST.=   27.5393 F=   10.8848
---------------------------------------------------------------------------------------------
ACTUAL AND ESTIMATED VALUES OF DEPENDENT VARIABLE
      ACTUAL            ESTIMATED          RESIDUAL
    230.00000           224.76929           5.23071
     81.00000           114.54561         -33.54561
    100.00000           153.57324         -53.57324
    212.00000           212.71362          -.71362
    216.00000           176.05005          39.94995
    156.00000           171.47192         -15.47192
    201.00000           175.96729          25.03271
    194.00000           177.30029          16.69971
    164.00000           182.36011         -18.36011
    166.00000           171.47192          -5.47192
    146.00000           175.50049         -29.50049
    196.00000           172.02173          23.97827
    202.00000           187.08911          14.91089
    203.00000           183.85864          19.14136
    201.00000           200.32690            .67310
    195.00000           177.46582          17.53418
    180.00000           193.54980         -13.54980
    174.00000           212.08105         -38.08105
    120.00000           148.76184         -28.76184
    198.00000           159.64990          38.35010
    189.00000           158.38501          30.61499
    184.00000           159.80054          24.19946
    174.00000           153.97217          20.02783
    168.00000           168.15894          -.15894
    143.00000           133.39302           9.60698
    131.00000           127.16570           3.83430
    130.00000           182.59375         -52.59375
```

chapter 5

Correlation, Analysis of Variance, and Covariance

In the previous chapter we discussed the basic elements of the linear regression model. In this chapter we shall extend our discussion of this model, treating in greater detail the subject of analysis of variance and discussing some relationships between analysis of variance and correlation.

5.1 simple, multiple, and partial correlation

Simple Correlation. The simple correlation coefficient between the variables X and Y is defined for the sample as [see Eq. (2-33)][1]

$$r = \frac{\sum xy}{\sqrt{\sum x^2}\sqrt{\sum y^2}}, \qquad x = X - \bar{X}, y = Y - \bar{Y} \tag{5-1}$$

If we square the correlation coefficient, we have

$$r^2 = \frac{\sum xy}{\sum x^2}\frac{\sum xy}{\sum y^2}$$

[1]Simple correlation as defined here is sometimes called *product moment* correlation or *Pearsonian* correlation.

Now the slope of the regression line between Y and X is $\hat{\beta}_1 = \sum xy/\sum x^2$; therefore

$$r^2 = \frac{\hat{\beta}_1 \sum xy}{\sum y^2} \tag{5-2}$$

From our previous discussion concerning analysis of variance, we know that if we carry out the regression leading to $\hat{\beta}_1$, then $\sum y^2$ is the total variation in the dependent variable, Y, to be explained by the regression. Furthermore, $\hat{\beta}_1 \sum xy$ is the variation explained by the regression [see Eq. (4-54)], so that we may state Eq. (5-2) verbally as

$$r^2 = \frac{\text{explained variation}}{\text{total variation}}$$

and

$$1 - r^2 = \frac{\text{unexplained variation}}{\text{total variation}}$$

The coefficient r^2 is often called the *coefficient of determination*; it is the fraction of total variation of the dependent variable explained by the regression. The coefficient $1 - r^2$ is often called the *coefficient of nondetermination*; it is the fraction of total variation of the dependent variable not explained by the regression.

In the case of simple regression, the F ratio of Eq. (4-55) becomes

$$F = \frac{\hat{\beta}_1 \sum xy/1}{\mathbf{e}'\mathbf{e}/(n-2)}$$

since $K = 1$. We may express this ratio verbally as

$$F = \frac{\text{explained variation}/1}{\text{unexplained variation}/(n-2)}$$

or as

$$F = \frac{r^2}{(1-r^2)/(n-2)} \tag{5-3}$$

Therefore the hypothesis that the population slope is zero can be tested by use of an F ratio that has been directly calculated from the correlation coefficient. Such a computing method is often convenient, although some accuracy will generally be lost by use of this method rather than by direct use of Eq. (4-55).

Let us further clarify the relationship between the hypothesis test that the slope is zero and the correlation coefficient. In the case where there is but a single independent variable

$$\mathbf{S}_{\beta} = \frac{\hat{\sigma}^2}{n \sum x^2} \begin{bmatrix} \sum X^2 & -\sum X \\ -\sum X & n \end{bmatrix}$$

by Eq. (4-44) and the definition of $(\mathbf{X}'\mathbf{X})^{-1}$. Therefore the estimated standard error for $\hat{\beta}_1$ is the square root of the southeast element of this matrix:

$$\sqrt{s_{\beta_{11}}} = \sqrt{\frac{\hat{\sigma}^2}{\sum x^2}}$$

The t statistic of Eq. (4-45) may be written (*under the hypothesis* that the population slope is zero) as

$$t = \frac{\hat{\beta}_1}{\sqrt{s_{\beta_{11}}}} = \frac{\hat{\beta}_1}{\sqrt{\hat{\sigma}^2 / \sum x^2}}$$

Since $\hat{\sigma}^2 = \mathbf{e}'\mathbf{e}/(n-2)$, upon squaring t, we have

$$t^2 = \frac{(\hat{\beta}_1)^2 \sum x^2 (n-2)}{\mathbf{e}'\mathbf{e}}$$

and upon further simplification,[2]

$$t^2 = \frac{r^2(n-2)}{1-r^2} \tag{5-4}$$

Since $t^2 = F$ (with 1 degree of freedom in the numerator of the F ratio) we see that Eqs. (5-3) and (5-4) are identical.

The population correlation coefficient, ρ, will be zero if and only if the population slope, β_1 is zero. Thus the hypothesis test that the population slope is zero is equivalent to the hypothesis test that the population correlation coefficient is zero and vice versa. Therefore we need not test both hypotheses. In the case where it is more convenient to test the correlation coefficient, the F ratio of Eq. (5-3) may be used with 1 and $n - 2$ degrees of freedom, or the square root of Eq. (5-4)

$$t = r\sqrt{\frac{n-2}{1-r^2}} \tag{5-4a}$$

may be used in conjunction with the t distribution with $n - 2$ degrees of freedom.

---

[2]Divide the numerator and denominator by $\sum y^2$ and recall that $\mathbf{e}'\mathbf{e}/\sum y^2 = 1 - r^2$ (i.e., unexplained variation/total variation).

We shall illustrate this by the example of the last chapter concerning the ability to set rivets. As reported in the Appendix to Chap. 4, the simple correlation matrix is

$$\mathbf{R} = \begin{bmatrix} r_{00} & r_{01} & r_{02} \\ r_{10} & r_{11} & r_{12} \\ r_{20} & r_{21} & r_{22} \end{bmatrix} = \begin{bmatrix} 1.0000 & 0.6482 & 0.5966 \\ 0.6482 & 1.0000 & 0.6417 \\ 0.5966 & 0.6417 & 1.0000 \end{bmatrix}$$

The subscript zero represents the dependent variable, Y.

To test whether the dependent variable is significantly correlated with, say, the first independent variable, we form the hypothesis

$$H_0: \quad \rho_{01} = 0$$
$$H_1: \quad \rho_{01} > 0$$

The alternative hypothesis need not be one-sided, but, as we have mentioned before, it generally is since in economic research we usually have some a priori notion about the sign of the slope and hence about the sign of the correlation coefficient. The t statistic of Eq. (5-4a) becomes

$$t = 0.6482\sqrt{\frac{27 - 2}{1 - (0.6482)^2}} = 4.3$$

In the t distribution at $\alpha = 0.05$ and 25 degrees of freedom we find that the critical value of t is $t_{0.05;\,25} = 1.708$. Since the calculated value of t exceeds the critical value, we reject the null hypothesis and, of course, we have at the same time rejected the null hypothesis at $\alpha = 0.05$ that the simple regression slope connecting these two variables is zero.[3]

As a general rule, one wishes, in multiple regression, a set of independent variables that are highly correlated with the dependent variable but not correlated between each other. That is, we would like a simple correlation matrix with values close to ± 1.0 on the first row (and column) and values close to zero off the main diagonal.[4]

In the case of two independent variables, the effect of high correlation between two independent variables is easy to visualize. As the correlation

---

[3] In some cases, fairly rare in economic research, one wishes to test the hypothesis that the population correlation coefficient is a nonzero constant. In this case one must use another method such as Fisher's z transformation [see Croxton, Cowden, and Bolch (1969)]. Alternatively, one may use David's *Tables* (1938).

[4] The presence of association between independent variables, whether theoretical or actual, is often called the problem of *multicollinearity*. The term is somewhat unfortunate since it imparts more mathematical precision to the problem than is warranted. We do not deal with theoretical multicollinearity (the Frish case). See Johnston (1963).

between the two variables increases, the joint confidence ellipse, such as the one shown in Figure 4.5, tends to elongate. That is, the major axis of the ellipse increases relative to the minor axis of the ellipse for a given level of confidence. Thus the estimated standard errors of the coefficients tend to increase. This result accords with common sense—as the correlation between the two variables increases, it becomes more difficult to separate out the independent contribution of the variables. Therefore it becomes more difficult to reject the univariate hypothesis that the population slope is zero.

There is some indication that this problem is present in the rivet-setting equation. From the correlation matrix given above we see that variable 2 is more highly correlated with variable 1 than it is with the dependent variable. Furthermore, variable 2 does not have associated with it a slope in the multiple regression equation which is twice its estimated standard error.

There are several methods of dealing with the problem of high correlation between independent variables. The most obvious cure for the problem is the simple expedient of dropping one or more variables from the equation. In the rivet-setting example we should probably drop variable X_2 and utilize the simple regression between Y and X_1. Another approach is to transform the independent variables in such a way as to cause them to have zero correlation among themselves. This transformation may be accomplished by the use of principal components analysis, to be discussed in Chap. 7. Furthermore, other transformation techniques are sometimes used to reduce the correlation among the independent variables while not completely eliminating it. Thus the "trend" is often removed from economic time series prior to carrying out a regression analysis.[5]

Multiple Correlation. The multiple correlation coefficient, $r_{0(1\ 2\ \cdots\ K)}$ is a straightforward extension of simple correlation. It is the maximum correlation between the dependent variable zero (0) and all the independent variables ($1\ 2\ \cdots\ K$). For a regression with an arbitrary number of independent variables

$$\begin{aligned} r^2_{0(1\ 2\ \cdots\ K)} &= \frac{\text{explained variation}}{\text{total variation}} \\ &= \frac{(\hat{\boldsymbol{\beta}}^*)'\mathbf{x}'\mathbf{y}}{\mathbf{y}'\mathbf{y}} \end{aligned} \tag{5-5}$$

For the rivet-setting regression of the last chapter

$$\begin{aligned} r^2_{0(1\ 2)} &= \frac{16506.07}{34708.07} \\ &= 0.4756 \end{aligned}$$

[5]In general, time series are often *filtered* for trend and seasonality.

Therefore the multiple regression explains about 48 percent of the variation in the dependent variable. Notice that if we drop the variable X_2 from the equation and use only X_1, then from the simple correlation matrix $r_{01}^2 = (0.6482)^2 = 0.4202$. Here we use the notation r_{01}^2 rather than $r_{0(1)}^2$ when there is a single independent variable. Apparently there is only a marginal improvement in the percentage of variation in Y explained when both X_1 and X_2 are used over the percentage explained when X_1 is used alone. Again, we reinforce our contention that X_2 should be dropped from the regression.

An important hypothesis test concerns the significance of the multiple correlation coefficient. To test the hypothesis that the population multiple correlation coefficient is *zero* against a two-sided alternative, we appeal directly to Eq. (5-3) and use

$$F = \frac{r_{0(1\ 2\ \cdots\ K)}^2/K}{(1 - r_{0(1\ 2\ \cdots\ K)}^2)/(n - K - 1)} \tag{5-6}$$

with K and $n - K - 1$ degrees of freedom. This is the same F ratio as that given by Eq. (4-55), and if we reject the null hypothesis that the population multiple correlation coefficient is zero, we also reject the null hypothesis that all population slopes are jointly zero.

Sometimes we have at our disposal only a simple correlation matrix and we wish to derive the multiple correlation coefficient. The following relationship exists between the multiple correlation coefficient and the simple correlation coefficients:

$$r_{0(1\ 2\ \cdots\ K)}^2 = 1 - \frac{|\mathbf{R}|}{|\mathbf{R}^*|} \tag{5-7}$$

The numerator in the ratio directly above is the determinant of the entire simple correlation matrix, and the denominator is the determinant of a submatrix of the simple correlation matrix which contains all the elements of the full matrix *except* the first row and column of the full matrix.

Although the sample multiple (and simple) correlation coefficients are maximum likelihood estimators of the corresponding population coefficients, they are biased (but consistent) estimators of their corresponding population parameters. To adjust for this bias, these estimates are often "adjusted" for degrees of freedom. Write the adjusted coefficient of nondetermination as

$$1 - \bar{r}_{0(1\ 2\ \cdots\ K)}^2 = \frac{\mathbf{e}'\mathbf{e}/(n - K - 1)}{\mathbf{y}'\mathbf{y}/(n - 1)} \tag{5-8}$$

Notice that $\mathbf{e}'\mathbf{e}$ and $\mathbf{y}'\mathbf{y}$ have each been divided by their corresponding numbers of degrees of freedom. From Eq. (5-8), the adjusted coefficient of determination is

$$\bar{r}^2_{0(1\ 2 \cdots K)} = 1 - [1 - r^2_{0(1\ 2 \cdots K)}]\frac{n-1}{n-K-1} \qquad \text{(5-8a)}$$

The adjusted coefficient tends to decrease, other things the same, as the number of independent variables increases. The adjusted coefficient may become negative, in which case it should be considered to be zero.

Partial Correlation. The partial correlation $r_{ij \cdot k}$ is the correlation between the variables i and j "holding constant" the effect of the variable k. Technically, the population partial correlation coefficient $\rho_{ij \cdot k}$ is the correlation coefficient associated with the conditional covariance matrix, that is, the covariance matrix associated with the conditional density $f(X_i, X_j \mid X_k)$. It is in this sense that the variable X_k has been held constant.

Another way to look at the partial correlation coefficient is as follows. For every slope in any regression equation there is a corresponding correlation coefficient. In the case of simple regression a simple correlation coefficient corresponds to the regression slope. If we have the equation $\hat{Y} = \hat{\beta}_0 + \hat{\beta}_1 X_1 + \hat{\beta}_2 X_2$, the slope $\hat{\beta}_1$ is the partial derivative $\partial \hat{Y}/\partial X_1$. That is, $\hat{\beta}_1$ is the slope of a side of a hyperplane which relates $\hat{Y}$ to X_1 "holding X_2 constant." Thus the correlation coefficient $r_{01 \cdot 2}$ corresponds to $\hat{\beta}_1$ and the correlation coefficient $r_{02 \cdot 1}$ corresponds to $\hat{\beta}_2$. It is our practice to use the dot in the subscript to separate the *active* variables from the *passive* variables, or variables held constant. Thus $r^2_{01 \cdot 2}$ will tell us the variation in Y explained by X_1 while holding X_2 *statistically constant.*

Partial correlation coefficients can be calculated from simple correlation coefficients. For first-order coefficients (coefficients with one passive variable) we use simple (or zero-order) coefficients:

$$r_{ij \cdot k} = \frac{r_{ij} - r_{ik} r_{jk}}{\sqrt{1 - r^2_{ik}}\sqrt{1 - r^2_{jk}}}$$

Thus for the rivet-setting example

$$r_{01 \cdot 2} = \frac{r_{01} - r_{02} r_{12}}{\sqrt{1 - r^2_{02}}\sqrt{1 - r^2_{12}}} = \frac{0.6482 - (0.5966)(0.6417)}{\sqrt{1 - (0.5966)^2}\sqrt{1 - (0.6417)^2}} = 0.431$$

$$r_{02 \cdot 1} = \frac{r_{02} - r_{01} r_{21}}{\sqrt{1 - r^2_{01}}\sqrt{1 - r^2_{21}}} = \frac{0.5966 - (0.6482)(0.6417)}{\sqrt{1 - (0.6482)^2}\sqrt{1 - (0.6417)^2}} = 0.310$$

Notice again that there is little correlation between the dependent variable and X_2 when account is taken of X_1. Again, the merit of dropping X_2 from the regression is pointed out.

Second-order coefficients may be calculated from first-order coefficients,

$$r_{ij\cdot km} = \frac{r_{ij\cdot k} - r_{ik\cdot m}r_{jk\cdot m}}{\sqrt{1 - r_{ik\cdot m}^2}\sqrt{1 - r_{jk\cdot m}^2}} \tag{5-9}$$

and so forth for higher orders.

In general, if we have $p + 1$ variables and wish to hold $p - 1$ variables constant, then, for example,

$$r_{01\cdot 23\cdots p} = \frac{-C_{01}}{(C_{00}C_{11})^{1/2}}$$

where C_{ij} is the cofactor associated with r_{ij} in the correlation matrix, **R**, for the $p + 1$ variables. For our example if we wish $r_{01\cdot 2}$, we find from the correlation matrix that

$$C_{00} = \begin{vmatrix} r_{11} & r_{12} \\ r_{21} & r_{22} \end{vmatrix}, \qquad C_{01} = -\begin{vmatrix} r_{10} & r_{12} \\ r_{20} & r_{22} \end{vmatrix}, \qquad C_{11} = \begin{vmatrix} r_{00} & r_{02} \\ r_{20} & r_{22} \end{vmatrix}$$

Upon expanding each of these cofactors (for example, $C_{00} = 1 - r_{12}^2$ since $r_{11} = r_{22} = 1$), it is easy to see that

$$r_{01\cdot 2} = \frac{-C_{01}}{(C_{00}C_{11})^{1/2}} = \frac{r_{10} - r_{20}r_{12}}{\sqrt{(1 - r_{12}^2)(1 - r_{02}^2)}}$$

which agrees with our previous calculation formula.

The coefficients can also be constructed from their corresponding regression slopes and estimated standard errors. Let t_j be the ratio of the corresponding slope to its standard error (i.e., $t_j = \hat{\beta}_j/s_{\hat{\beta}_j}$); then

$$r_{0j\cdot km\cdots} = \frac{t_j}{\sqrt{t_j^2 + (n - K - 1)}} \tag{5-10}$$

For our rivet example, if we desire to compute $r_{01\cdot 2}$, we calculate $t_1 = 1.4985/0.64 = 2.3414$ and

$$r_{01\cdot 2} = \frac{2.3414}{\sqrt{(2.3414)^2 + 24}} = 0.431$$

which is the same result as previously obtained.

To test the hypothesis that the population partial correlation coefficient is zero, we form

$$t = r_{ij\cdot km\cdots}\sqrt{\frac{n - P - 2}{1 - r_{ij\cdot km\cdots}^2}} \tag{5-11}$$

where P is the number of passive variables. This ratio follows the t distribution with $n - P - 2$ degrees of freedom. In the case of simple correlation, $P = 0$, and Eq. (5-11) reduces to Eq. (5-4a). As for simple correlation the test may be one-sided or two-sided and a rejection of the null hypothesis implies a rejection of the null hypothesis that the corresponding slope is zero.

Multiple-Partial Correlation. Suppose that we have a regression with K independent variables. Then the unexplained variation is

$$(\mathbf{e}'\mathbf{e})_K = \mathbf{y}'\mathbf{y}(1 - r^2_{0(1\ 2\ \cdots\ K)})$$

from Eq. (5-5). Now suppose that we have a regression with P independent variables, where $P > K$, and the K variables are a subset of the P variables. Then the unexplained variation after regressing Y on the P variables is

$$(\mathbf{e}'\mathbf{e})_P = \mathbf{y}'\mathbf{y}(1 - r^2_{0(1\ 2\ \cdots\ P)})$$

The reduction in unexplained variation associated with the use of the additional variables is

$$(\mathbf{e}'\mathbf{e})_K - (\mathbf{e}'\mathbf{e})_P = \mathbf{y}'\mathbf{y}(r^2_{0(1\ 2\ \cdots\ P)} - r^2_{0(1\ 2\ \cdots\ K)})$$

and the proportional reduction in an unexplained variation is

$$r^2_{0(K+1\ K+2\ \cdots\ P)\cdot(1\ 2\ \cdots\ K)} = \frac{r^2_{0(1\ 2\ \cdots\ P)} - r^2_{0(1\ 2\ \cdots\ K)}}{1 - r^2_{0(1\ 2\ \cdots\ K)}} \tag{5-12}$$

This coefficient is called the *multiple-partial coefficient of determination.* It is the correlation between the dependent variable and the variables $K + 1$, $K + 2, \ldots, P$ holding constant (or accounting for) the variables $1, 2, \ldots, K$.

The significance of the multiple-partial correlation coefficient may be tested by use of the ratio

$$F = \frac{((\mathbf{e}'\mathbf{e})_K - (\mathbf{e}'\mathbf{e})_P)/(P - K)}{(\mathbf{e}'\mathbf{e})_P/(n - P - 1)} \tag{5-13}$$

which follows the F distribution with $P - K$ and $n - P - 1$ degrees of freedom.[6] This F ratio compares the additional variation explained by the use of the $K + 1, K + 2, \ldots, P$ variables with the variation unexplained by all P variables.

5.2 analysis of variance—one basis of classification

In Sec. 3.3 we tested the equality of two means. In this section we shall generalize this test to include the case where the simultaneous equality of several means is tested. The test is closely related to the hypothesis test

[6]See Draper and Smith (1966), p. 68, and also Cowden (1952).

concerning the joint equality of several regression coefficients, which was discussed in Sec. 4.6.

Notation. Table 5.1 shows a typical data layout for analysis of variance with one basis of classification. Each column represents one set of sample values from a given classification. For each Y_{ij} observation, the subscript i represents the item number (or row) and the subscript j represents the classification (or column). In Table 5.1 each of the columns has the same number of observations. Equal numbers of observations are not always the case in this type of problem, and so we shall carry out our analysis in a manner that will allow for unequal numbers of observations, just as we did in Sec. 3.3 where two means were treated.

For economy of notation let

$$Y_{\cdot j} = \sum_{i=1}^{n_j} Y_{ij}, \qquad j = 1, 2, \ldots, K \tag{5-14}$$

That is, the sum of any column, j, is $Y_{\cdot j}$, where n_j is the number of observations in the column. The $\cdot j$ subscript indicates that the summation is over "all rows of the jth column." Furthermore, we denote the sum of all observations as

$$Y_{\cdot\cdot} = \sum_{j=1}^{K} \sum_{i=1}^{n_j} Y_{ij} \tag{5-15}$$

The $\cdot\cdot$ subscript thus indicates that the summation is over "all rows and all columns." Corresponding to the column sums will be *column means*:

$$\bar{Y}_{\cdot j} = \frac{Y_{\cdot j}}{n_j} \tag{5-16}$$

Similarly, the grand *mean*, $\bar{Y}_{\cdot\cdot}$, will correspond to the sum of all the observations. Therefore

$$\bar{Y}_{\cdot\cdot} = \frac{Y_{\cdot\cdot}}{N} \tag{5-17}$$

where $N = n_1 + n_2 + \cdots + n_K$, or, if all columns have the same number of observations, $N = nK$, where n is the number of observations in each column and K is the number of columns.

For any observation, the following equality obviously holds:

$$(Y_{ij} - \bar{Y}_{\cdot\cdot}) = (\bar{Y}_{\cdot j} - \bar{Y}_{\cdot\cdot}) + (Y_{ij} - \bar{Y}_{\cdot j}) \tag{5-18}$$

Equation (5-18) decomposes the total deviation of any Y_{ij} about the grand mean into two parts, a part associated with the deviation of the column

Table 5.1 Data Layout for Analysis of Variance—One Basis of Classification

Classification (Columns) / Item Number (Rows)	1	2	3	...	K
1	Y_{11}	Y_{12}	Y_{13}	...	Y_{1K}
2	Y_{21}	Y_{22}	Y_{23}	...	Y_{2K}
3	Y_{31}	Y_{32}	Y_{33}	...	Y_{3K}
.	.	.	.	.	.
.	.	.	.	.	.
.	.	.	.	.	.
n	Y_{n1}	Y_{n2}	Y_{n3}	...	Y_{nK}
Column Mean	$\bar{Y}_{.1}$	$\bar{Y}_{.2}$	$\bar{Y}_{.3}$	...	$\bar{Y}_{.K}$
Grand Mean					$\bar{Y}_{..}$

mean about the grand mean (*column deviation*) and a part associated with the deviation of the Y_{ij}s about its column mean (*error deviation*).[7] Thus Eq. (5-18) may be stated verbally as

Total deviation = column deviation + error deviation

If we square the deviations of Eq. (5-18) and sum them so that there are N deviations for each term, then we convert these measures of deviation into measures of *variation*. If each column has the same number of observations, then

$$\sum_{j=1}^{K}\sum_{i=1}^{n}(Y_{ij} - \bar{Y}_{..})^2 = n\sum_{j=1}^{K}(\bar{Y}_{.j} - \bar{Y}_{..})^2 + \sum_{j=1}^{K}\sum_{i=1}^{n}(Y_{ij} - \bar{Y}_{.j})^2 \qquad (5\text{-}19)$$

If the columns have differing numbers of observations, then

$$\sum_{j=1}^{K}\sum_{i=1}^{n_j}(Y_{ij} - \bar{Y}_{..})^2 = \sum_{j=1}^{K}n_j(\bar{Y}_{.j} - \bar{Y}_{..})^2 + \sum_{j=1}^{K}\sum_{i=1}^{n_j}(Y_{ij} - \bar{Y}_{.j})^2 \qquad (5\text{-}19a)$$

Stated verbally,

Total variation = column variation + error variation

Equation (5-19a) is called the *fundamental equality of variation*; it partitions the total variation of Y_{ij} into the variation due to the column classification and the variation due to the random error.

[7]Column deviation is sometimes called *among-column* deviation. Error deviation is sometimes called *within-column* deviation.

Let us illustrate these calculations by use of the data in Table 5.2. In this table, rates of return for sample corporations in three industrial classifications are given. There are 32 observations in all.

Table 5.2 Five-Year Rate of Return on Equity for Firms in Three Industries

	Steel	*Electronics*	*Energy*
	14.0	33.6	18.6
	12.0	25.1	16.3
	10.5	20.6	14.4
	9.9	19.6	13.3
	9.3	16.7	12.7
	8.7	16.0	12.0
	8.7	14.1	10.9
	7.7	13.1	10.2
	6.6	11.6	9.5
	6.2	10.3	—
	5.7	8.8	—
	1.1	—	—
Mean	8.3667	17.2273	13.1000
Grand Mean			12.7438

Source: *Forbes*, Jan. 1, 1970.

We may find the total variation indirectly by use of (see footnote 3 in Chap. 4)

$$\sum_{j=1}^{K}\sum_{i=1}^{n_j}(Y_{ij} - \bar{Y}_{..})^2 = \sum_{j=1}^{K}\sum_{i=1}^{n_j} Y_{ij}^2 - \frac{\left(\sum_{j=1}^{K}\sum_{i=1}^{n_j} Y_{ij}\right)^2}{N}$$

or directly by calculation of the sum of the squared values of $(Y_{ij} - \bar{Y}_{..})$. Again, the indirect method is preferred for desk calculator use, and the direct method is preferred for a digital computer. From Table 5.2

$$\sum_{j=1}^{K}\sum_{i=1}^{n_j} Y_{ij}^2 = (14.0)^2 + (12.0)^2 + \cdots + (10.2)^2 + (9.5)^2 = 6366.3$$

and

$$\sum_{j=1}^{K}\sum_{i=1}^{n_j} Y_{ij} = 14.0 + 12.0 + \cdots + 10.2 + 9.5 = 407.8$$

so that

$$\sum_{j=1}^{K}\sum_{i=1}^{n_j}(Y_{ij} - \bar{Y}_{..})^2 = 6366.3 - \frac{(407.8)^2}{32} = 1169.4$$

Column variation is generally easy to calculate by the direct method:

$$\sum_{j=1}^{3} n_j(\bar{Y}_{.j} - \bar{Y}_{..})^2 = 12(-4.3771)^2 + 11(4.4835)^2 + 9(0.3562)^2 = 452.2$$

Error variation is obtained by subtraction:

$$\sum_{j=1}^{K}\sum_{i=1}^{n_j}(Y_{ij} - \bar{Y}_{.j})^2 = \text{total variation} - \text{column variation}$$
$$= 1169.4 - 452.2 = 717.2$$

Model. We assume that within each column, the Y_{ij} values may be represented as

$$Y_{ij} = \mu_{.j} + \epsilon_{ij} \tag{5-20}$$

That is, each Y_{ij} within a column is given by the column mean plus an error, ϵ_{ij}, which will be assumed to be distributed $N_1(0, \sigma^2)$ for any column with $\text{cov}(\epsilon_{ij}, \epsilon_{ij'}) = 0$, $j \neq j'$. Furthermore, we assume for the model that each column mean in the population may be represented as

$$\mu_{.j} = \mu_{..} + \tau_j, \qquad j = 1, 2, \ldots, K \tag{5-21}$$

where $\mu_{..}$ is the grand population mean estimated by $\bar{Y}_{..}$ and τ_j is the *column effect*. By the definition in Eq. (5-21), $\tau_j = \mu_{.j} - \mu_{..}$; the column effect is the difference between the column mean and the grand mean. The sum of the column effects, that is, the sum of all τ_j, $j = 1, 2, \ldots, K$, is assumed to be zero in the population, which is nothing more than stating that the grand mean is the average of the column means. Combining Eqs. (5-20) and (5-21), we have

$$Y_{ij} = \mu_{..} + \tau_j + \epsilon_{ij} \tag{5-20a}$$

Under the model, it can be shown that the least-squares estimates of the $\mu_{.j}$ in Eq. (5-20) are[8]

$$\hat{\mu}_{.j} = \bar{Y}_{.j}, \qquad j = 1, 2, \ldots, K \tag{5-22}$$

and hence

$$Y_{ij} = \bar{Y}_{.j} + e_{ij}, \qquad j = 1, 2, \ldots, K \tag{5-23}$$

where e_{ij} is the least-squares residual. The residual sum of squares for any particular column j is therefore

$$\sum_{i=1}^{n_j} e_{ij}^2 = \sum_{i=1}^{n_j}(Y_{ij} - \bar{Y}_{.j})^2$$

[8]The least-squares estimates of the parameters in Eq. (5-20a), under the assumptions of the model, are

$$\hat{\mu}_{..} = \bar{Y}_{..} \quad \text{and} \quad \hat{\tau}_j = \bar{Y}_{.j} - \bar{Y}_{..}$$

with $n_j - 1$ degrees of freedom, since one mean is estimated from the n_j observations. Hence the total residual sum of squares of Eq. (5-23) for *all* column classifications is

$$S_1 = \sum_{j=1}^{K} \sum_{i=1}^{n_j} e_{ij}^2 = \sum_{j=1}^{K} \sum_{i=1}^{n_j} (Y_{ij} - \bar{Y}_{.j})^2 \tag{5-24}$$

with $\sum_{j=1}^{K} (n_j - 1) = N - K$ degrees of freedom.

Referring to Eq. (5-19a), we call S_1 the error variation. It is essentially the variation unexplained by the column classification. For our example $S_1 = 717.2$, as previously calculated.

Hypothesis. We desire to test whether or not the mean rates of return for each classification are jointly equal. That is, we wish to test

$$H_0: \quad \mu_{.1} = \mu_{.2} = \mu_{.3} = \mu_{..}$$

against

$$H_1: \quad \text{Not all column means are equal}$$

This test is equivalent to one where the null hypothesis is that the column effects, τ_j, are all zero against the alternative that they are not all zero. Under the hypothesis, $\mu_{..}$ replaces each $\mu_{.j}$, and the model of Eq. (5-20) becomes

$$Y_{ij} = \mu_{..} + \epsilon_{ij} \tag{5-25}$$

Moreover, under the hypothesis, it can be shown that the least-squares estimate of $\mu_{..}$ is

$$\hat{\mu}_{..} = \bar{Y}_{..}$$

and hence that

$$Y_{ij} = \bar{Y}_{..} + e_{ij} \tag{5-26}$$

where e_{ij} is the least-squares residual. Thus, under the hypothesis H_0, the total residual sum of squares of Eq. (5-26) for all column classifications is

$$S_0 = \sum_{j=1}^{K} \sum_{i=1}^{n_j} e_{ij}^2 = \sum_{j=1}^{K} \sum_{i=1}^{n_j} (Y_{ij} - \bar{Y}_{..})^2 \tag{5-27}$$

with $N - 1$ degrees of freedom.[9]

Referring to Eq. (5-19a), we call S_0 the total variation.[10] It is $S_0 = 1169.4$, as previously calculated.

---

[9]Since we are estimating a single parameter, we lose a single degree of freedom.

[10]Technically, we may call S_0 the unexplained variation about the grand mean, rather than total variation.

Hypothesis Test. Let us now develop the hypothesis test intuitively. Suppose that the hypothesis that all column means are equal to a common mean, $\mu_{..}$, is true. Then the specifications of Eqs. (5-25) and (5-20) are the same except for sampling error. In other words, Eq. (5-25) will fit the data just as well as Eq. (5-20). More specifically, the total variation, S_0, due to the hypothesis will be almost equal to the error variation, S_1, due to the model. The discrepancy, $S_0 - S_1$, will be attributable to sampling error *if* the hypothesis is true. The discrepancy is equal to

$$\begin{aligned} S_C &= S_0 - S_1 \\ &= \sum_{j=1}^{K}\sum_{i=1}^{n_j}(Y_{ij} - \bar{Y}_{..})^2 - \sum_{j=1}^{K}\sum_{i=1}^{n_j}(Y_{ij} - \bar{Y}_{.j})^2 = \sum_{j=1}^{K} n_j(\bar{Y}_{.j} - \bar{Y}_{..})^2 \end{aligned} \tag{5-28}$$

or

$$S_C = 1169.4 - 717.2 = 452.2$$

from the fundamental equality, Eq. (5-19a). Therefore S_C is referred to as column variation. If the hypothesis is true, the column variation, S_C, will be fairly small relative to S_1. Since S_0 and S_1 are associated, respectively, with $N - 1$ and $N - K$ degrees of freedom, $S_C = S_0 - S_1$ is associated with $K - 1 = (N - 1) - (N - K)$ degrees of freedom. The F ratio will then compare S_C and S_1, each being adjusted for its respective numbers of degrees of freedom:

$$F = \frac{S_C/(K-1)}{S_1/(N-K)} = \frac{(\text{column variation})/(K-1)}{(\text{error variation})/(N-K)} \tag{5-29}$$

This ratio follows the F distribution with $K - 1$ and $N - K$ degrees of freedom. That is, if column variation is large relative to error variation (after dividing each by the associated number of degrees of freedom), we reason that some means are statistically different. For our example

$$F = \frac{452.2/2}{717.2/29} = \frac{226.1}{24.7} = 9.15$$

In the F table at $\alpha = 0.05$, with 2 and 29 degrees of freedom, we find that the critical F is approximately 3.32. Since the calculated F exceeds the critical F, given alpha, we reject the null hypothesis.

As a matter of presentation, the analysis of variance results are often given by an analysis of variance table. Such a table is shown as Table 5.3 and is given by the computer program in the Appendix to this chapter. The column in the table entitled "mean square" is obtained by dividing the column entitled "amount of variation" by the associated number of degrees of freedom. The mean square of the error variation is an estimate[11] of the variance of ϵ_{ij}, i.e., σ^2. Therefore

$$\hat{\sigma}^2 = \frac{S_1}{N-K} \tag{5-30}$$

which, for our example, is

$$\hat{\sigma}^2 = \frac{717.2}{29} \doteq 24.7$$

Table 5.3 Analysis of Variance Table—One Basis of Classification

Source of Variation	*Amount of Variation*	*Degrees of Freedom**	*Mean Square*	*F Ratio*
Total	$S_0 = \sum_{j=1}^{K}\sum_{i=1}^{n_j}(Y_{ij} - \bar{Y}_{..})^2$	$N-1$		
Column	$S_C = \sum_{j=1}^{K} n_j(\bar{Y}_{.j} - \bar{Y}_{..})^2$	$K-1$	$\frac{S_C}{K-1}$	$F = \frac{S_C/(K-1)}{S_1/(N-K)}$
Error	$S_1 = \sum_{j=1}^{K}\sum_{i=1}^{n_j}(Y_{ij} - \bar{Y}_{.j})^2$	$N-K$	$\frac{S_1}{N-K}$	

*$N = n_1 + n_2 + \cdots + n_K$.

Linear Contrasts. When more than two means are involved, rejection of the null hypothesis that all means are equal does not necessarily offer any insight as to which one(s) are causing the rejection. One way to handle this problem is to use Scheffé's method of linear contrasts.[12]

By definition, a linear contrast, L, is a linear combination of, say, the parameters $\beta_1, \beta_2, \ldots, \beta_K$ using weights which sum to zero. Thus $L = \sum_{i=1}^{K} c_i\beta_i$, where $\sum_{i=1}^{K} c_i = 0$. If $\hat{\beta}_i$ is an estimate of β_i with estimated variance s_i^2 (i.e., s_i^2 = estimated variance of $\hat{\beta}_i$), then it is reasonable that the estimate of the linear contrast is

$$\hat{L} = \sum_{i=1}^{K} c_i\hat{\beta}_i \tag{5-31}$$

with estimated variance

$$s_{\hat{L}}^2 = \sum_{i=1}^{K} c_i^2 s_i^2 \tag{5-32}$$

[11] The mean square of the column variation is also an estimate of σ^2 *if* the null hypothesis is true.

[12] Scheffé (1959). Another method has been suggested by Tukey and is described by Morrison (1967).

if $\hat{\beta}_1, \ldots, \hat{\beta}_K$ are mutually independent. A $100(1-\alpha)$ percent confidence interval for the linear contrast, L, is given by

$$\hat{L} \pm s_{\hat{L}}\sqrt{(K-1)F_{\alpha;\,K-1,N-K}} \tag{5-33}$$

where $\hat{L}$ is given by Eq. (5-31).

For our example, under the model of Eq. (5-20), the sample column mean $\bar{Y}_{\cdot j}$ is an estimate of the population column mean $\mu_{\cdot j}$ with variance

$$s_j^2 = \text{estimated variance of } \bar{Y}_{\cdot j} = \frac{\hat{\sigma}^2}{n_j}$$

where $\hat{\sigma}^2$ is an estimate of σ^2 [see Eq. (5-30)]. Moreover, $\bar{Y}_{\cdot 1}, \ldots, \bar{Y}_{\cdot K}$ are mutally independent, so that the estimate of the linear contrast of the column means is equal to

$$\hat{L} = \sum_{j=1}^{K} c_j \bar{Y}_{\cdot j} \tag{5-31a}$$

with estimated variance

$$s_{\hat{L}}^2 = \hat{\sigma}^2 \sum_{j=1}^{K} \frac{c_j^2}{n_j} \tag{5-32a}$$

Returning to the numerical example of Table 5.2, we find that the sample column means are

$$[\bar{Y}_{\cdot 1} \quad \bar{Y}_{\cdot 2} \quad \bar{Y}_{\cdot 3}] = [8.3667 \quad 17.2273 \quad 13.1000]$$

with

$$[n_1 \quad n_2 \quad n_3] = [12 \quad 11 \quad 9]$$

and $\hat{\sigma}^2 = 24.7$ from Eq. (5-30). Three simple linear contrasts are possible. That is, we may test whether the rejection of the null hypothesis is because $\mu_{\cdot 1} \neq \mu_{\cdot 2}$, or because $\mu_{\cdot 2} \neq \mu_{\cdot 3}$, or because $\mu_{\cdot 1} \neq \mu_{\cdot 3}$. To test

$$H_0\colon \quad \mu_{\cdot 1} - \mu_{\cdot 2} = 0$$

against a two-sided alternative, let

$$[c_1 \quad c_2 \quad c_3] = [1 \quad -1 \quad 0]$$

Similarly, to test

$$H_0\colon \quad \mu_{\cdot 2} - \mu_{\cdot 3} = 0$$

against a two-sided alternative, let

$$[c_1 \quad c_2 \quad c_3] = [0 \quad 1 \quad -1]$$

and to test

$$H_0: \ \mu_{\cdot 1} - \mu_{\cdot 3} = 0$$

against a two-sided alternative, let

$$[c_1 \quad c_2 \quad c_3] = [1 \quad 0 \quad -1]$$

Choosing $\alpha = 0.05$, we find that $F_{0.05;\,2,29} = 3.3$, and for the first contrast $\hat{L} = 8.3667 - 17.2273 = -8.8606$ and $s_{\hat{L}}^2 = 24.7\ (1/12 + 1/11) = 4.3038$. Thus from Eq. (5-33)

$$-14.19 \leq \mu_{\cdot 1} - \mu_{\cdot 2} \leq -3.53$$

Since this interval does not cover zero, we conclude that $\mu_{\cdot 1}$ and $\mu_{\cdot 2}$ *are* significantly different. Proceeding in a similar way, we find the other contrasts:

$$-1.69 \leq \mu_{\cdot 2} - \mu_{\cdot 3} \leq 9.95$$
$$-10.36 \leq \mu_{\cdot 1} - \mu_{\cdot 3} \leq 0.89$$

Since these two intervals do cover zero, we conclude that these means are not significantly different from each other at the stated level of significance. We conclude that the difference between the means of the first two classifications is causing the rejection of the joint null hypothesis.

5.3 analysis of variance—two bases of classification

Analysis of variance with two bases of classification is an extension of the analysis of variance with a single basis of classification. In the current case the Y_{ij} observations are classified both by rows and by columns. Table 5.4 shows a typical data layout where there is only one Y_{ij} observation in each of the cells created by the classification scheme. We shall discuss shortly the case where there are several items in each cell.

The notation is the same as that of the last section. For any Y_{ij}, i is the row location and j is the column location. The row means $\bar{Y}_{i\cdot}$ are given by

$$\bar{Y}_{i\cdot} = \frac{Y_{i\cdot}}{K}, \qquad i = 1, 2, \ldots, R$$

where $Y_{i\cdot}$ is the sum of the Y_{ij} for any row i and K is the number of columns

Table 5.4 Data Layout for Analysis of Variance—Two Bases of Classification

Row Classification	*Column Classification*					*Row Mean*
	1	*2*	*3*	$\cdots$	*K*	
1	Y_{11}	Y_{12}	Y_{13}	$\cdots$	Y_{1K}	$\bar{Y}_{1\cdot}$
2	Y_{21}	Y_{22}	Y_{23}	$\cdots$	Y_{2K}	$\bar{Y}_{2\cdot}$
3	Y_{31}	Y_{32}	Y_{33}	$\cdots$	Y_{3K}	$\bar{Y}_{3\cdot}$
$\vdots$	$\vdots$	$\vdots$	$\vdots$	$\vdots$	$\vdots$	$\vdots$
R	Y_{R1}	Y_{R2}	Y_{R3}	$\cdots$	Y_{RK}	$\bar{Y}_{R\cdot}$
Column Mean	$\bar{Y}_{\cdot 1}$	$\bar{Y}_{\cdot 2}$	$\bar{Y}_{\cdot 3}$	$\cdots$	$\bar{Y}_{\cdot K}$	
Grand Mean						$\bar{Y}_{\cdot\cdot}$

(number of Y_{ij}s in any row). Similarly, the column means are given by

$$\bar{Y}_{\cdot j} = \frac{Y_{\cdot j}}{R}, \qquad j = 1, 2, \ldots, K$$

where R is the number of rows and $Y_{\cdot j}$ has the same definition as that given in the last section. Also, as in the last section, $\bar{Y}_{\cdot\cdot}$ is the grand mean.

Table 5.5 affords an illustration. In this table the maximum rates paid on time deposits, open account, for several sizes of banks (row classification) and three Federal Reserve Districts (column classification) are shown. We wish to know if there is any difference between the means by districts and if there is any difference between the means by size of bank.

Table 5.5 Highest Rate* Paid on Time Deposits, Open Account, July 31, 1967 for Three Federal Reserve Districts

District / *Size of Bank*†	*New York*	*Atlanta*	*San Francisco*	*Row Mean*
Less than 10	4.65	4.55	4.44	4.547
10– 50	4.76	4.40	4.99	4.717
50–100	4.60	4.92	5.00	4.840
100–500	4.83	4.70	4.97	4.833
500 and over	4.99	4.46	4.93	4.793
Column Mean	4.766	4.606	4.866	
Grand Mean				4.746

*Weighted Average.
†Total deposits, millions of dollars.
Source: *Federal Reserve Bulletin*, Sept. 1967, p. 1508.

For any observation, the following equality obviously holds:

$$(Y_{ij} - \bar{Y}_{..}) = (\bar{Y}_{i.} - \bar{Y}_{..}) + (\bar{Y}_{.j} - \bar{Y}_{..}) + (Y_{ij} - \bar{Y}_{i.} - \bar{Y}_{.j} + \bar{Y}_{..}) \tag{5-34}$$

Stated verbally,

Total deviation = row deviation + column deviation + error deviation

Since each of the columns (or rows) must have the same number of observations, upon summing the squared elements, we have

$$\sum_{j=1}^{K}\sum_{i=1}^{R}(Y_{ij} - \bar{Y}_{..})^2 = K\sum_{i=1}^{R}(\bar{Y}_{i.} - \bar{Y}_{..})^2 + R\sum_{j=1}^{K}(\bar{Y}_{.j} - \bar{Y}_{..})^2 + \sum_{j=1}^{K}\sum_{i=1}^{R}(Y_{ij} - \bar{Y}_{i.} - \bar{Y}_{.j} + \bar{Y}_{..})^2 \tag{5-35}$$

or stated verbally,

Total variation = row variation + column variation + error variation

Equation (5-35) is the fundamental equality of variation with two bases of classification and a single observation in a cell. It partitions the total variation of Y_{ij} about the grand mean into the variation due to the row classification, the column classification, and a random error.

Calculation is straightforward. Using the data of Table 5.5 we find, by the same method as used in the last section, that total variation is

$$\sum_{j=1}^{K}\sum_{i=1}^{R}(Y_{ij} - \bar{Y}_{..})^2 = 0.6714$$

Row variation is

$$K\sum_{i=1}^{R}(\bar{Y}_{i.} - \bar{Y}_{..})^2 = 3[(-0.199)^2 + (-0.029)^2 + \cdots + (0.047)^2] = 0.1779$$

and column variation is

$$R\sum_{j=1}^{K}(\bar{Y}_{.j} - \bar{Y}_{..})^2 = 5[(0.020)^2 + (-0.140)^2 + (0.120)^2] = 0.1720$$

Then, error variation is found by subtraction:

$$\sum_{j=1}^{K}\sum_{i=1}^{R}(Y_{ij} - \bar{Y}_{i.} - \bar{Y}_{.j} + \bar{Y}_{..})^2 = \text{total variation} - \text{row variation} - \text{column variation} = 0.3215$$

Model. Each cell in the data layout is assumed to have a population mean μ_{ij}. When there is only one observation in a cell, the observation in the ith row and jth column can be denoted by Y_{ij} with the random error ϵ_{ij}. Thus the model is

$$Y_{ij} = \mu_{ij} + \epsilon_{ij} \tag{5-36}$$

Let us define the column effect, τ_j, as the difference between the column mean and the grand mean,

$$\tau_j = \mu_{.j} - \mu_{..}, \qquad j = 1, 2, \ldots, K \tag{5-37}$$

as in the last section, and the row effect, ρ_i, as the difference between the row mean and the grand mean,

$$\rho_i = \mu_{i.} - \mu_{..}, \qquad i = 1, 2, \ldots, R \tag{5-38}$$

Each cell mean is assumed to be representable by the sum of the grand mean, a column effect, and a row effect:

$$\mu_{ij} = \mu_{..} + \tau_j + \rho_i, \qquad i = 1, 2, \ldots, R \quad j = 1, 2, \ldots, K \tag{5-39}$$

It is obvious that the implicit assumption in Eq. (5-39) is that the deviation of the cell mean about the grand mean can be expressed as

$$\mu_{ij.} - \mu_{..} = \tau_j + \rho_i \tag{5-39a}$$

i.e., the deviation of a cell mean from the grand mean is explainable by the sum of the column effect and the row effect. In other words, there is no *interaction* of row and column effects. This assumption, that the row effects are constant from column to column and that the column effects are constant from row to row, is called the *assumption of lack of interaction.*

Substituting Eq. (5-39) into Eq. (5-36), we have a model for analysis of variance with two bases of classification and no interaction:

$$Y_{ij} = \mu_{..} + \rho_i + \tau_j + \epsilon_{ij} \tag{5-40}$$

It can be shown that, under the model, the least-squares estimates of the parameters of Eq. (5-40) are

$$\hat{\mu}_{..} = \bar{Y}_{..}, \qquad \hat{\rho}_i = \bar{Y}_{i.} - \bar{Y}_{..}, \qquad \hat{\tau}_j = \bar{Y}_{.j} - \bar{Y}_{..} \tag{5-41}$$

Therefore

$$Y_{ij} = \bar{Y}_{..} + (\bar{Y}_{i.} - \bar{Y}_{..}) + (\bar{Y}_{.j} - \bar{Y}_{..}) + e_{ij} \tag{5-42}$$

where e_{ij} is the least-squares residual. The residual sum of squares of Eq. (5-42) for all row and column classifications is therefore

$$S_1 = \sum_{j=1}^{K}\sum_{i=1}^{R} e_{ij}^2 = \sum_{j=1}^{K}\sum_{i=1}^{R} (Y_{ij} - \bar{Y}_{i.} - \bar{Y}_{.j} + \bar{Y}_{..})^2 \tag{5-43}$$

Referring to Eq. (5-35), the fundamental equality, we call S_1 the error variation, and $S_1 = 0.3215$, as previously calculated. If $N = RK$ is the total number of Y_{ij} observations in the data layout, then the number of degrees of freedom associated with S_1 is $N - R - K + 1 = (R - 1)(K - 1)$.

Hypotheses. The two null hypotheses are that the row means are equal and that the column means are equal. Thus for the rows

$$H_0\colon \quad \mu_{1.} = \mu_{2.} = \cdots = \mu_{R.} = \mu_{..}$$
$$H_1\colon \quad \text{Not all row means are equal}$$

and for the columns

$$H_0\colon \quad \mu_{.1} = \mu_{.2} = \cdots = \mu_{.K} = \mu_{..}$$
$$H_1\colon \quad \text{Not all column means are equal}$$

The hypothesis that all row means (column means) are equal is equivalent to the hypothesis that all row effects (column effects) are equal to zero. For the rows

$$H_0\colon \quad \rho_1 = \rho_2 = \cdots = \rho_R = 0$$
$$H_1\colon \quad \text{Not all row effects are equal to zero}$$

and for the columns

$$H_0\colon \quad \tau_1 = \tau_2 = \cdots = \tau_K = 0$$
$$H_1\colon \quad \text{Not all column effects are equal to zero.}$$

Under the first hypothesis (that all row effects are equal to zero) the row effects drop out of Eq. (5-40) and the model becomes

$$Y_{ij} = \mu_{..} + \tau_j + \epsilon_{ij} \tag{5-44}$$

Since the least-squares estimates of the parameters are $\hat{\mu}_{..} = \bar{Y}_{..}$ and $\hat{\tau}_j = \bar{Y}_{.j} - \bar{Y}_{..}$, we have

$$Y_{ij} = \bar{Y}_{..} + (\bar{Y}_{.j} - \bar{Y}_{..}) + e_{ij} \tag{5-45}$$

The residual sum of squares of Eq. (5-45) for all rows and columns under the hypothesis is

$$S_0 = \sum_{j=1}^{K}\sum_{i=1}^{R} e_{ij}^2 = \sum_{j=1}^{K}\sum_{i=1}^{R} (Y_{ij} - \bar{Y}_{.j})^2 \tag{5-46}$$

with $N - K = K(R - 1)$ degrees of freedom. It can be shown that[13]

$$S_0 = \sum_{j=1}^{K}\sum_{i=1}^{R} (Y_{ij} - \bar{Y}_{..})^2 - R\sum_{j=1}^{K} (\bar{Y}_{.j} - \bar{Y}_{..})^2 \tag{5-47}$$

From the fundamental equality of Eq. (5-35), the difference between S_0 (given above) and S_1 [given in Eq. (5-43)], under the model is equal to

$$S_R = S_0 - S_1 = K\sum_{i=1}^{R} (\bar{Y}_{i.} - \bar{Y}_{..})^2 \tag{5-48}$$

with $R - 1 = K(R - 1) - (R - 1)(K - 1)$ degrees of freedom. S_R is therefore referred to as a row variation.

Under the second hypothesis (that all column effects are equal to zero), a similar argument can be made. Under this hypothesis, the model [Eq. (5-40)] becomes

$$Y_{ij} = \mu_{..} + \rho_i + \epsilon_{ij} \tag{5-49}$$

and the least-squares estimates of the parameters are $\hat{\mu}_{..} = \bar{Y}_{..}$ and $\hat{\rho}_i = \bar{Y}_{i.} - \bar{Y}_{..}$. Thus

$$S_0 = \sum_{j=1}^{K}\sum_{i=1}^{R} e_{ij}^2 = \sum_{j=1}^{K}\sum_{i=1}^{R} (Y_{ij} - \bar{Y}_{i.})^2 \tag{5-50}$$

with $N - R = R(K - 1)$ degrees of freedom. From the fundamental equality of variation, the difference between S_0 and S_1 is

$$S_C = S_0 - S_1 = R\sum_{j=1}^{K} (\bar{Y}_{.j} - \bar{Y}_{..})^2 \tag{5-51}$$

with $K - 1 = R(K - 1) - (R - 1)(K - 1)$ degrees of freedom. Therefore S_C is referred to as the column variation.

Hypotheses Tests. The hypotheses tests with two bases of classification are similar to the hypothesis test with one basis of classification. We argue that if the hypothesis is true, then all row (column) effects are equal to zero, and Eq. (5-44) [Eq. (5-49)] will fit the data just as well as Eq. (5-40). That is,

---

13 $$\sum_{j=1}^{K}\sum_{i=1}^{R} (Y_{ij} - \bar{Y}_{.j})^2 = \sum_{j=1}^{K}\sum_{i=1}^{R} [(Y_{ij} - \bar{Y}_{..}) - (\bar{Y}_{.j} - \bar{Y}_{..})]^2$$

Expansion of the right-hand expression proves our contention.

the residual sum of squares S_0 due to the hypothesis will be almost equal to the residual sum of squares S_1 due to the model. The difference $S_R(S_C)$ will be attributable to sampling error. The discrepancies are $S_R = 0.1779$ and $S_C = 0.1720$, as previously calculated. Upon adjusting for the numbers of degrees of freedom, we find that the F ratio to test the hypothesis concerning the rows

$$H_0\colon \quad \mu_{1\cdot} = \mu_{2\cdot} = \cdots = \mu_{R\cdot} = \mu_{\cdot\cdot}$$

or

$$H_0\colon \quad \rho_1 = \rho_2 = \cdots = \rho_R = 0$$

is

$$F = \frac{S_R/(R-1)}{S_1/(R-1)(K-1)} = \frac{(\text{row variation})/(R-1)}{(\text{error variation})/(R-1)(K-1)} \tag{5-52}$$

which, for our example, comes to

$$F = \frac{0.1779/4}{0.3215/8} = 1.11$$

with $R - 1$ and $(R - 1)(K - 1)$ degrees of freedom. This value does not exceed $F_{0.05;\,4,8} = 3.84$, so that we cannot reject the hypothesis. For the test of the hypothesis concerning the columns

$$H_0\colon \quad \mu_{\cdot 1} = \mu_{\cdot 2} = \cdots = \mu_{\cdot K} = \mu_{\cdot\cdot}$$

or

$$H_0\colon \quad \tau_1 = \tau_2 = \cdots = \tau_K = 0$$

the F ratio is

$$F = \frac{S_C/(K-1)}{S_1/(R-1)(K-1)} = \frac{(\text{column variation})/(K-1)}{(\text{error variation})/(R-1)(K-1)} \tag{5-53}$$

which, for our example, is

$$F = \frac{0.1720/2}{0.3215/8} = 2.14$$

with $K - 1$ and $(R - 1)(K - 1)$ degrees of freedom. This F value does not exceed $F_{0.05;\,2,8} = 4.46$. Therefore we find no significant difference between either the row or the column means. In other words, the row effects and the column effects do not differ from zero at the stated level of significance.

An analysis of variance table with two bases of classification is given in Table 5.6. Again the mean square of the error variation is an estimate[14] of

---

[14]Under the hypothesis of zero row (column) effect, the mean square of the row (column) variation is also an estimate of σ^2.

Table 5.6 Analysis of Variance Table—Two Bases of Classification

Source of Variation	*Amount of Variation*	*Degrees of Freedom**	*Mean Square*	*F Ratio*
Total	$S_0 = \sum_{j=1}^{K} \sum_{i=1}^{R} (Y_{ij} - \bar{Y}_{..})^2$	$N - 1$		
Column	$S_C = R \sum_{j=1}^{K} (\bar{Y}_{.j} - \bar{Y}_{..})^2$	$K - 1$	$\frac{S_C}{K-1}$	$F = \frac{S_C/(K-1)}{S_1/(R-1)(K-1)}$
Row	$S_R = K \sum_{i=1}^{R} (\bar{Y}_{i.} - \bar{Y}_{..})^2$	$R - 1$	$\frac{S_R}{R-1}$	$F = \frac{S_R/(R-1)}{S_1/(R-1)(K-1)}$
Error	$S_1 = \sum_{j=1}^{K} \sum_{i=1}^{R} (Y_{ij} - \bar{Y}_{i.} - \bar{Y}_{.j} + \bar{Y}_{..})^2$	$(R-1)(K-1)$	$\frac{S_1}{(R-1)(K-1)}$	

*$N = RK$.

the variance of ϵ_{ij}, i.e., σ^2,

$$\hat{\sigma}^2 = \frac{S_1}{(R-1)(K-1)} \tag{5-54}$$

Linear Contrasts. Had we rejected either of the null hypotheses, we would have probably been interested in detecting which row(s), or column(s), was causing the rejection. For the rows, the sample row mean $\bar{Y}_{i.}$ is an estimate of the row mean $\mu_{i.}$ with estimated variance $s_i^2 = \hat{\sigma}^2/K$. For the columns, the sample column mean $\bar{Y}_{.j}$ is an estimate of the population mean $\mu_{.j}$ with estimated variance $s_j^2 = \hat{\sigma}^2/R$. The estimate of the row linear contrast is then $\hat{L} = \sum_{i=1}^R c_i\bar{Y}_{i.}$ with estimated variance $s_{\hat{L}}^2 = s_i^2 \sum_{i=1}^R c_i^2$. The estimate of the column linear contrast is $\hat{L} = \sum_{j=1}^K c_j\bar{Y}_{.j}$ with estimated variance $s_{\hat{L}}^2 = s_j^2 \sum_{j=1}^K c_j^2$. Again the contrast coefficients c_i (or c_j) are assumed to sum to zero. A $100(1-\alpha)$ percent confidence interval for any row contrast is given by Eq. (5-33) with $R-1$ and $(R-1)(K-1)$ degrees of freedom for F, so that $R-1$ replaces $K-1$ in Eq. (5-33). A $100(1-\alpha)$ percent confidence interval for any column contrast is similarly given by Eq. (5-33) with $K-1$ and $(R-1)(K-1)$ degrees of freedom for F.

Several Items in a Cell. Analysis of variance with two bases of classification and several items in a cell can be carried out. We restrict ourself in this book to the case where all cells have the same number of observations.[15] Table 5.7 shows some hypothetical weight gains for yearling bulls of three

Table 5.7 Weight Gain (pounds) for Yearling Bulls

Feed \ Breed	Hereford		Angus		Charolais		Row Means
A	100 93	96.5	97 107	102.0	108 96	102.0	100.17
B	100 101	100.5	95 100	97.5	105 109	107.0	101.67
C	100 104	102.0	93 106	99.5	96 107	101.5	101.00
Column means	99.67		99.67		103.5		
Grand mean							100.95

Source: Hypothetical data. Cell means are in boxes.

[15]For the case where the numbers of items in a cell are not equal, see Kendall and Stuart (1968). Some simplifications are possible if the numbers of items are proportional to each other.

breeds. Two bulls of each breed were randomly placed on three different feed mixes for a given period of time. We wish to know if there is a significant difference in weight gain among breeds and among feeds.

Within each cell created by the row and column classification there are now two (or, in general, n) observations. Each observation will be designated as Y_{ijt}, where i and j represent the row and column created by the classification and t represents the item number within the cell. For each cell we have also enclosed within a box the mean of the observations in that cell. Call each of these *cell means* $\bar{Y}_{ij}$ for the sample. Then the cell means are given by

$$\bar{Y}_{ij} = \sum_{t=1}^{n} \frac{Y_{ijt}}{n}, \qquad \begin{aligned} i &= 1, 2, \ldots, R \\ j &= 1, 2, \ldots, K \end{aligned}$$

Similarly, the row means are given by

$$\bar{Y}_{i\cdot} = \sum_{j=1}^{K} \sum_{t=1}^{n} \frac{Y_{ijt}}{nK}, \qquad i = 1, 2, \ldots, R$$

and column means are given by

$$\bar{Y}_{\cdot j} = \sum_{i=1}^{R} \sum_{t=1}^{n} \frac{Y_{ijt}}{nR}, \qquad j = 1, 2, \ldots, K$$

Finally, the grand mean is given by

$$\bar{Y}_{\cdot\cdot} = \sum_{i=1}^{R} \sum_{j=1}^{K} \sum_{t=1}^{n} \frac{Y_{ijt}}{N}$$

where $N = nRK$ is the total number of observations in the table.

Again, let us partition the total deviation, $Y_{ijt} - \bar{Y}_{\cdot\cdot}$, in order to obtain the fundamental equality for variation. For any observation, the following equality obviously holds:

$$(Y_{ijt} - \bar{Y}_{\cdot\cdot}) = (\bar{Y}_{i\cdot} - \bar{Y}_{\cdot\cdot}) + (\bar{Y}_{\cdot j} - \bar{Y}_{\cdot\cdot}) + (\bar{Y}_{ij} - \bar{Y}_{i\cdot} - \bar{Y}_{\cdot j} + \bar{Y}_{\cdot\cdot}) + (Y_{ijt} - \bar{Y}_{ij}) \tag{5-55}$$

Stated verbally,

Total deviation = row deviation + column deviation
+ interaction deviation + error deviation

Notice that when there is only one observation in a cell that $Y_{ijt} = \bar{Y}_{ij} = Y_{ij}$ and that Eq. (5-55) reduces to Eq. (5-34). That is, when there is a single observation in a cell, we cannot distinguish between the interaction effect and the error effect. In Eq. (5-34) the term $Y_{ij} - \bar{Y}_{i\cdot} - \bar{Y}_{\cdot j} + \bar{Y}_{\cdot\cdot}$ plays the part of the error term only under the assumption that there is no interaction.

Since each cell must have the same number of observations, by summing the squared elements, we have the fundamental equality

$$\sum_{i=1}^{R}\sum_{j=1}^{K}\sum_{t=1}^{n}(Y_{ijt} - \bar{Y}_{\cdot\cdot})^2 = nK\sum_{i=1}^{R}(\bar{Y}_{i\cdot} - \bar{Y}_{\cdot\cdot})^2 + nR\sum_{j=1}^{K}(\bar{Y}_{\cdot j} - \bar{Y}_{\cdot\cdot})^2$$
$$+ n\sum_{i=1}^{R}\sum_{j=1}^{K}(\bar{Y}_{ij} - \bar{Y}_{i\cdot} - \bar{Y}_{\cdot j} + \bar{Y}_{\cdot\cdot})^2$$
$$+ \sum_{i=1}^{R}\sum_{j=1}^{K}\sum_{t=1}^{n}(Y_{ijt} - \bar{Y}_{ij})^2 \tag{5-56}$$

or, stated verbally,

Total variation = row variation + column variation
+ interaction variation + error variation

Calculation is once again straightforward. Using the data of Table 5.7, we may obtain total variation indirectly by

$$\sum_{j=1}^{K}\sum_{i=1}^{R}\sum_{t=1}^{n}(Y_{ijt} - \bar{Y}_{\cdot\cdot})^2 = \sum_{j=1}^{K}\sum_{i=1}^{R}\sum_{t=1}^{n}Y_{ijt}^2 - \frac{(\sum_{j=1}^{K}\sum_{i=1}^{R}\sum_{t=1}^{n}Y_{ijt})^2}{N}$$

which for our example is

$$\sum_{j=1}^{K}\sum_{i=1}^{R}\sum_{t=1}^{n}(Y_{ijt} - \bar{Y}_{\cdot\cdot})^2 = 183885.00 - 183416.06$$
$$= 468.94$$

Total variation may also be obtained directly, as usual.

Row variation is easy to obtain directly:

$$nK\sum_{i=1}^{R}(\bar{Y}_{i\cdot} - \bar{Y}_{\cdot\cdot})^2 = (2)(3)[(-0.78)^2 + (0.72)^2 + (0.05)^2]$$
$$= 6.78$$

Column variation is

$$nR\sum_{j=1}^{K}(\bar{Y}_{\cdot j} - \bar{Y}_{\cdot\cdot})^2 = (2)(3)[(-1.28)^2 + (-1.28)^2 + (2.55)^2]$$
$$= 58.77$$

Interaction variation is

$$n\sum_{j=1}^{K}\sum_{i=1}^{R}(\bar{Y}_{ij} - \bar{Y}_{i\cdot} - \bar{Y}_{\cdot j} + \bar{Y}_{\cdot\cdot})^2 = (2)[(-2.39)^2 + \cdots + (-2.05)^2]$$
$$= 82.89$$

Error variation is obtained by subtraction:

$$\sum_{j=1}^{K}\sum_{i=1}^{R}\sum_{t=1}^{n}(Y_{ijt} - \bar{Y}_{ij})^2 = \text{total variation} - \text{row variation}$$
$$- \text{column variation} - \text{interaction variation}$$
$$= 320.50$$

Model. Each of the observations within a cell may be specified by its cell mean plus an error:

$$Y_{ijt} = \mu_{ij} + \epsilon_{ijt} \tag{5-57}$$

In addition to the definition of the row and column effects, we shall also give a definition of the interaction effect. Each cell mean is assumed to be representable by the sum of the grand mean, a row effect, a column effect, and an interaction effect, I_{ij},

$$\mu_{ij} = \mu_{..} + \rho_i + \tau_j + I_{ij} \tag{5-58}$$

It is obvious that the interaction effect is just the difference between the deviation $(\mu_{ij} - \mu_{..})$ and the row and column effects:

$$I_{ij} = (\mu_{ij} - \mu_{..}) - (\rho_i + \tau_j) \tag{5-59}$$

Substituting Eq. (5-58) into Eq. (5-57), we have

$$Y_{ijt} = \mu_{..} + \rho_i + \tau_j + I_{ij} + \epsilon_{ijt} \tag{5-60}$$

The least-squares estimates of the parameters of Eq. (5-60) are

$$\hat{\mu}_{..} = \bar{Y}_{..}, \qquad \hat{\rho}_i = \bar{Y}_{i.} - \bar{Y}_{..}, \qquad \hat{\tau}_j = \bar{Y}_{.j} - \bar{Y}_{..} \tag{5-61}$$

and

$$\hat{I}_{ij} = \bar{Y}_{ij} - \bar{Y}_{i.} - \bar{Y}_{.j} + \bar{Y}_{..}$$

Therefore

$$Y_{ijt} = \bar{Y}_{..} + (\bar{Y}_{i.} - \bar{Y}_{..}) + (\bar{Y}_{.j} - \bar{Y}_{..}) + (\bar{Y}_{ij} - \bar{Y}_{i.} - \bar{Y}_{.j} + \bar{Y}_{..})$$
$$+ e_{ijt} \tag{5-62}$$

where $e_{ijt} = Y_{ijt} - \bar{Y}_{ij}$ is the least-squares residual. Thus, under the model of Eq. (5-60), the total residual sum of squares is

$$S_1 = \sum_{i=1}^{R}\sum_{j=1}^{K}\sum_{t=1}^{n} e_{ijt}^2 = \sum_{i=1}^{R}\sum_{j=1}^{K}\sum_{t=1}^{n}(Y_{ijt} - \bar{Y}_{ij})^2 \tag{5-63}$$

with $N - RK$ degrees of freedom. Referring to the fundamental equality,

Eq. (5-56), we call S_1 error variation. In our example, $S_1 = 320.50$, as previously calculated.

Hypotheses. There are three null hypotheses to be tested in the model. For the rows

$$H_0\colon\ \mu_{1\cdot} = \mu_{2\cdot} = \cdots = \mu_{R\cdot} = \mu_{\cdot\cdot}$$

or

$$H_0\colon\ \rho_1 = \rho_2 = \cdots = \rho_R = 0$$

and

H_1: Not all row means are equal

or

H_1: Not all row effects are zero

For the columns

$$H_0\colon\ \mu_{\cdot 1} = \mu_{\cdot 2} = \cdots = \mu_{\cdot K} = \mu_{\cdot\cdot}$$

or

$$H_0\colon\ \tau_1 = \tau_2 = \cdots = \tau_K = 0$$

and

H_1: Not all column means are equal

or

H_1: Not all column effects are zero

For interaction

$$H_0\colon\ I_{11} = I_{12} = \cdots = I_{RK} = 0$$

H_1: Not all interaction effects are zero

Under the null hypotheses, we can calculate the residual sum of squares S_0 for each of three hypotheses, and hence, from the fundamental equality, it can be shown that the differences between S_1 and S_0 are

$$S_R = nK \sum_{i=1}^{R} (\bar{Y}_{i\cdot} - \bar{Y}_{\cdot\cdot})^2 \tag{5-64}$$

$$S_C = nR \sum_{j=1}^{K} (\bar{Y}_{\cdot j} - \bar{Y}_{\cdot\cdot})^2 \tag{5-65}$$

$$S_I = n \sum_{i=1}^{R} \sum_{j=1}^{K} (\bar{Y}_{ij} - \bar{Y}_{i\cdot} - \bar{Y}_{\cdot j} + \bar{Y}_{\cdot\cdot})^2 \tag{5-66}$$

Here S_I is called the interaction variation with $N - R - K + 1 = (R - 1)(K - 1)$ degrees of freedom. In our example, $S_R = 6.78$, $S_C = 58.77$, and $S_I = 82.89$, as previously calculated.

Hypotheses Tests. By now the general method for testing the hypotheses should be clear. We first test the significance of the interaction effect, and we form the F ratio

$$F = \frac{S_I/(R-1)(K-1)}{S_1/(N-RK)} = \frac{(\text{interaction variation})/(R-1)(K-1)}{(\text{error variation})/(N-RK)} \tag{5-67}$$

For our example

$$F = \frac{82.89/4}{320.50/9} = 0.58$$

which is less than $F_{0.05;\,4,9} = 3.5$. Therefore we do not reject the null hypothesis that the interaction effect is zero. If interaction had been found to be significant, we would be placed in the situation of not being able to separate out the independent contributions of the row and column effects. For our example, some one(s) of the feeds might be specifically better when used with one breed than with another. Such information is sometimes important and can be investigated further. However, the analysis to follow may be completely misleading if interaction is present.[16]

Under the assumption that the row and column effects are fixed variates,[17] both row variation and column variation are compared to error variation to test the hypotheses that the row mean and column means are respectively equal. For rows

$$F = \frac{S_R/(R-1)}{S_1/(N-RK)} = \frac{(\text{row variation})/(R-1)}{(\text{error variation})/(N-RK)} \tag{5-68}$$

which follows the F ratio with $R-1$ and $N-RK$ degrees of freedom. For our example

$$F = \frac{6.78/2}{320.50/9} = 0.10$$

which does not exceed $F_{0.05;\,2,9} = 4.3$. We cannot reject the null hypothesis that all row means are equal at the stated level of significance.

---

[16]Scheffé (1959), pp. 94 ff., gives a mathematical discussion of the problem. However, as Scheffé points out, the discussion is of little practical significance. We content ourselves in this book with the advice that the analysis should be terminated when significant interaction is discovered.

[17]Three different models are sometimes discussed: the fixed model, which we assume here; the random model, which assumes that the row and column effects are random variables; and the mixed model, which assumes that the row (column) effects are fixed while the column (row) effects are random. See Li (1964), p. 59, for the differences these assumptions make in hypothesis testing.

For columns

$$F = \frac{S_C/(K-1)}{S_1/(N-RK)} = \frac{(\text{column variation})/(K-1)}{(\text{error variation})/(N-RK)} \tag{5-69}$$

which follows the F ratio with $K - 1$ and $N - RK$ degrees of freedom. For our example

$$F = \frac{58.77/2}{320.50/9} = 0.83$$

and we cannot reject the null hypothesis that the column means are equal.

The analysis of variance table with n observations in each cell is shown in Table 5.8. Again the mean square of the error variation is an estimate[18] of the variance of ϵ_{ijt}, i.e., σ^2.

Linear contrasts are difficult to form under the current model because of the complicating influence of interaction. The interested reader is referred to Scheffé.[19]

5.4 analysis of covariance

Analysis of covariance is a technique that combines the features of analysis of variance and regression. The technique offers an analysis of data which have been classified, and, as well, it allows for the inclusion of information obtained from ordinary regression analysis.[20]

Let us illustrate a typical problem that may be handled with this technique by reference to the data of Table 4.4 of Chap. 4. In that table are shown the selling price, Y, of residential dwellings. We may wonder whether or not the mean selling price differs among the condition classifications. That is, other things equal, does a house rated excellent tend to sell for a higher price than one rated poor? Obviously there is a significant difference between the selling price column means, and an analysis of variance done on the Y observations

---

[18]If we *assume* that all interaction effects are zero in the model of Eq. (5-60), S_I may be merged with S_1 to form a pooled error sum of squares with $N - R - K + 1$ degrees of freedom. Thus an estimate of σ^2 is

$$\hat{\sigma}^2 = \frac{S_I + S_1}{N - R - K + 1}$$

This estimate is then used as the denominator of the F ratio in testing the row and column effects.

[19]Scheffé (1959), pp. 109–110.

[20]A well-known review of the many uses of analysis of covariance can be found in Cochran (1957). A standard reference is Snedecor and Cochran (1967), and a method for use with ranked data is discussed by Quade (1967).

Table 5.8 Analysis of Variance Table—Two Bases of Classification and n Items Per Cell

Source of Variation	*Amount of Variation*	*Degrees of Freedom**	*Mean Square*	*F Ratio*
Total	$S_0 = \sum_{i=1}^{R} \sum_{j=1}^{K} \sum_{t=1}^{n} (Y_{ijt} - \bar{Y}_{..})^2$	$N - 1$		
Column	$S_C = nR \sum_{j=1}^{K} (\bar{Y}_{.j} - \bar{Y}_{..})^2$	$K - 1$	$\frac{S_C}{K - 1}$	$F = \frac{S_C/(K-1)}{S_1/(N - RK)}$
Row	$S_R = nK \sum_{i=1}^{R} (\bar{Y}_{i.} - \bar{Y}_{..})^2$	$R - 1$	$\frac{S_R}{R - 1}$	$F = \frac{S_R/(R-1)}{S_1/(N - RK)}$
Interaction	$S_I = n \sum_{i=1}^{R} \sum_{j=1}^{K} (\bar{Y}_{ij} - \bar{Y}_{i.} - \bar{Y}_{.j} + \bar{Y}_{..})^2$	$(R - 1)(K - 1)$	$\frac{S_I}{(R-1)(K-1)}$	$F = \frac{S_I/(R-1)(K-1)}{S_1/(N - RK)}$
Error	$S_1 = \sum_{i=1}^{R} \sum_{j=1}^{K} \sum_{t=1}^{n} (Y_{ijt} - \bar{Y}_{ij})^2$	$N - RK$	$\frac{S_1}{N - RK}$	

$^*N = nRK$.

will bear out this judgment. Recall, however, that we showed in Table 4.4 that dwellings rated fair tend to have relatively small floor space measurements and that dwellings rated good tend to have relatively large floor space measurements. Thus the variable X, often called a *concomitant variable*, may in reality be of greater importance in explaining differences between the mean selling prices than is the condition classification.

With this in mind let us propose the model

$$Y_{ij} = \mu_{.j} + \beta(X_{ij} - \bar{X}_{..}) + \epsilon_{ij}, \qquad j = 1, 2, \ldots, K \tag{5-70}$$

Here, apart from the error term, each Y_{ij} observation is representable by its population column mean, $\mu_{.j}$, plus an effect stemming from the fact that the associated concomitant variable, X_{ij}, differs from the grand mean of the concomitant variable, $\bar{X}_{..}$. The regression coefficient, β, indicates the importance of the influence of the concomitant variable upon Y_{ij}. Recalling that our main task is to compare the selling price means among condition classifications, it would seem reasonable to adjust for the floor space variable. Thus we wish to do an analysis of variance using the selling price column means after they have been "adjusted" for the effect of floor space. The adjusted selling price column means, $\bar{Y}^*_{.j}$, are defined as

$$\bar{Y}^*_{.j} = \bar{Y}_{.j} - \hat{\beta}(\bar{X}_{.j} - \bar{X}_{..}), \qquad j = 1, 2, \ldots, K \tag{5-71}$$

Let us inquire for a moment into the meaning of Eq. (5-71). Suppose that $\hat{\beta}$ is positive, which means that X is positively associated with Y. Then, if a given $\bar{X}_{.j}$ is greater than the grand mean $\bar{X}_{..}$, the adjusted column mean $\bar{Y}^*_{.j}$ will be smaller than the unadjusted column mean $\bar{Y}_{.j}$. Thus Eq. (5-71) will adjust the column mean of Y downward to take into account the fact that this column mean is associated with a large column mean for the concomitant variable. The opposite argument can be made for a negative $\hat{\beta}$ or a column mean $\bar{X}_{.j}$ which is smaller than $\bar{X}_{..}$.

It should now be clear why analysis of covariance combines regression analysis and analysis of variance. Regression analysis will provide us with the estimated slope $\hat{\beta}$. Then, after adjusting for the effect of the concomitant variable, we shall do an analysis of variance on the adjusted Y variables. The final F ratio will be of the form

$$F = \frac{(\text{adjusted column variation})/(K-1)}{(\text{adjusted error variation})/(N-K-1)} \tag{5-72}$$

Equation (5-70) implicitly assumes that the coefficient β is constant for all classifications. That is, the effect of floor space upon selling price is the same regardless of the condition classification.[21] This assumption of lack of

[21] We assume in this book that the concomitant variable is nonstochastic.

interaction between the concomitant variable and the classification scheme is vital to the analysis to follow. If it is found that interaction is present, the reasons for the interaction are generally explored, but the analysis of covariance usually terminates. Thus before we can do the analysis of covariance, we should test the validity of the assumption of a common coefficient. This test consists of fitting a regression line of the form

$$Y_{ij} = \beta_{0j} + \beta_{1j}X_{ij} + \epsilon_{ij}, \qquad j = 1, 2, \ldots, K$$

for each of the classifications. Then the hypothesis is tested that all $\beta_{1j}, j = 1, 2, \ldots, K$, are equal. If the hypothesis is accepted, then the K individual regression slopes are pooled to form the estimate $\hat{\beta}$.

In Sec. 4.6 we carried out the calculation of the three regression lines for our example [see Eqs. (4-68)]. In that section we also tested the hypothesis that all three slopes were equal, and we did not reject that hypothesis. Therefore we are justified in pooling these three slopes into a common slope which will estimate β of Eq. (5-70). This pooled slope may be obtained by an evaluation of Eq. (4-82) or by the simple technique of forming an average of the three individual regression slopes, so that the estimate is often called the *average* slope.

Before we form this average of the individual regression slopes, we introduce some new notation that will be helpful in the discussion to follow. For any two variables X and Y let us denote the sums of squares and products in the following way[22]:

$$E_{yx} = \sum_{j=1}^{K}\sum_{i=1}^{n_j}(Y_{ij} - \bar{Y}_{.j})(X_{ij} - \bar{X}_{.j})$$

$$E_{yy} = \sum_{j=1}^{K}\sum_{i=1}^{n_j}(Y_{ij} - \bar{Y}_{.j})^2$$

$$E_{xx} = \sum_{j=1}^{K}\sum_{i=1}^{n_j}(X_{ij} - \bar{X}_{.j})^2$$

$$T_{yx} = \sum_{j=1}^{K}\sum_{i=1}^{n_j}(Y_{ij} - \bar{Y}_{..})(X_{ij} - \bar{X}_{..})$$

$$T_{yy} = \sum_{j=1}^{K}\sum_{i=1}^{n_j}(Y_{ij} - \bar{Y}_{..})^2$$

$$T_{xx} = \sum_{j=1}^{K}\sum_{i=1}^{n_j}(X_{ij} - \bar{X}_{..})^2$$

Notice that the Es denote sums of squares and products about column means, while the Ts denote sums of squares and products about the grand means.

---

[22]This notation follows Snedecor and Cochran (1967). No confusion should arise between our use of E here and the previous use of E as the expectation operator.

The average (or pooled) estimate, $\hat{\beta}$, can now be obtained by evaluating[23]

$$\hat{\beta} = \frac{E_{yx}}{E_{xx}} \tag{5-73}$$

As the reader can easily verify by calculation of E_{yx} and E_{xx}, $\hat{\beta} = 13.247$ for our example (the calculation is given by the computer program in the Appendix). The adjusted column means are, by Eq. (5-71), $\bar{Y}^*_{.1} = 15{,}820.86$, $\bar{Y}^*_{.2} = 15{,}405.92$, and $\bar{Y}^*_{.3} = 19{,}880.55$. Note that the adjusted means do not differ among each other nearly so much as the unadjusted means. In fact, the adjusted mean selling price for houses rated fair and good are nearly equal.

Model. As we have already explained, the model is given by Eq. (5-70) with β estimated by Eq. (5-73) and the adjusted means estimated by Eq. (5-71). The sample least-squares residual of Eq. (5-70) is

$$e_{ij} = (Y_{ij} - \hat{\mu}_{.j}) - \hat{\beta}(X_{ij} - \bar{X}_{..})$$

However, the least-squares estimate of any $\mu_{.j}$ is the adjusted sample column mean

$$\hat{\mu}_{.j} = \bar{Y}^*_{.j} = \bar{Y}_{.j} - \hat{\beta}(\bar{X}_{.j} - \bar{X}_{..})$$

Therefore, upon substitution, the sample least-squares residual becomes

$$e_{ij} = (Y_{ij} - \bar{Y}_{.j}) - \hat{\beta}(X_{ij} - \bar{X}_{.j}) \tag{5-74}$$

---

[23]If we write Eq. (5-73) in extensive form, we have

$$\hat{\beta} = \frac{\sum_{i=1}^{n_1}(Y_{i1} - \bar{Y}_{.1})(X_{i1} - \bar{X}_{.1}) + \cdots + \sum_{i=1}^{n_K}(Y_{iK} - \bar{Y}_{.K})(X_{iK} - \bar{X}_{.K})}{\sum_{i=1}^{n_1}(X_{i1} - \bar{X}_{.1})^2 + \cdots + \sum_{i=1}^{n_K}(X_{iK} - \bar{X}_{.K})^2}$$

Since each of the elements in this equation represents sums of squares and products for each column, let us abbreviate further and write

$$\hat{\beta} = \frac{\sum y_{i1}x_{i1} + \cdots + \sum y_{iK}x_{iK}}{\sum x_{i1}^2 + \cdots + \sum x_{iK}^2}$$

If we multiply and divide each element in the numerator of the equation directly above (i.e., $\sum y_{ij}x_{ij}$) by the corresponding sum of squares of x(i.e., $\sum x_{ij}^2$), we have

$$\hat{\beta} = \frac{\hat{\beta}_{11}\sum x_{i1}^2 + \cdots + \hat{\beta}_{1K}\sum x_{iK}^2}{\sum x_{i1}^2 + \cdots + \sum x_{iK}^2}$$

Clearly, $\hat{\beta}$ is an average of the $\hat{\beta}_{1j}$s with the $\sum x_{ij}^2$s serving as weights.

The residual sum of squares is

$$S_1^* = \sum_{j=1}^{K} \sum_{i=1}^{n_j} e_{ij}^2 = \sum_{j=1}^{K} \sum_{i=1}^{n_j} (Y_{ij} - \bar{Y}_{.j})^2 - \hat{\beta} \sum_{j=1}^{K} \sum_{i=1}^{n_j} (Y_{ij} - \bar{Y}_{.j})(X_{ij} - \bar{X}_{.j})$$

or, using Eq. (5-73) and the new notation,

$$S_1^* = E_{yy} - \hat{\beta} E_{yx} = S_1 - \hat{\beta} E_{yx} \tag{5-75}$$

The quantity S_1^* is called the *adjusted error variation*; it may be interpreted as the error variation in Eq. (5-24) [the error variation under Eq. (5-20)] adjusted by the amount of variation which is "explained" by the concomitant variable (i.e., $\hat{\beta} E_{yx}$). Since E_{yy} is associated with $N - K$ degrees of freedom and explained variation is associated with 1 degree of freedom, S_1^* is associated with $N - K - 1$ degrees of freedom. In our example (see the Appendix)

$$S_1^* = 184{,}241{,}700$$

with 18 degrees of freedom.

Hypothesis. We desire to test whether or not the column means, $\mu_{.j}$, $j = 1, 2, \ldots, K$, which are estimated by $\bar{Y}_{.j}^*$, are equal. That is, we wish to test for the Y variable

$$H_0\colon \quad \mu_{.1} = \mu_{.2} = \cdots = \mu_{.K} = \mu_{..}$$

against

$$H_1\colon \quad \text{Not all column means are equal}$$

Under the hypothesis we ignore the column classification, and the model of Eq. (5-70) becomes

$$Y_{ij} = \mu_{..} + \beta_G(X_{ij} - \bar{X}_{..}) + \epsilon_{ij} \tag{5-76}$$

and the least-squares estimates of $\mu_{..}$ and β_G are

$$\hat{\beta}_G = \frac{T_{yx}}{T_{xx}} \tag{5-77}$$

and

$$\hat{\mu}_{..} = \bar{Y}_{..} \tag{5-78}$$

For our example

$$\hat{\beta}_G = 15.998 \quad \text{and} \quad \bar{Y}_{..} = 17{,}146.23$$

Notice that in Eq. (5-76) β_G is simply the slope of the regression of the pooled Y observations on the pooled X observations. Thus β_G is often called the *grand* regression slope since it is found by regressing all Y and X values. The least-squares residuals for Eq. (5-76) are

$$e_{ij}^* = (Y_{ij} - \bar{Y}_{..}) - \hat{\beta}_G(X_{ij} - \bar{X}_{..})$$

and the sum of the squared values of these residuals, using Eq. (5-77), is

$$\begin{aligned} S_0^* &= \sum_{j=1}^{K} \sum_{i=1}^{n_j} e_{ij}^2 = T_{yy} - \hat{\beta}_G T_{yx} \\ &= S_0 - \hat{\beta}_G T_{yx} \end{aligned} \tag{5-79}$$

where S_0^* is called the adjusted error variation under the hypothesis; it may be interpreted as the error variation in Eq. (5-27) [the error variation under Eq. (5-25)] adjusted by the variation which is "explained" by the concomitant variable, i.e., $\hat{\beta}_G T_{yx}$. Since S_0 is associated with $N - 1$ degrees of freedom and explained variation is associated with 1 degree of freedom, S_0^* is associated with $N - 2$ degrees of freedom.

Hypothesis Test. The discrepancy between S_0^* and S_1^* should be small if the null hypothesis is true. Define the adjusted column variation as

$$S_C^* = S_0^* - S_1^*$$

Thus the F ratio will be

$$F = \frac{S_C^*/(K-1)}{S_1^*/(N-K-1)} \tag{5-80}$$

which is the F ratio stated verbally in Eq. (5-72). For our example

$$S_C^* = 60{,}110{,}590$$

and

$$F = \frac{60{,}110{,}590/2}{184{,}241{,}700/18} = 2.94$$

which does not exceed $F_{0.05;\,2,18} = 3.5$. Thus we do not reject the null hypothesis that the adjusted column means are equal. Apparently, the condition classification is of little influence upon the selling price when floor space is taken into account. Table 5.9 summarizes the discussion. Again,

$$\hat{\sigma}^2 = \frac{S_1^*}{N - K - 1} \tag{5-81}$$

is an estimate of the variance of ϵ_{ij}.

Linear Contrasts. If the null hypothesis that all adjusted column means are equal is rejected, we may be interested in knowing which one(s) of the means is causing this rejection. Let the contrast be defined as

$$L = \sum_{j=1}^{K} c_j \mu_{\cdot j} \qquad \text{with} \sum_{j=1}^{K} c_j = 0$$

Then the estimate of the contrast is

$$\hat{L} = \sum_{j=1}^{K} c_j \bar{Y}_{\cdot j}^* = \sum_{j=1}^{K} c_j [\bar{Y}_{\cdot j} - \hat{\beta}(\bar{X}_{\cdot j} - \bar{X}_{\cdot\cdot})]$$

from Eq. (5-71). It can be shown that the estimated variance of $\hat{L}$ is[24]

$$s_{\hat{L}}^2 = \hat{\sigma}^2 \left\{ \sum_{j=1}^{K} \frac{c_j^2}{n_j} + \frac{[\sum_{j=1}^{K} c_j(\bar{X}_{\cdot j} - \bar{X}_{\cdot\cdot})]^2}{E_{xx}} \right\}$$

Therefore the $100(1 - \alpha)$ percent confidence interval for the linear contrast is given by Eq. (5-33) with $K - 1$ and $N - K - 1$ degrees of freedom.

Two Concomitant Variables. Sometimes we wish to adjust for more than one concomitant variable. In the housing example of this section we might, for example, have wished to adjust for both floor space and the median income of the neighborhood in which the house was located. If we wish to adjust for two variables, say X and Z, the model is

$$Y_{ij} = \mu_{\cdot j} + \beta_1(X_{ij} - \bar{X}_{\cdot\cdot}) + \beta_2(Z_{ij} - \bar{Z}_{\cdot\cdot}) + \epsilon_{ij} \tag{5-82}$$

Under the model, the least-squares estimates of β_1 and β_2 are found by solving the normal equations

$$E_{xx}\hat{\beta}_1 + E_{xz}\hat{\beta}_2 = E_{yx}$$
$$E_{zx}\hat{\beta}_1 + E_{zz}\hat{\beta}_2 = E_{yz}$$

In matrix notation we may write the solution as

$$\begin{bmatrix} \hat{\beta}_1 \\ \hat{\beta}_2 \end{bmatrix} = \begin{bmatrix} E_{xx} & E_{xz} \\ E_{zx} & E_{zz} \end{bmatrix}^{-1} \begin{bmatrix} E_{yx} \\ E_{yz} \end{bmatrix}$$

---

[24] $s_{\hat{L}}^2 = \sum_{j=1}^{K} c_j^2 s_{\bar{Y}\cdot j}^2 + [\sum_{j=1}^{K} c_j(\bar{X}_{\cdot j} - \bar{X}_{\cdot\cdot})]^2 s_{\hat{\beta}}^2$ since $\bar{Y}_{\cdot j}$ and $\hat{\beta}$ are independent under the model of Eq. (5-70). Also, $s_{\bar{Y}\cdot j}^2 = \hat{\sigma}^2/n_j$ and $s_{\hat{\beta}}^2 = \hat{\sigma}^2/E_{xx}$.

Table 5.9 Analysis of Covariance

Source of Variation	*Sums of Squares and Products*			*Regression Coefficient*	*Explained by Concomitant Variable*	*Adjusted Variation*	*Degrees of Freedom*	*F Ratio*
Total	T_{yx}	T_{xx}	T_{yy}	$\hat{\beta}_G = \frac{T_{yx}}{T_{xx}}$	$\hat{\beta}_G T_{yx}$	$S_0^* = T_{yy} - \hat{\beta}_G T_{yx}$	$N - 2$	
Error	E_{yx}	E_{xx}	E_{yy}	$\hat{\beta} = \frac{E_{yx}}{E_{xx}}$	$\hat{\beta} E_{yx}$	$S_1^* = E_{yy} - \hat{\beta} E_{yx}$	$N - K - 1$	
Column						$S_C^* = S_0^* - S_1^*$	$K - 1$	$F = \frac{S_C^*/(K-1)}{S_1^*/(N-K-1)}$

which may be abbreviated to read

$$\hat{\boldsymbol{\beta}} = \mathbf{E}_{xx}^{-1}\mathbf{E}_{yy}$$

The adjusted column means are

$$\bar{Y}^*_{\cdot j} = \bar{Y}_{\cdot j} - \hat{\beta}_1(\bar{X}_{\cdot j} - \bar{X}_{\cdot\cdot}) - \hat{\beta}_2(\bar{Z}_{\cdot j} - \bar{Z}_{\cdot\cdot})$$

and the adjusted error variation, which corresponds to Eq. (5-75), is

$$S_1^* = S_1 - \hat{\boldsymbol{\beta}}\mathbf{E}_{yx}$$

where

$$S_1 = \sum_{j=1}^{K}\sum_{i=1}^{n_j}(Y_{ij} - \bar{Y}_{\cdot j})^2$$

Under the hypothesis

$$H_0\colon \quad \mu_{\cdot 1} = \mu_{\cdot 2} = \cdots = \mu_{\cdot K} = \mu_{\cdot\cdot}$$

the model of Eq. (5-82) becomes

$$Y_{ij} = \mu_{\cdot\cdot} + \beta_{1G}(X_{ij} - \bar{X}_{\cdot\cdot}) + \beta_{2G}(Z_{ij} - \bar{Z}_{\cdot\cdot}) + \epsilon_{ij}$$

and the least-squares estimates of β_{1G} and β_{2G} are found by solving

$$T_{xx}\hat{\beta}_{1G} + T_{xz}\hat{\beta}_{2G} = T_{yx}$$
$$T_{zx}\hat{\beta}_{1G} + T_{zz}\hat{\beta}_{2G} = T_{yz}$$

These normal equations may be written as

$$\mathbf{T}_{xx}\hat{\boldsymbol{\beta}}_G = \mathbf{T}_{yx}$$

and their solution is

$$\hat{\boldsymbol{\beta}}_G = \mathbf{T}_{xx}^{-1}\mathbf{T}_{yx}$$

The adjusted error variation under the hypothesis is

$$S_0^* = S_0 - \hat{\boldsymbol{\beta}}_G\mathbf{T}_{yx}$$

where $S_0 = \sum_{j=1}^{K}\sum_{i=1}^{n_j}(Y_{ij} - \bar{Y}_{\cdot\cdot})^2$ and $\hat{\boldsymbol{\beta}}_G$ is the vector of grand regression slopes found by regressing all values of Y on all values of X and Z. If we define the adjusted column variation as $S_C^* = S_0^* - S_1^*$, then the F ratio becomes

$$F = \frac{S_{\bar{C}}^*/(K-1)}{S_{\bar{1}}^*/(N-K-2)}$$

For linear contrasts we note that $\hat{\sigma}^2$ is still given by Eq. (5-81). Define the covariance matrix of the estimated $\hat{\beta}_i$ $(i = 1, 2)$ as

$$\begin{bmatrix} s_{11} & s_{12} \\ s_{21} & s_{22} \end{bmatrix} = \hat{\sigma}^2 \mathbf{E}_{xx}^{-1}$$

The estimate of the linear contrast is

$$\hat{L} = \sum_{j=1}^{K} c_j \hat{\mu}_{.j} = \sum_{j=1}^{K} c_j [\bar{Y}_{.j} - \hat{\beta}_1(\bar{X}_{.j} - \bar{X}_{..}) - \hat{\beta}_2(\bar{Z}_{.j} - \bar{Z}_{..})]$$

with variance

$$s_{\hat{L}}^2 = \hat{\sigma}^2 \sum_{j=1}^{K} \frac{c_j^2}{n_j} + s_{11}[\sum_{j=1}^{K} c_j(\bar{X}_{.j} - \bar{X}_{..})]^2 + s_{22}[\sum_{j=1}^{K} c_j(\bar{Z}_{.j} - \bar{Z}_{..})]^2$$
$$+ 2s_{12}[\sum_{j=1}^{K} c_j(\bar{X}_{.j} - \bar{X}_{..})][\sum_{j=1}^{K} c_j(\bar{Z}_{.j} - \bar{Z}_{..})]$$

A $100(1 - \alpha)$ percent confidence interval can be obtained from Eq. (5-33) with $K - 1$ and $N - K - 2$ degrees of freedom for F.

Several Concomitant Variables. The formulas above are easily extended to the case where there are several concomitant variables. As an exercise, the student should generalize these results to three concomitant variables. The computer program in the Appendix will handle the case of one concomitant variable since in the authors' experience this is the case most often encountered. Programs exist which will handle several concomitant variables.[25] Again, the equality of the slopes for the individual columns should be tested as a prelude to the analysis.

Finally, we wish to mention that analysis of variance and analysis of covariance (as well as a host of other techniques) may be viewed as special cases of regression analysis when *dummy* variables are used to indicate classifications. Regrettably, lack of space prevents us from presenting this theoretically interesting (but computationally inefficient) way of looking at analysis of variance and covariance. The interested reader is referred to Bottenberg and Ward (1963), Graybill (1961), Gujarati (1970), Mendenhall (1968), and Skvarcius and Cromer (1971).

---

[25] See Dixon (1967). Analysis of covariance can also be extended to two bases of classification. See Scheffé (1959), pp. 209–213.

Questions and Problems

Sec. 5.1

1. Calculate $r^2_{0(12)}$ using Eq. (5-7) and the correlation matrix of this section.
2. Calculate $\bar{r}^2_{0(12)}$ using Eq. (5-8a).
3. Does $r^2_{0(1).2}$ differ from $r^2_{01.2}$? Why, or why not?

Sec. 5.2

1. The following data give the cost of a market basket of items purchased in three types of stores in the Nashville, Tennessee area in 1971. Each store is located in a middle-income white neighborhood. Test for equality of mean cost of the market basket among the three types of stores. Use $\alpha = 0.05$ and linear contrasts if necessary.

Market Type		
Drive-In	*Independent*	*Chain*
\$20.36	\$17.61	\$17.96
19.78	18.10	17.74
18.82	18.02	18.00
20.47	17.74	17.71
18.78	17.69	17.58
20.62	18.42	17.42
21.05	18.15	17.62
19.70	18.30	17.50
20.20	18.25	17.16
21.68	18.36	16.77
—	18.26	16.27
—	—	16.71

Source: Data courtesy of William F. Steel.

2. In the same study, Professor Steel did an analysis of variance on prices charged by chain stores classified by the income of the neighborhood in which they were located (low, medium, high). He could not reject the hypothesis (at $\alpha = 0.01$) that the mean price of the market basket did not differ among income areas. He also found that chain stores tended to locate in upper-middle-class and high-income neighborhoods, while independent stores tended to locate in low-income neighborhoods. What can you say about the thesis "The poor pay more?"

Sec. 5.3

1. The following table shows the median salary (thousands of dollars) in 1966 for economists with three different degrees classified by years of work experience. Test for differences in salary among degrees and among years of work experience. Use $\alpha = 0.05$ and round out your analysis by use of linear contrasts if appropriate.

Years of Experience \ *Degree*	*Ph.D.*	*Masters*	*Bachelors*
Under 2	9.8	8.0	9.0
2– 4	10.0	8.8	8.9
5– 9	11.5	10.5	10.6
10–14	13.0	12.3	13.0
15–19	15.0	15.0	15.6

Source: Tolles and Melichar (1968), p. 92.

2. The following table shows inventory turnover for three types of stores (grocery, department, and variety) for 4 years (1, 2, 3, and 4). The sample for each year was selected independently in that different stores were used each year. Use $\alpha = 0.05$ to test for significance of difference in inventory turnover among years and among store types if such a test appears valid in view of possible interaction effects.

Type of Store \ *Year*	*1*	*2*	*3*	*4*
Grocery	7.4 7.7 7.2	9.1 8.8 7.0	10.1 8.7 7.4	7.2 7.1 7.0
Department	6.5 6.6 7.0	6.7 7.2 6.4	6.7 7.0 7.0	6.3 6.6 6.6
Variety	3.1 6.2 5.8	6.1 4.2 7.0	4.9 5.5 6.1	5.3 3.3 3.2

Source: Data courtesy Dudley J. Cowden.

Sec. 5.4

1. Cowden (1957) presents some data illustrating the use of analysis of

covariance in industrial quality control. Suppose that a machine produces output which has a quality measurement Y. For example, Y may measure the diameter of a drill hole which is, say, 0.0100 inches. The measurement may be coded by multiplying it by 1000 to give a measurement of, say, 10. Let X denote five machine settings represented by the integers 0, 1, 2, 3, and 4. Samples were taken for three different runs of the machine. Do an analysis of covariance to test for equality of mean diameter among runs taking into account machine settings. Use $\alpha = 0.05$.

First Run		*Second Run*		*Third Run*	
Y	*X*	*Y*	*X*	*Y*	*X*
4	0	6	0	5	0
6	0	7	0	7	0
6	2	10	0	9	0
10	2	8	1	8	2
12	2	11	1	12	2
14	2	11	3	12	2
11	3	14	3	14	2
13	3	15	3	16	4
13	4	18	3	19	4
14	4	15	4	20	4
15	4	17	4	—	—
17	4	—	—	—	—

APPENDIX

In this appendix we shall give three subroutines which are useful in both analysis of variance and covariance. We shall then give main programs for analysis of variance and covariance. More general programs for both analysis of variance and covariance are available.[1]

A5.1 subroutine GRAND

A. Description. This subroutine computes the grand mean and the total sum of squares from a data layout.

[1]See Dixon (1967).

B. Limitation. The data layout can have at most 10 columns with 50 observations per column. This restriction may be changed by altering the DIMENSION statement.

C. Use. We assume that the following items are in storage:

1. The data layout *X*.
2. The number of observations in each column of X, N (which is an array).
3. The total number of observations, NT.
4. The number of columns in X, NC.
5. The number of rows in X, NR. (Note that if the data layout is two bases with equal numbers of observations in a cell or one basis with equal numbers of observations in a column, NR = N.)
6. The number of observations in a cell, NCE.
7. The model type, MOD.
   a. If MOD = 0, there are unequal numbers of observations in a column.
   b. If MOD is positive, there are equal numbers of observations in a column.

Upon receipt of these items, GRAND calculates the grand mean, GM, and the total sum of squares, SST, for any of the data layouts discussed in Secs. 5.2 and 5.3. The statement

CALL GRAND (X, N, NT, NC, NR, NCE, GM, SST, MOD)

will cause entry into the subroutine. The subroutine is shown below.

```
      SUBROUTINE GRAND(X,N,NT,NC,NR,NCE,GM,SST,MOD)
      DIMENSION X(50,10),N(10)
      GM=0.0
      DO 10 J=1,NC
      IF(MOD)2,2,4
2     NR=N(J)
4     NF=NCE*NR
      DO 10 I=1,NF
10    GM=GM+X(I,J)
      GM=GM/FLOAT(NT)
      SST=0.0
      DO 20 J=1,NC
      IF(MOD)13,13,14
13    NR=N(J)
14    NF=NCE*NR
      DO 20 I=1,NF
20    SST=SST+(X(I,J)-GM)**2
      RETURN
      END
```

A5.2 subroutine COLS

A. Description. This subroutine calculates the column means, XBC, and the column sum of squares, SSC.

B. Limitations. Same as for GRAND.

C. Use. X, N, NC, NR, NCE, and MOD are assumed to be in storage as they were for GRAND. In addition the grand mean, GM, is assumed to have been calculated and placed in storage by GRAND. Then COLS calculates the column means, XBC, and the column sum of squares, SSC. The statement

CALL COLS (X, XBC, N, NC, NR, NCE, GM, SSC, MOD)

will cause entry into the subroutine. The subroutine is shown below.

```
      SUBROUTINE COLS(X,XBC,N,NC,NR,NCE,GM,SSC,MOD)
      DIMENSION X(50,10),XBC(10),N(10)
      DO 10 J=1,NC
      XBC(J)=0.0
      IF(MOD)2,2,4
2     NR=N(J)
4     NF=NCE*NR
      DO 5 I=1,NF
5     XBC(J)=XBC(J)+X(I,J)
10    XBC(J)=XBC(J)/FLOAT(NF)
      SSC=0.0
      DO 15 J=1,NC
      IF(MOD)13,13,14
13    NR=N(J)
14    NF=NCE*NR
15    SSC=SSC+FLOAT(NF)*(XBC(J)-GM)**2
      RETURN
      END
```

A5.3 subroutine FCAL

A. Description. This subroutine receives two sums of squares and two numbers of degrees of freedom. It returns two mean squares and an *F* ratio.

B. Limitations. The denominator sum of squares or numbers of degrees of freedom must be positive.

C. Use. The numerator sum of squares, SSC, and the denominator sum of squares, SSE, are assumed to be in storage. Likewise, the numerator number of degrees of freedom, NTOP, and the denominator degrees of freedom, NBOT, are assumed to be in storage. The subroutine then returns the mean squares SMC = SSC/NTOP and SME = SSE/NBOT. It also returns the F ratio, F = SMC/SME. The statement

CALL FCAL (SSC, SSE, NTOP, NBOT, SMC, SME, F)

will cause entry into the subroutine. The subroutine is shown below.

```
SUBROUTINE FCAL(SSC,SSE,NTOP,NBOT,SMC,SME,F)
SMC=SSC/FLOAT(NTOP)
SME=SSE/FLOAT(NBOT)
F=SMC/SME
RETURN
END
```

A5.4 main for analysis of variance

A. Description. This program calculates an analysis of variance table for all cases discussed in this chapter. It also computes the row means, column means, cell means, and the grand mean as indicated by the type of problem.

B. Limitations. Same as for GRAND and FCAL.

C. Use. Four cases were discussed in Chap. 5. Each case is indicated by a numerical value for MOD

Value for MOD	*Case*
0	One basis of classification, unequal numbers of items in a column
1	One basis of classification, equal numbers of items in a column
2	Two bases of classification, one item in a cell
3	Two bases of classification, several (but equal) numbers of observations in a cell

The first data card contains the value of MOD (i.e., 0, 1, 2, or 3) punched in column 3.

If MOD = 0, the next data card contains the number of columns in the data layout punched in columns 1–3, right-adjusted with no decimal point. The third card contains the number of observations in each column punched in FORMAT 10I3. That is, in columns 1–3 we punch right-adjusted without the decimal point, the number of observations in the first column of the data layout. In a similar way, punch the number of observations in the second column of the data layout in columns 4–6 and so forth.

If MOD = 1 or MOD = 2, the second card contains the number of columns in the data layout punched in columns 1–3 and the number of rows in the data layout punched in columns 4–6. Both of these figures are right-adjusted and punched without a decimal point.

If MOD = 3, punch on the second card the number of columns in the data layout in columns 1–3 and the number of rows in columns 4–6 in the same way as described in the previous paragraph. In addition, punch the number of observations in each cell in columns 7–9 (right-adjusted with no decimal point).

Having punched the first two (in the case of MOD $\neq$ 0) or three (in the case of M = 0) data cards, we are now ready to punch the data layout. The data layout is punched the same way in all four cases. Column 1 of the data layout is punched across the card in eight fields of width 10, continuing for as many cards as are necessary until all the elements are punched. Column 2 is begun on a fresh card, and so forth. These figures are all punched with decimal points. This is the same format followed in previous chapters.

Shown below is the output corresponding to the data of Tables 5.2, 5.5, and 5.7, respectively. The program follows these output examples.

SOURCE OF VARIATION	AMOUNT OF VARIATION	DEGREES OF FREEDOM	MEAN SQUARE	F
TOTAL	.1169396E+04	31		
COLUMN	.4521697E+03	2	.2260848E+03	9.14
ERROR	.7172266E+03	29	.2473195E+02	

```
GRAND MEAN
     12.744
COLUMN MEANS
      8.367( 1)          17.227( 2)          13.100( 3)
```

Data of Table 5.2

SOURCE OF VARIATION	AMOUNT OF VARIATION	DEGREES OF FREEDOM	MEAN SQUARE	F
TOTAL	.6713601E+00	14		
COLUMN	.1720014E+00	2	.8600068E-01	2.14
ROW	.1778938E+00	4	.4447344E-01	1.11
ERROR	.3214650E+00	8	.4018312E-01	

```
GRAND MEAN
     4.746
COLUMN MEANS
     4.766( 1)          4.606( 2)          4.866( 3)
ROW MEANS
     4.547( 1)          4.717( 2)          4.840( 3)          4.833( 4)          4.793( 5
```

Data of Table 5.5

```
-----------------------------------------------------------------------------
SOURCE OF         AMOUNT OF       DEGREES OF          MEAN              F
VARIATION         VARIATION        FREEDOM           SQUARE
-----------------------------------------------------------------------------
TOTAL             .4689441E+03       17
COLUMN            .5877806E+02        2         .2938902E+02          .83
ROW               .6777783E+01        2         .3388891E+01          .10
INTERACTION       .8288882E+02        4         .2072220E+02          .58
ERROR             .3204993E+03        9         .3561102E+02
-----------------------------------------------------------------------------
GRAND MEAN
    100.944
COLUMN MEANS
     99.667( 1)         99.667( 2)        103.500( 3)
ROW MEANS
    100.167( 1)        101.667( 2)        101.000( 3)
CELL MEANS
     96.500( 1 1)        102.000( 1 2)        102.000( 1 3)
    100.500( 2 1)         97.500( 2 2)        107.000( 2 3)
    102.000( 3 1)         99.500( 3 2)        101.500( 3 3)
```

Data of Table 5.7

```
C      GRAND, COLS, AND FCAL NEEDED
       DIMENSION X(50,10),N(10),XBC(10),XBR(10),XBCE(10,10)
       READ(5,5)MOD
 5     FORMAT(10I3)
       IF(MOD)10,10,15
 10    READ(5,5)NC
       READ(5,5)(N(J),J=1,NC)
       NT=0.0
       DO 12 J=1,NC
 12    NT=NT+N(J)
       NCE=1
       GO TO 40
 15    CONTINUE
       IF(MOD-1)20,20,25
 20    READ(5,5)NC,NR
       NCE=1
       NT=NC*NR
       GO TO 40
 25    READ(5,5)NC,NR,NCE
       IF(NCE) 30,30,35
 30    NCE=1
 35    NT=NC*NR*NCE
 40    DO 60 J=1,NC
       IF(MOD)45,45,50
 45    NR=N(J)
 50    NF=NCE*NR
 60    READ(5,65)(X(I,J),I=1,NF)
 65    FORMAT(8F10.0)
       CALL GRAND(X,N,NT,NC,NR,NCE,GM,SST,MOD)
       CALL COLS(X,XBC,N,NC,NR,NCE,GM,SSC,MOD)
       IF(MOD)80,80,70
 70    F1=NCE*NC
       DO 75 I=1,NR
       XBR(I)=0.0
       NST=NCE*(I-1)+1
       NF=NST+NCE-1
       DO 72 K=NST,NF
       DO 72 J=1,NC
 72    XBR(I)=XBR(I)+X(K,J)
 75    XBR(I)=XBR(I)/F1
       SSR=0.0
       DO 78 I=1,NR
```

```
78    SSR=SSR+F1*(XBR(I)-GM)**2
80    CONTINUE
      IF(MOD-2)90,90,81
81    DO 85 I=1,NR
      NST=NCE*(I-1)+1
      NF=NST+NCE-1
      DO 85 J=1,NC
      XBCE(I,J)=0.0
      DO 83 K=NST,NF
83    XBCE(I,J)=XBCE(I,J)+X(K,J)
85    XBCE(I,J)=XBCE(I,J)/FLOAT(NCE)
      SSI=0.0
      DO 87 I=1,NR
      DO 87 J=1,NC
87    SSI=SSI+FLOAT(NCE)*(XBCE(I,J)-XBC(J)-XBR(I)+GM)**2
90    WRITE(6,95)
95    FORMAT(' ------------------------------------------------------------
     1-----------')
      WRITE(6,100)
100   FORMAT(' SOURCE OF',6X,'AMOUNT OF',4X,'DEGREES OF',6X,'MEAN',15X,
     1'F',/,' VARIATION',6X,'VARIATION',6X,'FREEDOM',6X,'SQUARE')
      WRITE(6,95)
      NC1=NT-1
      WRITE(6,110)SST,NC1
110   FORMAT(' TOTAL',7X,E15.7,5X,I3)
      NTOP=NC-1
      IF(MOD-1)115,115,130
115   SSE=SST-SSC
      NBOT=NT-NC
      CALL FCAL(SSC,SSE,NTOP,NBOT,SMC,SME,F)
      WRITE(6,117)SSC,NTOP,SMC,F
117   FORMAT(' COLUMN ',5X,E15.7,5X,I3,4X,E15.7,F12.2)
      WRITE(6,118)SSE,NBOT,SME
118   FORMAT(' ERROR',7X,E15.7,5X,I3,4X,E15.7)
      GO TO 500
130   NTOP=NC-1
      IF(MOD-2)135,135,150
135   NBOT=(NR-1)*(NC-1)
      SSE=SST-SSC-SSR
      CALL FCAL(SSC,SSE,NTOP,NBOT,SMC,SME,F)
      WRITE(6,117)SSC,NTOP,SMC,F
      NTOP=NR-1
      CALL FCAL(SSR,SSE,NTOP,NBOT,SMC,SME,F)
      WRITE(6,140)SSR,NTOP,SMC,F
140   FORMAT(' ROW ',8X,E15.7,5X,I3,4X,E15.7,F12.2)
      WRITE(6,118)SSE,NBOT,SME
      GO TO 500
150   NBOT=NT-NR*NC
      SSE=SST-SSC-SSR-SSI
      CALL FCAL(SSC,SSE,NTOP,NBOT,SMC,SME,F)
      WRITE(6,117)SSC,NTOP,SMC,F
      NTOP=NR-1
      CALL FCAL(SSR,SSE,NTOP,NBOT,SMC,SME,F)
      WRITE(6,140) SSR,NTUP,SMC,F
      NTOP=(NR-1)*(NC-1)
      CALL FCAL(SSI,SSE,NTOP,NBOT,SMC,SME,F)
      WRITE(6,155)SSI,NTOP,SMC,F
155   FORMAT(' INTERACTION ' ,E15.7,5X,I3,4X,E15.7,F12.2)
      WRITE(6,118)SSE,NBOT,SME
500   WRITE(6,95)
      WRITE(6,505)GM
505   FORMAT(' GRAND MEAN',/,F12.3,/,' COLUMN MEANS')
      WRITE(6,510)(XBC(J),J,J=1,NC)
510   FORMAT(5(F12.3,'(',I2,')'))
      IF(MOD-1)550,550,520
```

```
520   WRITE(6,525)
525   FORMAT(' ROW MEANS')
      WRITE(6,510)(XBR(I),I,I=1,NR)
      IF(MOD-2)550,550,530
530   WRITE(6,535)
535   FORMAT(' CELL MEANS')
      DO 540 I=1,NR
540   WRITE(6,545)(XBCE(I,J),I,J,J=1,NC)
545   FORMAT(5(F12.3,'(',2I2,')'))
550   CONTINUE
      CALL EXIT
      END
```

A5.5 main for analysis of covariance

A. Description. This program calculates an analysis of covariance with one concomitant variable. It also computes the column and grand means for X and Y, the column slopes, the average slope, and the grand slope. The information of Table 5.9 is reproduced with the exception of the column labeled "explained by concomitant variable," which may be easily calculated from the rest of the output if desired. Three F ratios are produced: The first is used for testing the equality of the adjusted column means for Y, the second is used for testing the equality of the column slopes, and the third is for testing the equality of the unadjusted column means for Y. Thus the F ratios correspond to Eqs. (5-80), (4-83), and (5-29), respectively.

B. Limitations. There may be at most 10 column classifications for X and Y with at most 50 pairs of observations in each column. These restrictions may be changed by altering the DIMENSION statements in the programs.

C. Use. On the first card, punch MOD in column 3. MOD may be zero, in which case there are unequal numbers of observations in the classifications, or 1, in which case there are equal numbers of observations in the classifications.

If MOD = 0, the next data card contains the number of classifications (columns) punched in columns 1–3, right-adjusted with no decimal point. The third card contains the number of observations in each classification punched in columns 1–3, 4–6, 7–9, and so forth, for as many column classifications as are listed on data card 2. These numbers are right-adjusted and punched without a decimal point.

If MOD = 1, the second data card contains the number of classifications (columns) punched in columns 1–3 and the number of pairs of observations in each classification punched in columns 4–6, right-adjusted with no decimal point.

The data layout is then punched. The Y observations in the first classification are punched across the card in eight fields of width 10, continuing for as many cards as necessary. On a fresh card, punch the X observations in the first classification in the same way. Then, on a fresh card, punch the Y observations in the second classification, followed by the X observations in the second classification. Continue in this way, Y observations and then X observations for each classification, until all observations are punched.

The output for the problem of this chapter is shown below, followed by the program.

```
GRAND MEAN FOR X
  1326.727
COLUMN MEANS FOR X
    777.833( 1)     1350.125( 2)     1715.000( 3)
GRAND MEAN FOR Y
  17146.227
COLUMN MEANS FOR Y
   8549.664( 1)   15715.875( 2)    25024.000( 3)
AVERAGE SLOPE
     13.247
GRAND SLOPE
     15.998
COLUMN SLOPES
     12.451( 1)        10.914( 2)       14.945( 3)
----------------------------------------------------------------------------------------
SOURCE OF             SUMS OF SQUARES AND PRODUCTS              ADJUSTED     DEGREES OF
VARIATION                                                       VARIATION     FREEDOM
----------------------------------------------------------------------------------------
TOTAL              TYX             TXX             TYY
             .7289122E+08    .4556406E+07    .1410431E+10     .2443523E+09    20
ERROR              EYX             EXX             EYY
             .2037771E+08    .1538281E+07    .4541865E+09     .1842417E+09    18
COLUMN
                                                              .6011059E+08     2
----------------------------------------------------------------------------------------
F=       2.936 WITH  2 AND 18 DEGREES OF FREEDOM(ADJUSTED COLUMN VARIATION)
F=        .266 WITH  2 AND 16 DEGREES OF FREEDOM(EQUALITY OF COLUMN SLOPES)
F=      20.001 WITH  2 AND 19 DEGREES OF FREEDOM(ORDINARY AN. OF VAR. ON Y)
```

```
C      GRAND,COLS, AND FCAL NEEDED
       DIMENSION X(50,10),Y(50,10),N(10),XBCX(10),XBCY(10),B(10)
       READ(5,5)MOD
 5     FORMAT(10I3)
       IF(MOD)10,10,15
 10    READ(5,5)NC
       READ(5,5)(N(J),J=1,NC)
       NT=0.0
       DO 12 J=1,NC
 12    NT=NT+N(J)
       GO TO 20
 15    READ(5,5)NC,NR
       NT=NC*NR
 20    DO 25 J=1,NC
       IF(MOD)22,22,23
 22    NF=N(J)
       GO TO 24
 23    NF=NR
 24    READ(5,28)(Y(I,J),I=1,NF)
 25    READ(5,28)(X(I,J),I=1,NF)
 28    FORMAT(8F10.0)
       CALL GRAND(X,N,NT,NC,NR,1,GMX,TXX,MOD)
       CALL GRAND(Y,N,NT,NC,NR,1,GMY,TYY,MOD)
       CALL COLS(X,XBCX,N,NC,NR,1,GMX,SSC,MOD)
       CALL COLS(Y,XBCY,N,NC,NR,1,GMY,SSC,MOD)
```

```
      TS=0.0
      TYX=0.0
      EYX=0.0
      EXX=0.0
      EYY=0.0
      DO 40 J=1,NC
      TP=0.0
      BT=0.0
      IF(MOD) 32,32,33
32    NF=N(J)
      GO TO 34
33    NF=NR
34    DO 35 I=1,NF
      TYX=TYX+(X(I,J)-GMX)*(Y(I,J)-GMY)
      TP=TP+(X(I,J)-XBCX(J))*(Y(I,J)-XBCY(J))
      BT=BT+(X(I,J)-XBCX(J))**2
35    EYY=EYY+(Y(I,J)-XBCY(J))**2
      B(J)=TP/BT
      TS=TS+B(J)*TP
      EXX=EXX+BT
40    EYX=EYX+TP
      SLG=TYX/TXX
      SLA=EYX/EXX
      WRITE(6,50) GMX,(XBCX(J),J,J=1,NC)
50    FORMAT(' GRAND MEAN FOR X',/,F12.3,/,' COLUMN MEANS FOR X',/,
     15(F12.3,'(',I2,')'))
      WRITE(6,55)GMY,(XBCY(J),J,J=1,NC)
55    FORMAT(' GRAND MEAN FOR Y',/,F12.3,/,' COLUMN MEANS FOR Y',/,
     15(F12.3,'(',I2,')'))
      WRITE(6,60) SLA,SLG,(B(J),J,J=1,NC)
60    FORMAT(' AVERAGE SLOPE',/,F12.3,/,' GRAND SLOPE',/,F12.3,/,
     1' COLUMN SLOPES',/,5(F12.3,'(',I2')'))
      WRITE(6,70)
70    FORMAT(' -----------------------------------------------------------
     1---------------------------')
      WRITE(6,80)
80    FORMAT(' SOURCE OF',11X,'SUMS OF SQUARES AND PRODUCTS',11X ,
     1'ADJUSTED',4X,'DEGREES OF',/,' VARIATION',50X,'VARIATION',4X,
     2'FREEDOM')
      WRITE(6,70)
      NT=NT-2
      NTOP=NC-1
      NBOT=NT-NC+1
      SO=TYY-SLG*TYX
      S1=EYY-SLA*EYX
      SC=SO-S1
      WRITE(6,90) TYX,TXX,TYY,SO,NT,EYX,EXX,EYY,S1,NBOT,SC,NTOP
90    FORMAT(' TOTAL',12X,'TYX',12X,'TXX',12X,'TYY',/,12X,4E15.7,I5,/,
     1' ERROR',12X,'EYX',12X,'EXX',12X,'EYY',/,12X,4E15.7,I5,/,
     2' COLUMN',/,57X,E15.7,I5)
      WRITE(6,70)
      CALL FCAL(SC,S1,NTOP,NBOT,SMC,SME,F)
      WRITE(6,100)F,NTOP,NBOT
100   FORMAT(' F=',F12.3,' WITH ',I2,' AND ',I2,' DEGREES OF FREEDOM(ADJ
     1USTED COLUMN VARIATION)')
      SC=S1-EYY+TS
      S1=EYY-TS
      NBOT=NT+2-2*NC
      CALL FCAL(SC,S1,NTOP,NBOT,SMC,SME,F)
      WRITE(6,110) F,NTOP,NBOT
110   FORMAT(' F=',F12.3,' WITH ',I2,' AND ',I2,' DEGREES OF FREEDOM(EQU
     1ALITY OF COLUMN SLOPES)')
      S1=TYY-SSC
      NBOT=NT+2-NC
      CALL FCAL(SSC,S1,NTUP,NBOT,SMC,SME,F)
      WRITE(6,115)F,NTOP,NBOT
115   FORMAT(' F=',F12.3,' WITH ',I2,' AND ',I2,' DEGREES OF FREEDOM(ORD
     1INARY AN. OF VAR. ON Y)')
      CALL EXIT
      END
```

chapter 6

Nonlinear Regression

In Chap. 4 we restricted ourselves to linear regression. That is, the dependent variable was expressed as a linear combination of the independent variables and an error term. The exact specification was given in that chapter by Eq. (4-10). There are, of course, many ways in which Y may relate to the Xs and βs, and the particular relationship will depend on the exact specification of the regression function, Eq. (4-1). As shown in Figure 4.1, the dashed curve which connects the conditional means, $\bar{Y}_i$, is one reasonable estimate of the population regression function. Naturally, if this curve is more or less a straight line, linear regression should be acceptable; if not, nonlinear regression should be utilized.

6.1 the correlation ratio and a test of linearity

Table 6.1 shows some figures for the hypothetical average cost, Y, and output, X, for an industry. The arrangement of the table is similar to that of Table 4.1. The conditional means, $\bar{Y}_i$, are plotted in Figure 6.1 and are connected by a dashed curve.[1] Also plotted in Figure 6.1 are the linear

[1]We continue to assume that the conditional distributions of the dependent variable are normal and independent.

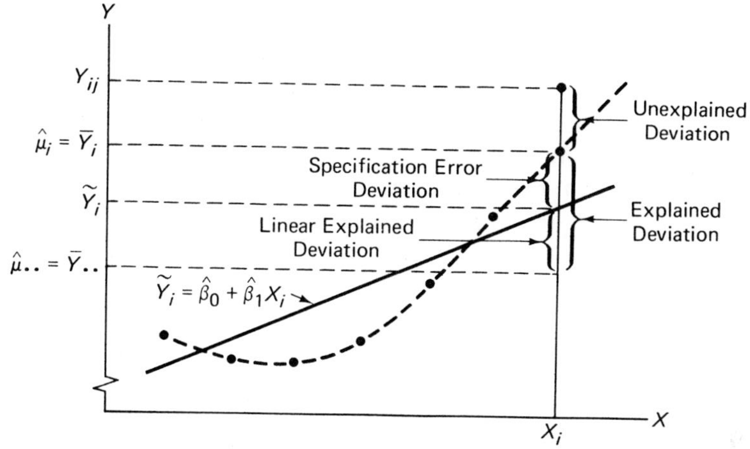

Figure 6.1. Conditional means, linear and nonlinear regression functions.

Source: Table 6.1

Table 6.1 Cost, Y, and Output, X, of an Industry

Output, X_i *(units)*	*Cost,* Y_{ij} (\$)	*Conditional Mean,* $\bar{Y}_i$ (\$)
11	37, 33, 34, 35, 34	34.6
12	33, 34, 32, 33, 31	32.6
13	33, 31, 32, 31, 30	31.4
14	31, 33, 32, 33, 31	32.0
15	35, 34, 36, 33, 34	34.4
16	39, 41, 39, 40, 40	39.8
17	46, 47, 45, 45, 46	45.8
18	57, 55, 56, 57, 54	55.8
19	63, 67, 67, 64, 68	65.8
20	80, 79, 81, 78, 77	79.0

regression estimates, $\tilde{Y}_i$. In this section we use only the symbol ~ to denote that these values are estimates of the conditional population means, μ_i, *if* the population regression relationship is linear.

Since the sample conditional mean $\bar{Y}_i$ and the grand mean $\bar{Y}..$ are unbiased estimates of the conditional population mean μ_i and the grand population mean $\mu..$, respectively,

$$(Y_{ij} - \bar{Y}..) = (\bar{Y}_i - \bar{Y}..) + (Y_{ij} - \bar{Y}_i) \qquad (6\text{-}1)$$

Stated verbally,

Total deviation = explained deviation + unexplained deviation

Recall that in Chap. 5 we called $\bar{Y}_i - \bar{Y}..$ column deviation because it was the deviation attributable to the column classifications. However, if we assume that the regression relation passes through each of the conditional means, this deviation can be attributable to the regression relationship. Hence we use the term *explained deviation* in this section. Explained deviation for one Y_{ij} point is indicated in Figure 6.1.

Upon taking the sum of squares of the elements of Eq. (6-1), we obtain

$$\sum_{i=1}^{K}\sum_{j=1}^{n_i}(Y_{ij} - \bar{Y}..)^2 = \sum_{i=1}^{K} n_i(\bar{Y}_i - \bar{Y}..)^2 + \sum_{i=1}^{K}\sum_{j=1}^{n_i}(Y_{ij} - \bar{Y}_i)^2 \qquad (6\text{-}2)$$

Here K is the number of observations (or classifications) for the X variable and the n_is are the number of Y observations for each X_i classification. For our example, $K = 10$ and all $n_i = 5$. Stated verbally,

Total variation = explained variation + unexplained variation

From Figure 6.1 it should be clear that if Y and X have no regression relationship, then for the sample we shall expect that the dashed curve will be horizontal and correspond to the line representing the grand sample mean. That is, $\bar{Y}_i = \bar{Y}..$, and explained variation will be zero. Thus the *determination ratio* (eta squared)

$$\hat{\eta}^2 = \frac{\text{explained variation}}{\text{total variation}} \qquad (6\text{-}3)$$

will indicate the degree of regressional relation between Y and X in the sample. The square root of Eq. (6-3) is called the *correlation ratio*:[2]

$$\hat{\eta} = \sqrt{\frac{\text{explained variation}}{\text{total variation}}} \qquad (6\text{-}3a)$$

Since total variation is equal to the sum of explained and unexplained variation, it is obvious that $0 \leq \hat{\eta} \leq 1$. If Y and X are strictly functionally related in the sample, then $\hat{\eta} = 1$. For our example, the elements of Eq. (6-2) are 12,375.28 = 12,305.28 + 70.00. Therefore, $\hat{\eta}^2 = 12{,}305.28/12{,}375.28 = 0.9943$ and $\hat{\eta} = 0.9971$.

Test of the Correlation Ratio. The test of the correlation ratio is similar to the test of the correlation coefficient as given by Eq. (5-3). The family of

[2]Unlike the correlation coefficient, the correlation ratio is always nonnegative. It cannot indicate whether the regression relationship is positive or negative. In Figure 6.1, over one region of the dashed curve, Y is negatively related to X; over another region of the curve, Y is positively related to X. We cannot say whether X and Y are positively or negatively related.

hypotheses is

$$H_0: \ \eta = 0, \qquad \text{i.e., there is no functional relationship}$$
$$H_1: \ \eta \neq 0, \qquad \text{i.e., there is a functional relationship}$$

where η is the population correlation ratio. Under the null hypothesis, the F ratio is

$$F = \frac{(\text{explained variation})/(K-1)}{(\text{unexplained variation})/(n-K)} = \frac{\hat{\eta}^2/(K-1)}{(1-\hat{\eta}^2)/(n-K)} \tag{6-4}$$

This ratio follows the F distribution with $K-1$ and $n-K$ degrees of freedom where $n = \sum_{i=1}^{K} n_i$. For our example

$$F = \frac{12{,}305.28/(10-1)}{70.00/(50-10)} = 781.3$$

This value exceeds the critical value $F_{0.05;\,9,40} = 2.12$. Therefore at the 0.05 level we reject the null hypothesis and conclude that a functional relationship exists.

Test for Linearity. We have explained the correlation ratio by use of the decomposition of total deviation in Eq. (6-1). In that equation the explained deviation is measured by the distance between the conditional mean and the grand mean (i.e., $\bar{Y}_i - \bar{Y}..$). Suppose, however, that we (incorrectly) specify that the regression relationship is linear. In this case, the difference between the conditional mean, $\bar{Y}_i$, and the linear estimated value may be called *specification error* deviation. This specification error deviation is shown in Figure 6.1 and results from our incorrect specification that the functional form is linear. We may now decompose total deviation as follows:

$$(Y_{ij} - \bar{Y}..) = (\tilde{Y}_i - \bar{Y}..) + (\bar{Y}_i - \tilde{Y}_i) + (Y_{ij} - \bar{Y}_i) \tag{6-5}$$

Stated verbally,

Total deviation = linear explained deviation + specification error deviation
+ unexplained deviation

From Figure 6.1 it is obvious that the specification error deviation will be zero if the true functional relationship is linear. Another way to look at specification error deviation is to notice from Figure 6.1 that specification error deviation will be zero if explained deviation is equal to linear explained deviation. In this case the linear relationship explains as much of the deviation of the Y_{ij} observation about the grand mean as does the nonlinear relationship; the proper specification is linear and specification error is zero.

Upon taking the sum of squares of the elements of Eq. (6-5), we have

$$\sum_{i=1}^{K}\sum_{j=1}^{n_i}(Y_{ij}-\bar{Y}..)^2=\sum_{i=1}^{K}n_i(\tilde{Y}_i-\bar{Y}..)^2+\sum_{i=1}^{K}n_i(\bar{Y}_i-\tilde{Y}_i)^2+\sum_{i=1}^{K}\sum_{j=1}^{n_i}(Y_{ij}-\bar{Y}_i)^2 \tag{6-6}$$

Stated verbally,

Total variation = linear explained variation + specification error variation + unexplained variation

The null hypothesis that we wish to test is that the relationship is linear. The alternative hypothesis is that the relationship is nonlinear. Under the null hypothesis the F ratio is given by

$$F=\frac{(\text{specification error variation})/(K-2)}{(\text{unexplained variation})/(n-K)} \tag{6-7}$$

This ratio follows the F distribution with $K-2$ and $n-K$ degrees of freedom.

For our present example, the linear estimates of the conditional means are given by the linear regression equation

$$\begin{aligned}\tilde{Y}_i&=\hat{\beta}_0+\hat{\beta}_1X_i\\&=-30.11+4.85X_i\end{aligned}$$

The decomposition of total variation according to Eq. (6-6) is given by $12{,}375.28 = 9716.37 + 2588.91 + 70.00$. The F ratio is then

$$F=\frac{2588.91/(10-2)}{70.00/(50-10)}=184.9$$

which exceeds the critical value $F_{0.05;\,8,40}=2.18$. Thus we reject the null hypothesis at the stated level of significance and conclude that the population regression relation is nonlinear.

Using the notation of the last chapter, we find that $\sum_{i=1}^{K} n_i(\tilde{Y}_i-\bar{Y}..)^2$ is what we called *explained variation* in Sec. 5.1 and that $\sum_{i=1}^{K}\sum_{j=1}^{n_i}(Y_{ij}-\bar{Y}..)^2$ is what we called *total variation*. Therefore the coefficient of determination is equal to the ratio of linear explained variation to total variation. That is,

$$r^2=\frac{\sum_{i=1}^{K}n_i(\tilde{Y}_i-\bar{Y}..)^2}{\sum_{i=1}^{K}\sum_{j=1}^{n_i}(Y_{ij}-\bar{Y}..)^2} \tag{6-8}$$

If we divide both sides of Eq. (6-6) by total variation and rearrange terms, we obtain

$$\hat{\eta}^2 = r^2 + \frac{\sum_{i=1}^{K} n_i(\bar{Y}_i - \tilde{Y}_i)^2}{\sum_{i=1}^{K}\sum_{j=1}^{n_i} (Y_{ij} - \bar{Y}..)^2} \tag{6-9}$$

Then the F ratio of Eq. (6-7) can be expressed as

$$F = \frac{(\hat{\eta}^2 - r^2)/(K-2)}{(1-\hat{\eta}^2)/(n-K)} \tag{6-10}$$

with $K - 2$ and $n - K$ degrees of freedom.

Let us denote the rightmost term of Eq. (6-9) by B^2. That is, let

$$B^2 = \frac{\sum_{i=1}^{K} n_i(\bar{Y}_i - \tilde{Y}_i)^2}{\sum_{i=1}^{K}\sum_{j=1}^{n_i} (Y_{ij} - \bar{Y}..)^2} \tag{6-11}$$

This term is the ratio of specification error variation to total variation and it is always nonnegative since it is a ratio of sums of squares. Notice that

$$B^2 = \hat{\eta}^2 - r^2 \geq 0 \tag{6-12}$$

so that B^2 indicates the degree of error in a linear specification of a regression relationship that is nonlinear. It is clear from Eq. (6-12) that the determination ratio is always greater than or equal to the coefficient of determination (i.e., $0 \leq r^2 \leq \hat{\eta}^2 \leq 1$). Similarly, in the population $0 \leq \rho^2 \leq \eta^2 \leq 1$. Thus the coefficient of determination is equal to the determination ratio only if the relationship between Y and X is exactly linear. Put another way, the determination ratio and the coefficient of determination will be equal only if linear explained variation is equal to explained variation.

The tests of this section do not work well unless there are several observations on Y for each value of X, or unless the sample is large enough so that the X values may be grouped in such a way as to yield several Y values for each group. When these conditions do not exist, another method, which is explained in Sec. 6.3, may be used.

6.2 estimation using simple transformations

In the previous section we concluded that the relationship between cost and output is a nonlinear one, so that we are now faced with the task of fitting such a nonlinear regression. Before we do this, let us discuss for a moment two different classes of nonlinear regressions. The first class is a class that can be handled by direct application of ordinary least-squares to the data after the data have undergone some simple transformation. Some writers call this class of regressions *intrinsically linear*. The second class of regressions, sometimes called *intrinsically nonlinear*, cannot be handled by ordinary least-squares without some approximation or iterative technique. We shall discuss two methods for dealing with this type of regression in the next sections.

Polynomial Regression. Recall that a polynomial is an equation of the form

$$Y = a + bX + cX^2 + dX^3 + \cdots$$

The *degree* of the polynomial is the highest exponent to which the independent variable is raised. Thus a straight line is a first-degree polynomial. Furthermore, just as a first-degree polynomial has no "bends," a second-degree polynomial has one bend, like the dashed curve in Figure 6.1. A third-degree polynomial has two bends, and, in the authors' experience, is the highest-degree polynomial that is needed in routine statistical analysis.

Without loss of generality, let us assume that there is one observation on Y for each observation on X (i.e., assume that $j = 1$). In this case we can write the statistical specification of a Kth-degree polynomial as

$$Y_i = \beta_0 + \beta_1 X_i + \beta_2 X_i^2 + \cdots + \beta_K X_i^K + \epsilon_i, \qquad i = 1, 2, \ldots, n \tag{6-13}$$

Estimation of the coefficients of this equation is easily done by use of the regression program in the Appendix to Chap. 4. The first independent variable consists of the values X_i, the second independent variable consists of the squared values of X_i, and so forth for as high a degree of polynomial as is desired. More formally, let the observed variable X be transformed as follows:

$$Z_{mi} = X_i^m, \qquad m = 1, 2, \ldots, K \tag{6-14}$$

Then, Eq. (6-13) becomes

$$Y_i = \beta_0 + \beta_1 Z_{1i} + \beta_2 Z_{2i} + \cdots + \beta_K Z_{Ki} + \epsilon_i \tag{6-13a}$$

which is exactly the form given in Chap. 4. Thus the material of Chap. 4 concerning estimation and hypothesis testing applies to this equation if the error term satisfies the requirements laid down in Chap. 4.

We offer an example of polynomial regression by fitting the data of Table 6.1. Let $Z_{1i} = X_i$ and $Z_{2i} = X_i^2$. If we fit a first-degree polynomial (a straight line) through the data, we get $\hat{Y} = -30.11 + 4.85Z_1$. If we fit a second-degree polynomial, we get $\hat{Y} = 199.23 - 25.79Z_1 + 0.99Z_2$. The coefficient of determination for the linear fit is 0.7851, and the multiple coefficient of determination for the second-degree fit is 0.9993. The second-degree polynomial appears to give a nearly perfect fit to the observations.

Loglinear Regression. Another common type of equation encountered in economic and business research is one of the form

$$Y_i = AX_{1i}^{\beta_1}X_{2i}^{\beta_2} \cdots X_{Ki}^{\beta_K}\epsilon_i, \qquad i = 1, 2, \ldots, n \tag{6-15}$$

where ϵ_i is the error term. If we take the logarithms of all observations on all variables, i.e.,

$$\begin{aligned} Y_i^* &= \ln Y_i \\ X_{1i}^* &= \ln X_{1i} \\ &\vdots \\ X_{Ki}^* &= \ln X_{Ki} \end{aligned}$$

then Eq. (6-15) can be expressed in the linear form

$$Y_i^* = \beta_0 + \beta_1 X_{1i}^* + \cdots + \beta_K X_{Ki}^* + \epsilon_i^* \tag{6-15a}$$

where $\beta_0 = \ln A$ and $\epsilon_i^* = \ln \epsilon_i$. The fact that Eq. (6-15) can be written in linear form by taking the logarithms of all variables leads to the term *loglinear* regression.

By assuming that ϵ_i^*(not ϵ_i) behaves according to the conditions set down in Chap. 4, the techniques of that chapter apply. Again, the computer program of the Appendix to Chap. 4 may be used by entering the logarithms of the original data rather than the original data. One possible problem with loglinear regression is that all original observations must be positive, since the logarithm of a nonpositive number is not defined. In economics this is usually a minor restriction since, in practice, most series in economics (prices, quantities, index numbers, and so forth) are positive.

The least-squares estimates $\hat{\beta}_1, \ldots, \hat{\beta}_K$ calculated by use of Eq. (6-15a) will be unbiased estimates of the parameters $\beta_1, \ldots, \beta_K$ of Eq. (6-15). And, since $\beta_0 = \ln A$ then, $A = \exp(\beta_0)$ and we might think intuitively that we could estimate A by the simple expedient of computing

$$\hat{A} = \exp(\hat{\beta}_0) \tag{6-16}$$

However, it can be shown that $\hat{A}$ computed by this method is biased upward as an estimator of A. An improvement in the estimate of A can be obtained by adjusting for this bias by using[3]

$$\hat{\hat{A}} = \hat{A}\exp(-\tfrac{1}{2}s_{\beta_{00}}) \tag{6-16a}$$

where $s_{\beta_{00}}$ is the estimated variance of $\hat{\beta}_0$ obtained by estimating Eq. (6-15a).

By the same token, since $Y_i^* = \ln Y_i$, it would seem natural to consider

$$\begin{aligned}\hat{Y}_i &= \exp(\hat{Y}_i^*)\\ &= \hat{A}X_{1i}^{\beta_1}X_{2i}^{\beta_2}\cdots X_{Ki}^{\beta_K}\end{aligned} \tag{6-17}$$

as an estimate of the conditional population mean $E(Y_i)$. Again, it can be shown that such an estimator is not unbiased, and the bias can be adjusted for by use of[4]

$$\hat{\hat{Y}}_i = \hat{Y}_i\exp\{\tfrac{1}{2}[\hat{\sigma}^2 - s_{\hat{Y}_i^*}^2]\} \tag{6-17a}$$

where $s_{\hat{Y}_i^*}^2$ is the estimated variance of $\hat{Y}_i^*$ [see Eq. (4-57) and its estimate $s_{\hat{Y}_i^*}^2$, with X^* replacing X].

Other Transformations. There are a great many more transformations which are available. In this book we shall concentrate on polynomial and loglinear regression, while providing reference for the interested reader to a discussion of other types of transformations.[5]

6.3 estimation by power transformation

A recent method of estimation of nonlinear regression functions is the method of power transformation. This method is associated with the pioneering work of Box and Cox.[6]

Define the power transformation as follows,

$$\begin{aligned}Y_i^{(\lambda)} &= \frac{Y_i^\lambda - 1}{\lambda}, & \lambda \neq 0\\ &= \ln Y_i, & \lambda = 0\\ X_{ji}^{(\lambda)} &= \frac{X_{ji}^\lambda - 1}{\lambda}, & \lambda \neq 0\\ &= \ln X_{ji}, & \lambda = 0\end{aligned} \tag{6-18}$$

---

[3] $\hat{\hat{A}}$ is an asymptotically unbiased estimator of A. An exactly unbiased estimator is given by Goldberger (1968). The difference between the exact estimate and our Eq. (6-16a) is, in practice, generally small.

[4] Again, $\hat{\hat{Y}}_i$ is an asymptotically unbiased estimator of $E(Y_i)$. The exactly unbiased estimator is given by Goldberger (1968).

[5] Johnston (1963), Chap. 2, covers many cases. Deming (1938) is a classic reference.

[6] Box and Cox (1964). The method is also useful for analysis of variance problems. Zarembka (1968) gives an economic application.

recognizing that the superscript (λ) on the left-hand side of these equations denotes an operator, not an exponent. Then the power transformation of the linear regression model becomes

$$Y_i^{(\lambda)} = \beta_0 + \beta_1 X_{1i}^{(\lambda)} + \beta_2 X_{2i}^{(\lambda)} + \cdots + \beta_K X_{Ki}^{(\lambda)} + \epsilon_i \tag{6-19}$$

Equation (6-19) is a family of regression functions, each member of which depends on the value of λ. When $\lambda = 1$, we have the linear regression function of Eq. (4-10). As λ approaches zero, it can be shown that Eq. (6-19) approaches the loglinear form of Eq. (6-15a). Thus the power transformation is general enough to include both the linear and loglinear forms as well as other forms of equations.

Estimation of Parameters. Let the observation matrices of the power-transformed dependent and independent variables be given by

$$\mathbf{Y}^{(\lambda)} = \begin{bmatrix} Y_1^{(\lambda)} \\ \cdot \\ \cdot \\ \cdot \\ Y_n^{(\lambda)} \end{bmatrix} \quad \text{and} \quad \mathbf{X}^{(\lambda)} = \begin{bmatrix} 1 & X_{11}^{(\lambda)} & \cdots & X_{K1}^{(\lambda)} \\ \cdot & & & \cdot \\ \cdot & & & \cdot \\ \cdot & & & \cdot \\ 1 & X_{1n}^{(\lambda)} & \cdots & X_{Kn}^{(\lambda)} \end{bmatrix}$$

Then the sample regression equation corresponding to Eq. (6-19) can be written for some given λ as

$$\mathbf{Y}^{(\lambda)} = \mathbf{X}^{(\lambda)}\hat{\boldsymbol{\beta}} + \mathbf{e} \tag{6-20}$$

where $\hat{\boldsymbol{\beta}}$ and $\mathbf{e}$ have the usual connotation. It can be shown that under the assumption that the error term is normally and independently distributed, the logarithm of the likelihood function for Eq. (6-20) is (except for a constant)

$$\ln L(\lambda) = -\frac{1}{2} n \ln [\hat{\sigma}^2(\lambda)] + (\lambda - 1) \sum_{i=1}^{n} \ln Y_i \tag{6-21}$$

Here $\hat{\sigma}^2(\lambda) = \mathbf{e}'\mathbf{e}/n$ and n is the number of observations. Thus $\hat{\sigma}^2(\lambda)$ is the maximum likelihood estimate of σ^2 for a given λ. By allowing λ to vary over a suitable range, it is then possible to plot (or otherwise inspect) the values for $\ln L(\lambda)$ in order to determine the maximum of the logarithm of the likelihood function. Once λ is determined, its use in Eq. (6-20) will allow maximum likelihood estimation of Eq. (6-19).

Table 6.2 gives data used by Lawrence Klein in the estimation of a consumption function associated with a model which has come to be known as Klein's model I.[7] Let us specify the form of the consumption function as

---

[7]Klein (1950).

Table 6.2 Data for Consumption, Profits, and Wage Income (billions of 1934 dollars)

Observation Number	*Consumption,* Y_i	*Current Profits,* X_{1i}	*Lagged Profits,* X_{2i}	*Wages,* X_{3i}
1	41.9	12.4	12.7	28.2
2	45.0	16.9	12.4	32.2
3	49.2	18.4	16.9	37.0
4	50.6	19.4	18.4	37.0
5	52.6	20.1	19.4	38.6
6	55.1	19.6	20.1	40.7
7	56.2	19.8	19.6	41.5
8	57.3	21.1	19.8	42.9
9	57.8	21.7	21.1	45.3
10	55.0	15.6	21.7	42.1
11	50.9	11.4	15.6	39.3
12	45.6	7.0	11.4	34.3
13	46.5	11.2	7.0	34.1
14	48.7	12.3	11.2	36.6
15	51.3	14.0	12.3	39.3
16	57.7	17.6	14.0	44.2
17	58.7	17.3	17.6	47.7
18	57.5	15.3	17.3	45.9
19	61.6	19.0	15.3	49.4
20	65.0	21.1	19.0	53.0
21	69.7	23.5	21.1	61.8

Source: Klein (1950), p. 135.

follows:

$$Y_i^{(\lambda)} = \beta_0 + \beta_1 X_{1i}^{(\lambda)} + \beta_2 X_{2i}^{(\lambda)} + \beta_3 X_{3i}^{(\lambda)} + \epsilon_i$$

Using Klein's data, we chose values for λ ranging from -2.0 to $+1.0$. That is, λ was allowed to increment in steps of 0.01 from the lower bound of -2.0 to the upper bound of $+1.0$. At each step an ordinary least-squares regression was performed on the variables transformed according to Eq. (6-20) and $\ln L(\lambda)$ was calculated. The results of the calculation for $\ln L(\lambda)$ are shown in Figure 6.2 and in the Appendix to this chapter. Apparently the maximum of the logarithm of the likelihood function is in the neighborhood of $\lambda = -0.52$. Therefore the regression based upon the power transformation using $\lambda = -0.52$ is the maximum likelihood estimate of Eq. (6-20). The computer program in the Appendix to this chapter will carry out the necessary calculations in conjunction with the program in the Appendix to Chap. 4. It is clear that this technique of estimation is totally impractical without the use of a computer.

For purposes of comparison our computed linear form of Klein's con-

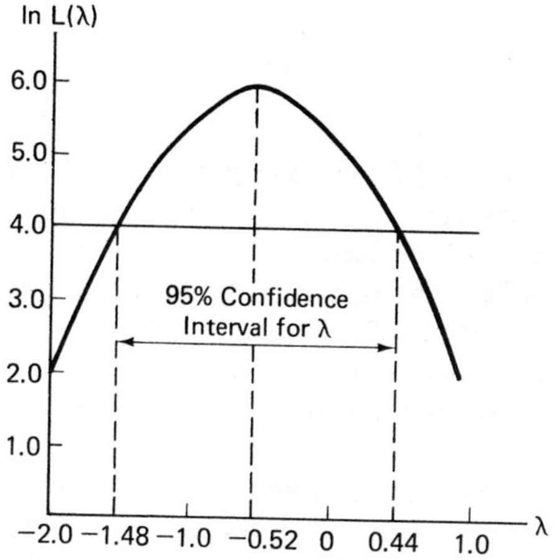

Figure 6.2. Logarithm of likelihood function of Eq. (6.21).
Source: Equations (6.20 and 6.21) and Table 6.2

sumption function is (standard errors in parentheses)

$$\hat{Y} = \underset{}{16.24} + \underset{(0.091)}{0.193X_1} + \underset{(0.091)}{0.090X_2} + \underset{(0.040)}{0.796X_3} \qquad r^2_{0(123)} = 0.981$$

The power-transformed estimate using $\lambda = -0.52$ is

$$\hat{Y}^{(-0.52)} = 0.715 + \underset{(0.00001)}{0.026X_1^{(-0.52)}} + \underset{(0.00001)}{0.008X_2^{(-0.52)}} + \underset{(0.00003)}{0.557X_3^{(-0.52)}}$$

$$r^2_{0(123)} = 0.987$$

It is interesting to note the substantial reduction in the standard errors when one uses the power-transformed variables.

Confidence Interval for λ and Test for Linearity. Call the value of λ that maximizes $\ln L(\lambda)$, $\hat{\lambda}$. In our example $\hat{\lambda} = -0.52$. Recalling Eq. (3-29), we find that the likelihood ratio

$$L^* = \frac{\max L_0}{\max L}$$

can be used to test hypotheses and set confidence limits for λ. For large samples, and making the full normality assumptions for the error term, we find that $-2 \ln L^*$ approaches the chi square distribution with 1 degree of freedom. Using this approximation, one can show that the $100(1 - \alpha)$ percent confidence interval for λ may be obtained from

$$-2 \ln L^* = -2\,[\ln L(\lambda) - \ln L(\hat{\lambda})] < \chi^2_{\alpha;\,1} \tag{6-22}$$

or

$$\ln L(\lambda) > \ln L(\hat{\lambda}) - \tfrac{1}{2}\chi^2_{\alpha;\,1}$$

To illustrate, in our example $\ln L(\hat{\lambda} = -0.52) \doteq 5.96$. This value may be read from Figure 6.2 or from the computer program in the Appendix. For a 95 percent confidence interval, $\frac{1}{2}\chi^2_{0.05;\,1} = 1.92$. Therefore $\ln L(\lambda) > 4.04$ by Eq. (6-22). In Figure 6.2 we have indicated 4.04 by a horizontal line. There are two places where the line cuts the $\ln L(\lambda)$ curve. Therefore we are approximately 95 percent confident that λ must lie in the region bounded by the two vertical lines dropped from the two points where the horizontal line cuts $\ln L(\lambda)$. This must be the confidence region for λ since within this region values for $\ln L(\lambda)$ exceed 4.04.

To test the hypothesis that the consumption function is linear, we recall that Eq. (6-19) will be linear when $\lambda = 1$. Since our confidence interval does not cover $\lambda = 1$, we reject the null hypothesis that the function is linear. However, the interval does cover zero, and so we conclude that the function is not significantly different from the loglinear type given by Eq. (6-17.)

6.4 estimation by taylor expansion

Suppose that Y is functionally related to X and three parameters α, β, and γ. Thus[8] let us write

$$Y = f(X;\, \alpha, \beta, \gamma) + \epsilon \tag{6-23}$$

where ϵ is the error term. One special example is a function relating the rate of change in money wage rates, Y, to the percentage of unemployment, X, for a historical period in England. The original form of the function as specified by Phillips[9] is

$$Y = \alpha + \beta X^{\gamma} + \epsilon \tag{6-24}$$

The general configuration of the curve estimated by Phillips is shown in Figure 6.3. This type of curve has come to be called a Phillips curve.

Notice that Eq. (6-24) is intrinsically nonlinear since there is no transformation which will convert it to an expression which is linear in the parameters α, β, and γ. Therefore we cannot resort to some simple transformation that will allow us to use ordinary least-squares estimation.

One method of estimating Eq. (6-24) is to linearize the equation by a Taylor series expansion and curtail the expansion at the first derivatives.

---

[8] A more general treatment of this subject is given by Draper and Smith (1966).

[9] See Phillips (1958).

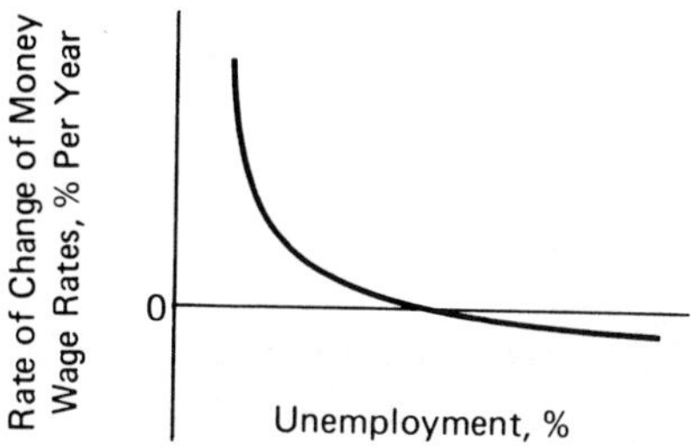

Figure 6.3. General shape of a Phililps Curve.
Source: Adapted from Phillips (1958).

Thus Eq. (6-23) can be approximated by

$$Y = f(X; \alpha_0, \beta_0, \gamma_0) + \frac{\partial f}{\partial \alpha}(\alpha - \alpha_0) + \frac{\partial f}{\partial \beta}(\beta - \beta_0) + \frac{\partial f}{\partial \gamma}(\gamma - \gamma_0) + \epsilon \qquad \text{(6-25)}$$

The approximation of Eq. (6-25) is exact if $\alpha_0 = \alpha$, $\beta_0 = \beta$, and $\gamma_0 = \gamma$, in which case the three terms containing derivatives vanish and Eq. (6-25) becomes Eq. (6-23). In Eq. (6-25) each of the partial derivatives is evaluated at the point where $\alpha = \alpha_0$, $\beta = \beta_0$, and $\gamma = \gamma_0$. The values of α_0, β_0, and γ_0 are educated guesses about the true values of the parameters α, β, and γ. We hope that our guesses are not far from correct and that they can be improved upon by the iterative procedure described below.

We begin by guessing that $\alpha_0 = 1$, $\beta_0 = 10$, and $\gamma_0 = -1$. Then by Eq. (6-24)

$$f(X; \alpha_0, \beta_0, \gamma_0) = 1 + 10X^{-1} = 1 + \frac{10}{X}$$

Furthermore, $\partial f/\partial \alpha = 1$; $\partial f/\partial \beta = X^\gamma$, but when $\gamma = -1$, $\partial f/\partial \beta = 1/X$; $\partial f/\partial \gamma = \beta X^\gamma \ln X$, but when $\beta = 10$ and $\gamma = -1$, $\partial f/\partial \gamma = 10 \ln X/X$. With these derivatives we can now write Eq. (6-25) as

$$Y = 1 + \frac{10}{X} + (\alpha - 1) + \frac{1}{X}(\beta - 10) + \frac{10 \ln X}{X}(\gamma + 1) + \epsilon$$

which reduces to

$$Y = \alpha + \beta\frac{1}{X} + (\gamma + 1)\frac{10 \ln X}{X} + \epsilon \qquad \text{(6-26)}$$

Equation (6-26) can now be estimated by ordinary least-squares. That is, let $\beta_1 = \beta$, $\beta_2 = (\gamma + 1)$, $Z_1 = 1/X$, and $Z_2 = (10 \ln X)/X$. Then

$$Y = \alpha + \beta_1 Z_1 + \beta_2 Z_2 + \epsilon \tag{6-26a}$$

Upon estimation of the parameters of Eq. (6-26a) we can obtain revised values for α_0, β_0, and γ_0. The entire process can then be repeated by using the revised estimates which replace the initial guesses. Hopefully, the process will converge in that each repetition will produce a smaller change in the parameters. Convergence, however, cannot be guaranteed, and the rather complicated computing procedure causes this estimation technique to be of limited usefulness.

6.5 additional remarks

No general rules can be given on the choice of a transformation to use in research involving regression analysis. The choice is made in practice by considerations of economic theory, by an inspection of the data, or both. Unfortunately, the choice is usually made by an inspection of the data since economic theory usually specifies only the existence of a functional relationship between variables—not the type of functional relationship.

Polynomial regression is widely used in time series analysis to approximate the "trend" in time series. In this case the X variable represents time, or the integrers $0, 1, 2, \ldots, n - 1$. The degree of the polynomial which gives a satisfactory representation of the trend is generally visually determined. A polynomial equation may be caused to fit a scatter of points as closely as is desired by the expedient of increasing the degree of the polynomial. Remember, however, that 1 degree of freedom is lost each time the degree of the polynomial is increased.

Most equations found in modern econometric models are either of the linear or loglinear form. These two types of equations predominate, no doubt because of their simplicity. But it should be remembered that while we assume that the error term is distributed normally in the linear model, we assume that the log of the error term is distributed normally in the loglinear model. This discussion leads to the more general assumption in the power transformation technique that the error term is distributed normally for *any* λ. Such an assumption is difficult to maintain unless it is believed that the λ which is associated with the maximum $\ln L(\lambda)$ will induce normality to the errors. Further problems with the power transformation technique include the accuracy with which to compute λ and the relationship between $E(Y)$ and $E(Y^{(\lambda)})$, which is, to our knowledge, as yet unknown.

Questions and Problems

Sec. 6.1

1. Show the following [Walker and Lev (1953), pp. 276–277]:
   a. If the conditional means, $\bar{Y}_i$, are equal, the correlation ratio will be zero.
   b. If the conditional means, $\bar{Y}_i$, are greatly different but the observations about each conditional mean have a very small variance, the correlation ratio will be close to unity.

Sec. 6.2

1. Three equations are given below. In each case discuss a simple transformation that will allow least-squares estimation. Also, draw a chart showing the general shape of the equation. More than one chart may be needed for each equation depending on the signs of (or relationships between) the parameters.
   a. $Y = \beta_0 \beta_1^X$.
   b. $Y = \beta_0 + \beta_1/X$.
   c. $Y = e^{(\beta_0 - \beta_1/X)}$.

Sec. 6.3

1. The following data were collected by Rolfe Wyer as reported in Croxton, Cowden and Bolch (1969), p. 725. The data relate to the so-called learning curve. This curve, now a standard tool of cost accounting, describes the fact that workers are able to produce more output per unit of time as they learn the job. This learning-by-doing effect appears in economic theory, especially in modern growth theory, where it is associated mainly with the economist K. Arrow. Let X represent cumulative units produced and Y represent cumulative hours of labor input per unit. The function is usually fit using the logarithms of both Y and X. Comment on the loglinear form using Wyer's data.

X	20	35	60	100	150	300	500	800	1500
Y	150	125	105	100	92	77	62	58	47

Sec. 6.4

1. Kementa (1967) discusses a C.E.S. production function which may be

written as

$$Y = [\beta_1 K^\lambda + \beta_2 L^\lambda]^{1/\lambda}$$

Here Y represents output and K and L represent inputs of capital and labor, respectively.

a. Show that this function may be converted to the form

$$Y^\lambda = \beta_1 K^\lambda + \beta_2 L^\lambda$$

and that in this form the function may be estimated by the Box-Cox method. Do not confuse the exponent λ in this equation with the operator (λ).

b. Using a Taylor transformation about $\lambda = 0$, Kementa is able to approximate the C.E.S. function (neglecting third- and higher-order terms) by

$$\ln Y = \beta_1 \ln K + \beta_2 \ln L + \tfrac{1}{2}\lambda\beta_1\beta_2[\ln K - \ln L]^2$$

Discuss the possibility of ordinary least-squares estimation of this equation.

c. In part b., what is the consequence of an estimated zero value for λ?

APPENDIX

A6.1 use of regression program of chapter 4 with transformations

The main for the regression program of Sec. A4.1 may be easily modified to handle the transformations discussed in this chapter. The modifications consist of adding statements immediately following statement 15 in that program. The location is indicated in that program by the comment card "TRANSFORMATIONS MAY BE ADDED HERE."

A. Polynomial regression. As we have noted, polynomial regression may be carried out by entering the values for X, X^2, X^3, and so forth in the usual way in the regression program. To avoid calculating and punching the powers of X, the following technique may be used.

On the first data card set $M = 2$ (two variables will be read in). Then punch the dependent variable and X_1 as usual. Finally, at the location of the comment card mentioned above, insert the following statements for, say, a *second*-degree polynomial:

```
      M = 3
      DO 17 I = 3, M
      I1 = I - 1
      DO 17 J = 1, N
17    X (I, J) = X (2, J)**I1
```

For a second-degree polynomial there will be three variables in all, and so M is set equal to 3. The only change in these five cards necessary for a higher-degree polynomial is in the first card. For a third-degree polynomial the first card should read $M = 4$, for a fourth-degree polynomial the first card should read $M = 5$, and so forth.

B. Loglinear regression. For this transformation it is only necessary to take the logarithms of all variables. To avoid having to look up and punch the logs of the variables, follow this procedure. Punch and enter the original data as usual. Then, at the location of the comment card, insert the following statements:

```
      DO 17 I = 1, M
      DO 17 J = 1, N
17    X (I, J) = ALOG(X(I, J))
```

The output of the program will now conform to Eq. (6-15a).

C. Power transformed regression. Once $\lambda \neq 0$, FL, is determined (see the following program), enter the variable as usual and insert the following at the location of the comment card for, say FL = −0.52:

```
      FL = -0.52
      DO 17 I = 1, M
      DO 17 J = 1, N
17    X (I, J) = (X(I, J)**FL - 1.0)/FL
```

In these cards the only change necessary for a different λ is in the first card, where FL is set equal to any desired nonzero value. If FL = 0, use the loglinear transformation.

A6.2 main for power transformed regression

A. Description. The following program calculates ln $L(\lambda)$ of Eq. (6-21). The program uses subroutine REG of Chap. 4, which in turn uses subroutine

MEAN, COVAR, and INVS. Thus for each value of λ, called FL in the program, a complete regression is calculated leading to $\ln L(\lambda)$.

B. Limitations. The limitations are the same as for subroutine REG. In principle one can choose any trial values for λ. In practice, however, the values for λ are limited. The programs in this book are all written in single precision arithmetic for fast execution and to cause the programs to be usable on small computers. Most computers can retain only eight significant digits with single precision arithmetic, and when raising modest-sized numbers to powers greater than 3 or so, eight digits are not enough to store the result. Thus, if λ is set too large or small in absolute value, the program will fail because the variables entering the regression program will appear to be essentially the same and the $\mathbf{x}'\mathbf{x}$ matrix will be considered singular.

For these reasons, we recommend that FL not exceed 3 in absolute value and that FL be allowed to increment in steps of 0.1 or 0.01.

C. Use. On the first card, enter the following data:

Columns	*Name*	*Meaning*
1– 3	M	Number of variables (same meaning as in main for regression); punch right-adjusted without decimal point
4– 6	N	Number of observations for each variable (same meaning as in main for regression); punch right-adjusted without decimal point
7–12	ST	The smallest value for λ; punch with decimal point; ST is assumed to be a whole number (i.e., −2.0, −1.0, etc.)
13–18	FIN	The largest value for λ; punch with decimal point; FIN is assumed to be a whole number (i.e., +1.0, +2.0, etc.)
19–24	STEP	The interval at which λ will be incremented; it is assumed that this parameter will be 0.1 or 0.01

On the remaining data cards, punch the dependent and independent variables in the same manner as they are punched for the main for regression program.

Given this information, the computer sets FL = ST, performs the power transformation, calls REG, and computes $\ln L(\lambda)$. It then sets FL = ST + STEP and repeats the process, until FL = FIN, where the program terminates. Once λ is determined, the main for regression may be used to estimate the equation. For the Klein consumption function example of this chapter, the first card appears as follows:

1	2	3	4	5	6	7	8	9	10	11	12	13	14	15	16	17	18	19	20	21	22	23	24	25
		4		2	1			−	2	.	0				1	.	0				.	0	1	

The program is shown below followed by the output for this problem. The maximum value of ln $L(\lambda)$ is circled on the output page. Notice that this value is located on the row −.600 and the column .08. Adding the two locations, we get $\lambda = -0.600 + 0.08 = -0.52$.

This is the only program in this book which took longer than 15 seconds to complete on an XDS Sigma 7 computer. For this program the time was about 20 seconds—it seems short when one considers that 300 regressions with four variables were calculated.

```
C       REG NEEDED
        DIMENSION X(10,100),Y(10,100),XBAR(10), S(10,10),
       1YHAT(100),YRES(100),SER(10),FLMAX(10),B(10),A(10,10)
        READ(5,5) M,N,ST,FIN,STEP
  5     FORMAT(2I3,3F6.0)
        DO 7 I=1,M
  7     READ(5,9)(Y(I,J),J=1,N)
  9     FORMAT(8F10.0)
        SLG=0.0
        DO 10 I=1,N
 10     SLG=SLG+ALOG(Y(1,I))
        WRITE(6,12)
 12     FORMAT(43X,'LOG L(LAMBDA)')
        WRITE(6,15)
 15     FORMAT(' ----------------------------------------------------------
       1-----------------------------------------')
        FLMAX(1)=0.0
        DO 20 K=2,10
 20     FLMAX(K)=FLMAX(K-1)+STEP
        WRITE(6,25)(FLMAX(K),K=1,10)
 25     FORMAT(' LAMBDA',10F9.5)
        WRITE(6,15)
 30     CONTINUE
        DO 50 K=1,10
        FL=ST+FLOAT(K-1)*STEP
        IF(ABS(FL)-0.0001)35,35,40
 35     DO 38 I=1,M
        DO 38 J=1,N
 38     X(I,J)=ALOG(Y(I,J))
        GO TO 45
 40     DO 43 I=1,M
        DO 43 J=1,N
 43     X(I,J)=(Y(I,J)**FL-1.0)/FL
 45     CALL REG(M,N,X,XBAR,S,A,AA,B,SEE,YHAT,YRES,R2,F,SER,SSE)
        SEE=SSE/FLOAT(N)
 50     FLMAX(K)=-0.5*FLOAT(N)*ALOG(SEE)+(FL-1.0)*SLG
        WRITE(6,60) ST,(FLMAX(K),K=1,10)
 60     FORMAT(F7.3,10F9.5)
        ST=ST+10.0*STEP
        IF(ST-FIN) 30,75,75
 75     CALL EXIT
        END
```

LOG L(LAMBDA)

LAMBDA	.00000	.01000	.02000	.03000	.04000	.05000	.06000	.07000	.08000	.09000
-2.000	1.95538	2.00929	2.05406	2.08473	2.18597	2.17227	2.26125	2.26357	2.32996	2.36264
-1.900	2.39395	2.46243	2.50447	2.51721	2.54939	2.58760	2.66570	2.69893	2.73360	2.78824
-1.800	2.81938	2.87521	2.90924	2.92598	2.97948	3.03850	3.05956	3.11812	3.13852	3.18176
-1.700	3.22046	3.28204	3.30791	3.35460	3.38167	3.42227	3.46480	3.51181	3.54344	3.58685
-1.600	3.62201	3.65588	3.69136	3.72731	3.75903	3.80716	3.84201	3.88373	3.91568	3.94482
-1.500	3.98286	4.02235	4.05942	4.09042	4.13286	4.15793	4.20189	4.22905	4.26308	4.29816
-1.400	4.32947	4.36206	4.39929	4.43494	4.46361	4.49573	4.52335	4.55739	4.59174	4.62314
-1.300	4.64931	4.68320	4.71338	4.74181	4.76823	4.80049	4.82973	4.85783	4.88818	4.91541
-1.200	4.94336	4.97018	5.00006	5.02351	5.04883	5.07773	5.10248	5.12978	5.15517	5.17805
-1.100	5.20505	5.22771	5.25214	5.27478	5.29832	5.32147	5.34366	5.36502	5.38721	5.40962
-1.000	5.42812	5.44865	5.47536	5.48273	5.51163	5.52356	5.55728	5.57083	5.57901	5.60252
-.900	5.62756	5.63965	5.65071	5.66101	5.68083	5.68948	5.70943	5.72748	5.74167	5.75140
-.800	5.76593	5.77887	5.78928	5.80258	5.81454	5.82726	5.83444	5.84409	5.85913	5.86665
-.700	5.87306	5.88110	5.89064	5.89806	5.90511	5.90918	5.91394	5.92111	5.92805	5.93501
-.600	5.93712	5.94206	5.94479	5.94598	5.94885	5.95137	5.95187	5.95406	5.95593	5.95265
-.500	5.95506	5.95456	5.95258	5.95007	5.94737	5.94788	5.94464	5.94138	5.93700	5.93196
-.400	5.92772	5.92224	5.91599	5.91039	5.90404	5.89566	5.88995	5.88170	5.87302	5.86398
-.300	5.85628	5.84453	5.83466	5.82544	5.81453	5.80304	5.78995	5.77832	5.76465	5.75198
-.200	5.73779	5.72444	5.71086	5.69511	5.67934	5.66399	5.64758	5.63126	5.61348	5.59641
-.100	5.57852	5.56076	5.54155	5.52251	5.50241	5.48294	5.46382	5.44257	5.42187	5.39856
.000	5.37755	5.35634	5.32768	5.31058	5.28400	5.26237	5.23685	5.21283	5.18666	5.16322
.100	5.13744	5.10931	5.08438	5.05821	5.03036	5.00302	4.97505	4.94702	4.91841	4.88983
.200	4.86011	4.83061	4.80029	4.76973	4.73943	4.70880	4.67822	4.64655	4.61327	4.58102
.300	4.54955	4.51654	4.48390	4.45038	4.41670	4.38254	4.34827	4.31389	4.27863	4.24321
.400	4.20779	4.17230	4.13692	4.10132	4.06528	4.02783	3.99022	3.95261	3.91489	3.87727
.500	3.83946	3.80188	3.76353	3.72491	3.68568	3.64612	3.60666	3.56691	3.52646	3.48720
.600	3.44707	3.40701	3.36665	3.32526	3.28355	3.24152	3.19969	3.15845	3.11682	3.07555
.700	3.03310	2.99011	2.94742	2.90442	2.86081	2.81754	2.77417	2.73152	2.68785	2.64394
.800	2.59972	2.55545	2.51076	2.46615	2.42159	2.37693	2.33226	2.28723	2.24232	2.19667
.900	2.15081	2.10490	2.05923	2.01342	1.96777	1.92177	1.87562	1.82891	1.78221	1.73582

chapter 7

Discrimination, Principal Components, and Canonical Correlations

Discriminant analysis treats the problem of attempting to differentiate between two or more classes of persons or objects. A banker may wish to classify a loan applicant into two categories, "good loan risk" or "bad loan risk," on the basis of the amount of loan equity, the monthly income of the applicant, and other measures thought to be relevant. Similarly, personnel officers wish to classify job applicants into "hire" or "no hire" categories on the basis of measures afforded by aptitude tests and the like. A more ominous, but nevertheless interesting, application of discriminant analysis is the use to which it is put by the Internal Revenue Service. Although the public is allowed to know very little about the details of the procedure, the Service uses discriminant analysis to select income tax returns for audit. The use of discriminant analysis in the Service promises "greater probability of error in the returns examined [i.e., once discriminant analysis selects a return for audit it stands a good chance of needing to be audited], improved taxpayer relationships due to the nonselection of cases which would result in little or no change, and reduced costs of technical and clerical operation."[1]

Principal component analysis and *canonical correlation analysis* both deal with the coordinate structure of multivariate observations. Both techniques

[1]From a letter by S. B. Wolf, Director, Audit Division, Internal Revenue Service, to all employees, Sept. 8, 1969.

are useful in exploratory statistical work, both are helpful in the task of reducing the number of measurements needed to describe a phenomenon, and both require a good deal of nonstatistical skill (or intuition) in their interpretation. Principal component analysis and its extension *factor analysis* have been little used in economics and business until recent years. However, these techniques have been recently used in such diverse problem areas as the discovery of "factors" associated with economic development, the study of money market conditions as a guide to monetary policy, and the study of technical problems associated with the use of certain simultaneous equation estimation techniques.[2]

In a similar way canonical correlation has been used to attack technical problems arising in simultaneous equation estimation, to study the associative nature of characteristics of primary and secondary goods, and to study the relationship between prices and quantities sold.[3]

Discriminant analysis has been developed along several lines by various writers.[4] Both principal component analysis and canonical correlation analysis are largely attributable to the work of Hotelling.[5]

7.1 linear discriminant function

Let us return to our example of Chap. 3 concerning the verbal and quantitative GRE scores for successful and unsuccessful graduate students in economics. We concluded in Chap. 3 that while the mean vectors of the GRE scores were significantly different, it seemed to be the quantitative scores that were causing that difference. Let us now take a fresh approach to this same problem using the technique of discriminant analysis.

The question that we ask in discriminant analysis is as follows: Can we set up an index which will classify a student as belonging to the successful or to the unsuccessful group of students? Since we have the GRE scores available, let us base our index upon these scores. We mention here that the analysis could probably be improved with the addition of other data, but to keep the calculation simple, we shall omit such other measures.

The index that we shall establish is a linear combination of the two GRE scores. In addition, a certain critical value for the index will be established

---

[2]See Adelman and Morris (1968), Anderson (1969), and Evans and Klein (1967). Leonall Anderson also makes an interesting application of discriminant analysis in the article cited.

[3]Hannan (1967) and Chow and Ray-Chaudhuri (1967) discuss the simultaneous equation application. The other applications mentioned are reviewed by Tintner (1946).

[4]It was apparently developed independently by Fisher, Hotelling, and Mahalanobis. Anderson (1958) gives a good historical perspective and is particularly strong on the Bayesian approach, which we do not treat here.

[5]Hotelling (1933) and (1936).

such that if the index value for a given student falls below the critical value, the student will be classified in one group; if the student's index value falls above the critical value, the student will be classified in the other group.

It is clear that if the two groups of students are fairly similar with respect to GRE scores (i.e., similar in mean vectors and covariance matrices) it may not be possible to classify the students satisfactorily because of the large amount of overlap between the groups. Figure 7.1 illustrates this point. In that figure let X_1 stand for the verbal GRE scores and X_2 stand for the quantitative GRE scores. Recall that we have assumed that the scores for each group were distributed normally with the same covariance matrices. Therefore, in Figure 7.1 let the two ellipses represent the same constant density for the two bivariate normal distritutions. Notice that there is some overlap between these ellipses. Clearly, the closer the mean vectors, the greater will be the overlap for any given constant density and the more difficult it will be to discriminate between the populations.

The procedure in linear discriminant analysis is to find a linear combination of the measures (X_1 and X_2) such that the distributions for the two groups will possess "little" overlap. The linear function

$$Y_{it} = \beta_1 X_{i1t} + \beta_2 X_{i2t}, \qquad i = 1, 2 \qquad t = 1, 2, \ldots, n_i \tag{7-1}$$

is called a linear discriminant function with unknown coefficients β. The subscript i represents the group, and the subscript t refers to the item number

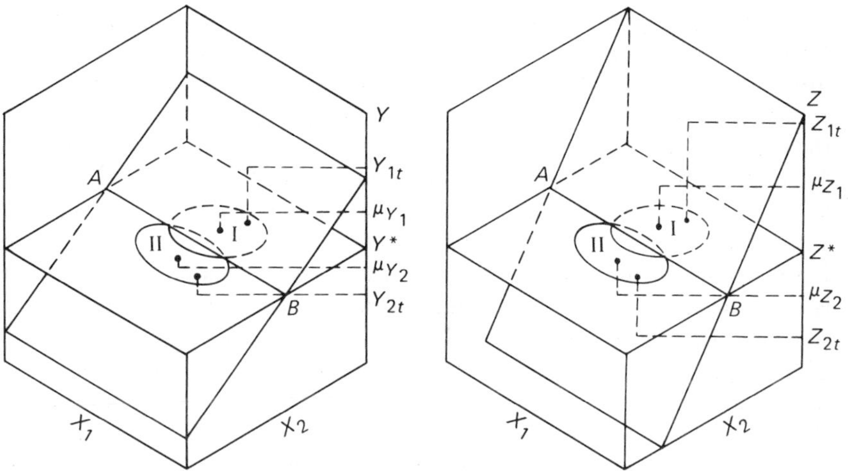

Figure 7.1. Constant density ellipses, discriminant planes, and projections.

of the observation within a group. Notice carefully that, unlike regression analysis, the variable Y is a result of combining the X variables—it is *not* a set of values to be fit by use of the X variables.

Geometrically, Eq. (7-1) defines a plane. The projection of X_{i1t} and X_{i2t} on the plane therefore transforms the two-dimensional GRE scores into a one-dimensional score, Y_{it}. Notice on the left-hand side of Figure 7.1 that the plane cuts the ellipses along the line AB which passes through their points of intersection. The projection of line AB is Y^*. Thus the plane cuts the ellipses with *most* of ellipse I being under the plane and most of ellipse II being above the plane. Students with GRE scores which project onto the Y axis above Y^* will be classified as belonging to group I. One such student's projection is Y_{1t}. Students with scores which project below Y^* will be classified as belonging to group II. One such student's projection is Y_{2t}. Similarly, the mean vectors of groups I and II project onto the Y axis as μ_{Y_1} and μ_{Y_2}. *Misclassification* occurs whenever an individual from group I projects below Y^* or whenever an individual from group II projects above Y^*. From Figure 7.1 the scores which will be misclassified (the area of misclassification) lie in the overlap of the two ellipses. Using the figure, the reader should try to convince himself that if the plane cuts the ellipses at their points of intersection, the area of misclassification will be smaller than for any other plane which does not cut the ellipses at their points of intersection. However, there are an infinite number of planes which pass through the points of intersection of the ellipses. For example, one such plane, with a steeper slope, is shown on the right-hand side of Figure 7.1.

Let us now compare the two discriminant planes in Figure 7.1. Let the squared distances $(\mu_{Y_1} - \mu_{Y_2})^2$ and $(\mu_{Z_1} - \mu_{Z_2})^2$ represent the *separation* of the two groups by the left and right discriminant planes, respectively. Each of these distances measures the variation between the means of the discriminant scores for each group, or, in short, each measures *between group variation*. From Figure 7.1 it is clear that this squared distance is greater for the Z axis than it is for the Y axis. In this respect the Z axis is superior to the Y axis since we desire to separate the groups, one from the other, as much as possible. On the other hand, notice that for, say, group I, the variation within the projections on the Y axis is less than the variation within the group I projections on the Z axis. Thus for any given Y_{1t} and Z_{1t}, $(Y_{1t} - \mu_{Y_1})^2$ is less than $(Z_{1t} - \mu_{Z_1})^2$. In this respect the Z axis is inferior to the Y axis since the *within-group variation* is larger for the Z axis than it is for the Y axis. It appears that the price that we must pay for greater separation of the means (increased between group variation) is greater within-group variation. Of course, large within-group variation is undesirable because any given distance between two means is less meaningful in a statistical sense the greater the variation of the distributions associated with each mean. Surely there is some optimal discriminant plane, and one method of seeking this optimum is to maximize the following ratio:

$$\lambda = \frac{\text{between-group variation}}{\text{within-group variation}} \tag{7-2}$$

We shall discuss the maximization of this ratio shortly.

So far we have discussed only the case where there are two variables, X_1 and X_2. It is easy to expand the discriminant function of Eq. (7-1) to include the case where there are p variables in each group. In this case the discriminant function defines a hyperplane. Thus

$$Y_{it} = \beta_1 X_{i1t} + \beta_2 X_{i2t} + \cdots + \beta_p X_{ipt} \qquad i = 1, 2, \quad t = 1, 2, \ldots, n_i \tag{7-1a}$$

where X_{ijt} represents the value of the jth variable for the tth item in the ith group. When there are only two groups, it is clear that the projections of the center (mean vectors) of the ellipsoids are given by

$$\mu_{Y_1} = \beta_1 \mu_{X_{11}} + \beta_2 \mu_{X_{12}} + \quad \cdots \quad + \beta_p \mu_{X_{1p}} \tag{7-3}$$

and

$$\mu_{Y_2} = \beta_1 \mu_{X_{21}} + \beta_2 \mu_{X_{22}} + \quad \cdots \quad + \beta_p \mu_{X_{2p}} \tag{7-4}$$

Here μ_{Y_1} and μ_{Y_2} are defined as they were for Figure 7.1. The values $\mu_{X_{ij}}$ represent the means for the jth variable in the ith group. Equations (7-3) and (7-4) must hold by definition since the variable Y is a linear combination of the X variables. Using the results of Eqs. (7-1a), (7-3), and (7-4), we find that the deviation of any Y_{it} about its mean μ_{Y_i} can be written (for $i = 1, 2$) as

$$Y_{it} - \mu_{Y_i} = \beta_1(X_{i1t} - \mu_{Xi1}) + \beta_2(X_{i2t} - \mu_{Xi2}) + \cdots + \beta_p(X_{ipt} - \mu_{Xip}) \tag{7-5}$$

The sum of squares of $Y_{it} - \mu_{Y_i}$ will then be a measure of *within-group variation.*

For the sample we replace all population means with sample means. Let μ_{Y_i} be estimated by $\bar{Y}_i = \sum_{t=1}^{n_i} Y_{it}/n_i$, $i = 1, 2$. Furthermore, estimate the population means for the X variable by use of $\bar{X}_{ij} = \sum_{t=1}^{n_i} X_{ijt}/n_i$ for $i = 1, 2$ and $j = 1, 2, \ldots, p$. Then, for the sample, within-group variation is, from Eq. (7-5), given by

$$\sum_{i=1}^{2} \sum_{t=1}^{n_i} (Y_{it} - \bar{Y}_i)^2 = \sum_{i=1}^{2} \sum_{t=1}^{n_i} [\beta_1(X_{i1t} - \bar{X}_{i1}) + \beta_2(X_{i2t} - \bar{X}_{i2}) + \cdots + \beta_p(X_{ipt} - \bar{X}_{ip})]^2 \tag{7-6}$$

Let us now simplify matters by introducing matrix notation. Define the matrix $\mathbf{X}'$ as

$$\mathbf{X}' = \left[\begin{array}{cccc|cccc} X_{111} & X_{112} & \cdots & X_{11n_1} & X_{211} & X_{212} & \cdots & X_{21n_2} \\ X_{121} & X_{122} & \cdots & X_{12n_1} & X_{221} & X_{222} & \cdots & X_{22n_2} \\ \cdot & \cdot & & \cdot & \cdot & \cdot & & \cdot \\ \cdot & \cdot & & \cdot & \cdot & \cdot & & \cdot \\ \cdot & \cdot & & \cdot & \cdot & \cdot & & \cdot \\ X_{1p1} & X_{1p2} & \cdots & X_{1pn_1} & X_{2p1} & X_{2p2} & \cdots & X_{2pn_2} \end{array}\right]$$

While the matrix $\mathbf{X}'$ appears somewhat formidable, it can be partitioned into two submatrices as indicated by the vertical dashed line in the $\mathbf{X}'$ matrix. Therefore, write

$$\mathbf{X}' = [\mathbf{X}_1' \mid \mathbf{X}_2']$$

where $\mathbf{X}_1'$ simply contains all the observations on all the p variables in group I, while $\mathbf{X}_2'$ contains all the observations on all the p variables in group II. Treated separately, $\mathbf{X}_1'$ and $\mathbf{X}_2'$ are the same as matrices of observations that we have used many times. Associated with the two matrices, we have the mean vectors

$$\bar{\mathbf{X}}_1' = [\bar{X}_{11} \quad \bar{X}_{12} \quad \cdots \quad \bar{X}_{1p}]$$
$$\bar{\mathbf{X}}_2' = [\bar{X}_{21} \quad \bar{X}_{22} \quad \cdots \quad \bar{X}_{2p}]$$

as defined by Eq. (1-33) and the covariation matrices $\mathbf{x}_1'\mathbf{x}_1$ and $\mathbf{x}_2'\mathbf{x}_2$ as defined by Eq. (2-31). Finally, let $\boldsymbol{\beta}' = [\beta_1 \quad \beta_2 \quad \ldots \quad \beta_p]$. Then Eq. (7-6) can be written as

$$\sum_{i=1}^{2}\sum_{t=1}^{n_i}(Y_{it} - \bar{Y}_i)^2 = \boldsymbol{\beta}'\mathbf{x}_1'\mathbf{x}_1\boldsymbol{\beta} + \boldsymbol{\beta}'\mathbf{x}_2'\mathbf{x}_2\boldsymbol{\beta}$$
$$= \boldsymbol{\beta}'(\mathbf{x}_1'\mathbf{x}_1 + \mathbf{x}_2'\mathbf{x}_2)\boldsymbol{\beta}$$

Recall, however, that in Eq. (3-21) the pooled covariance matrix is defined as

$$\mathbf{S}_* = \frac{1}{n_1 + n_2 - 2}(\mathbf{x}_1'\mathbf{x}_1 + \mathbf{x}_2'\mathbf{x}_2)$$

Therefore we may write within-group variation as

$$\sum_{i=1}^{2}\sum_{t=1}^{n_i}(Y_{it} - \bar{Y}_i)^2 = \boldsymbol{\beta}'[(n_1 + n_2 - 2)\mathbf{S}_*]\boldsymbol{\beta}$$

The numerator of Eq. (7-2), between-group variation, may be expressed for the sample as $(\bar{Y}_1 - \bar{Y}_2)^2$. However, if we appeal to Eqs. (7-3) and (7-4) and replace population means with sample means, we have

$$(\bar{Y}_1 - \bar{Y}_2)^2 = (\boldsymbol{\beta}'\bar{\mathbf{X}}_1 - \boldsymbol{\beta}'\bar{\mathbf{X}}_2)^2$$
$$= \boldsymbol{\beta}'(\bar{\mathbf{X}}_1 - \bar{\mathbf{X}}_2)(\bar{\mathbf{X}}_1 - \bar{\mathbf{X}}_2)'\boldsymbol{\beta} \qquad (7\text{-}7)$$

We are now in a position to write the sample analogue of Eq. (7-2),

$$L = \frac{\text{between-group variation}}{\text{within-group variation}} = \frac{1}{n_1 + n_2 - 2} \frac{\boldsymbol{\beta}'(\bar{\mathbf{X}}_1 - \bar{\mathbf{X}}_2)(\bar{\mathbf{X}}_1 - \bar{\mathbf{X}}_2)'\boldsymbol{\beta}}{\boldsymbol{\beta}'\mathbf{S}_*\boldsymbol{\beta}} \tag{7-8}$$

which is the function that we wish to maximize. Before carrying out the maximization, notice that we may maximize the following ratio

$$l = \frac{\boldsymbol{\beta}'(\bar{\mathbf{X}}_1 - \bar{\mathbf{X}}_2)(\bar{\mathbf{X}}_1 - \bar{\mathbf{X}}_2)'\boldsymbol{\beta}}{\boldsymbol{\beta}'\mathbf{S}_*\boldsymbol{\beta}} \tag{7-9}$$

and obtain the same result since l is proportional to L. Upon setting the first partial derivatives $\partial l/\partial\boldsymbol{\beta}$ equal to zero and doing some tedious algebra, which we shall not repeat here, we obtain

$$c(\bar{\mathbf{X}}_1 - \bar{\mathbf{X}}_2) - \mathbf{S}_*\boldsymbol{\beta} = \mathbf{0} \tag{7-10}$$

where $c = \boldsymbol{\beta}'(\bar{\mathbf{X}}_1 - \bar{\mathbf{X}}_2)/l$ is a nonzero constant. The estimated values of the coefficients are then given by $\hat{\boldsymbol{\beta}} = c\mathbf{S}_*^{-1}(\bar{\mathbf{X}}_1 - \bar{\mathbf{X}}_2)$. The solutions, $\hat{\boldsymbol{\beta}}$, depend on the nonzero constant c. From Eq. (7-1a) it is obvious that such a factor can be discounted because the two sets of Ys will differ only by this constant factor, and therefore the constant factor makes no difference in the ability to discriminate between the two groups. Usually, c is set equal to 1,[6] in which case the estimates of the coefficients become

$$\hat{\boldsymbol{\beta}} = \mathbf{S}_*^{-1}(\bar{\mathbf{X}}_1 - \bar{\mathbf{X}}_2) \tag{7-11}$$

If we substitute this solution for $\hat{\boldsymbol{\beta}}$ into Eq. (7-9), we can write

$$\hat{l} = (\bar{\mathbf{X}}_1 - \bar{\mathbf{X}}_2)'\mathbf{S}_*^{-1}(\bar{\mathbf{X}}_1 - \bar{\mathbf{X}}_2) = \mathbf{D}^2 \tag{7-12}$$

which is called *Mahalanobis' D^2*, or *generalized distance.*

Let us now calculate the discriminant function coefficients for the examination scores. From Sec. 3.4

$$\mathbf{S}_*^{-1} = \begin{bmatrix} 0.0001335 & -0.0000637 \\ -0.0000637 & 0.0002645 \end{bmatrix} \quad \text{and} \quad \bar{\mathbf{X}}_1 - \bar{\mathbf{X}}_2 = \begin{bmatrix} 22.461 \\ 114.769 \end{bmatrix}$$

Therefore

$$\hat{\boldsymbol{\beta}} = \mathbf{S}_*^{-1}(\bar{\mathbf{X}}_1 - \bar{\mathbf{X}}_2) = \begin{bmatrix} -0.0043 \\ 0.0289 \end{bmatrix}$$

[6]The constant is also often set so as to make $\hat{\beta}_1 = 1.0$.

and the estimated discriminant function of Eq. (7-1) becomes

$$Y_{it} = -0.0043X_{i1t} + 0.0289X_{i2t}$$

It can be shown[7] that if the costs of misclassification are equal for the two groups and if the probability that an observation comes from the two groups is equal, then Y^* will lie midway between μ_{Y_1} and μ_{Y_2}. For the sample, $Y^* = \frac{1}{2}(\bar{Y}_1 + \bar{Y}_2)$, but by using Eqs. (7-3) and (7-4) we may write

$$Y^* = \tfrac{1}{2}\hat{\boldsymbol{\beta}}'(\bar{\mathbf{X}}_1 + \bar{\mathbf{X}}_2) \tag{7-13}$$

For our example

$$Y^* = \frac{1}{2}[-0.0043 \quad 0.0289]\begin{bmatrix} 1{,}174.46 \\ 1{,}306.77 \end{bmatrix} = 16.36$$

The rule for discrimination can now be expressed in the following manner, under the assumptions that we have made:

$$\begin{array}{ll} \text{Classify as group I if} & Y_{it} \geq Y^* \\ \text{Classify as group II if} & Y_{it} < Y^* \end{array}$$

As a final step, define the vector $\mathbf{Y}'$ as

$$\mathbf{Y}' = [Y_{11} \quad Y_{12} \quad \cdots \quad Y_{1n_1} \quad Y_{21} \quad Y_{22} \quad \cdots \quad Y_{2n_2}]$$

Then the discriminant function of Eq. (7-1a) can be written for the sample as

$$\mathbf{Y} = \hat{\boldsymbol{\beta}}\mathbf{X}' \tag{7-14}$$

We can now write the discrimination rule in its final form by use of Eqs. (7-13) and (7-14):

$$\begin{array}{ll} \text{Classify group I if} & \hat{\boldsymbol{\beta}}\mathbf{X}' - \tfrac{1}{2}\hat{\boldsymbol{\beta}}'(\bar{\mathbf{X}}_1 + \bar{\mathbf{X}}_2) \geq 0 \\ \text{Classify group II if} & \hat{\boldsymbol{\beta}}\mathbf{X}' - \tfrac{1}{2}\hat{\boldsymbol{\beta}}'(\bar{\mathbf{X}}_1 + \bar{\mathbf{X}}_2) < 0 \end{array} \tag{7-15}$$

We shall illustrate the calculation with the first pair of scores in the first (successful) group. From Table 3.2, $X_1 = 750$ and $X_2 = 590$. Then

$$-0.0043(750) + 0.0289(590) - 16.36 = -2.5$$

This number is less than zero so that this individual has been *incorrectly* classified as belonging to group II. This same calculation has been carried out for the rest of the pairs of test scores of Table 3.2, and the results are

[7] Anderson (1958), Chap. 6.

presented by the computer program in the Appendix. Notice from the computer program that two individuals were incorrectly classified for group I and that two were incorrectly classified for group II. Thus of a total of 23 persons, 19 persons, or about 83 percent, were *correctly* classified.

In Chap. 3 we inferred that the verbal GRE scores were not significantly different between successful and unsuccessful students. As an exercise, the reader should repeat the entire discriminant analysis using only the GRE quantitative scores. If the reader does this analysis, he will find that the same four students are misclassified when using only the quantitative scores as were misclassified when using both the verbal and the quantitative scores. From the point of view of misclassification of this particular sample of students, the GRE verbal scores add nothing to the ability to discriminate between successful and unsuccessful students. The analysis of Chap. 3 is verified, and if we are satisfied with the ability of this function to discriminate between successful and unsuccessful students, we can presumably use it in the future to aid in the decision to admit or reject students.

Hypothesis Tests. There are three general types of tests that we shall mention here: (1) a test for the usefulness of the entire discriminant function, (2) a test for deciding whether a hypothetical discriminant function agrees with the discriminant function calculated from the data, and (3) a test for the inclusion or omission of a variable from the function.

1. If the two populations are homogeneous, there is no way that a discriminant function can separate one population from the other. If the two populations are not homogeneous, the discriminant function is, by definition, useful since it is based upon an optimum method of separating the populations. However, if both populations are normally distributed, with equal covariance matrices, nonhomogeneity can imply only inequality of the mean vectors of the two populations. Thus, to test the possible usefulness of the discriminant function, we use Hotelling's T^2 of Eq. (3-23) and compare it to the critical value of T^2 as given by Eq. (3-24). Since we have already rejected the hypothesis of equal mean vectors in Chap. 3, we have already confirmed the usefulness of the discriminant function.

We mention in passing that T^2 of Eq. (3-23) and Mahalanobis' D^2 of Eq. (7-12) are closely related. In fact, $T^2 = [n_1 n_2/(n_1 + n_2)]D^2$, so that T^2 may be considered, itself, to be a measure of distance.

2. To test whether a hypothetical discriminant function,

$$Y_{it} = c_1 X_{i1t} + c_2 X_{i2t} + \cdots + c_p X_{ipt}, \qquad i = 1, 2 \qquad t = 1, 2, \ldots, n_i$$

is in agreement with the discriminant function computed from the data, the first step is to compute the generalized distance, D_0^2, associated with the hypothetical discriminant function (i.e., the given discriminant function which holds under the null hypothesis)[8], namely,

$$D_0^2 = (\mathbf{C}' \, \Delta\bar{\mathbf{X}})'(\mathbf{C}'\mathbf{S}_*\mathbf{C})^{-1}(\mathbf{C}' \, \Delta\bar{\mathbf{X}})$$

where $\Delta\bar{\mathbf{X}} = \bar{\mathbf{X}}_1 - \bar{\mathbf{X}}_2$ and $\mathbf{C}' = [c_1 \quad c_2 \quad \cdots \quad c_p]$. We then compare D_0^2 with D^2 computed from the data. The hypothetical discriminant function is rejected as incompatible with the one computed from the data at the stated level of significance if

$$F = \frac{n_1 + n_2 - p - 1}{p - 1} \, \frac{m(D^2 - D_0^2)}{1 + mD_0^2} > F_{\alpha;\, p-1,\, n_1+n_2-p-1}$$

where $m = n_1 n_2/(n_1 + n_2)(n_1 + n_2 - 2)$.

3. We have already concluded that the GRE verbal scores are not aiding in discrimination between successful and unsuccessful students. Aside from recomputing the discriminant function with some variables missing, another way to detect redundant variables is to compare the coefficients with their asymptotic (large-sample) standard errors. The method is much like the one used for conducting univariate tests on regression coefficients.

There is some difficulty with this approach in that $\hat{\boldsymbol{\beta}}$ depends on the nonzero constant, c, of Eq. (7-10). In practice, we often compare the generalized distance D_p^2 based upon p variables with the generalized distance D_{p+q}^2 based upon $p + q$ variables. Generally $D_{p+q}^2 \geq D_p^2$, and the smaller the difference $D_{p+q}^2 - D_p^2$, the less the contribution attributable to the extra q variables. Thus the null hypothesis that the q extra variables do not increase the discriminating power of the function is rejected at the stated level of significance if

$$F = \frac{n_1 + n_2 - p - q - 1}{q} \, \frac{m(D_{p+q}^2 - D_p^2)}{1 + mD_p^2} > F_{\alpha;\, q,\, n_1+n_2-p-q-1}$$

where m is defined as before.

More than Two Groups. The analysis can be extended to an arbitrary number of groups. That is, we may wish to classify items into "poor," "average," and "good" categories or, for that matter, into K categories. This treatment is beyond the scope of this text.[9]

7.2 principal component analysis

Suppose that two random variables X_1 and X_2 are normally distributed with mean vector $\boldsymbol{\mu} = [\mu_1, \mu_2]$ and covariance matrix $\boldsymbol{\Sigma}$. Figure 7.2 shows

---

[8]Observe that D_0^2 is nothing more than a linear combination of D^2 of Eq. (7-12). See also Eq. (3-26).

[9]See Rao (1952) and Anderson (1958).

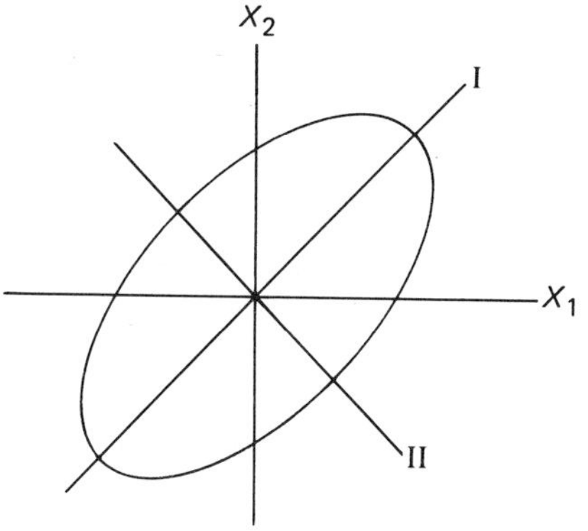

Figure 7.2. Constant density ellipse.

a particular constant density ellipse for this distribution. Recall that if X_1 and X_2 are perfectly positively correlated, the ellipse will degenerate into a straight line, part of which might be represented by line I in Figure 7.2. This line can be written as a linear combination of X_1 and X_2, say, $Y = \beta_1 X_1 + \beta_2 X_2$, and this linear combination can be used to represent the distribution of X_1 and X_2 just as well as the original X_1 and X_2 variables themselves.

In a less extreme case assume that the two variables are positively correlated. This is the case supposedly shown by the ellipse in Figure 7.2. In this case, the linear combination represented by line I gives only an approximation to the distribution of the two variables in that it captures much of the joint variability in X_1 and X_2. Clearly, as the correlation between the two variables increases, the concentration ellipse more closely approaches a straight line, and the degree of approximation of the straight line to the distribution of the two variables improves.

Generally, suppose that we have p random variables $\mathbf{X}' = [\mathbf{X}_1 \quad \mathbf{X}_2 \quad \cdots \quad \mathbf{X}_p]$. Principal component analysis seeks to make p linear combinations of the variables (called *principal components*) such that each of these linear combinations captures as much of the variation in $\mathbf{X}$ as possible while *at the same time* being linearly independent of all the other principal components. Let us now formalize these ideas. We do so in terms of the sample rather than in terms of the population.

A principal component Y_j is a linear combination of p variables. Thus

$$Y_j = \hat{\beta}_1 X_{1j} + \hat{\beta}_2 X_{2j} + \cdots + \hat{\beta}_p X_{pj}, \qquad j = 1, 2, \ldots, n \qquad (7\text{-}16)$$

is a principal component with unknown coefficients $\hat{\beta}_1, \hat{\beta}_2, \ldots, \hat{\beta}_p$. It will be convenient to express Eq. (7-16) in matrix notation. Let

$$\hat{\boldsymbol{\beta}} = \begin{bmatrix} \hat{\beta}_1 \\ \hat{\beta}_2 \\ \cdot \\ \cdot \\ \cdot \\ \hat{\beta}_p \end{bmatrix}, \quad \mathbf{Y} = \begin{bmatrix} Y_1 \\ Y_2 \\ \cdot \\ \cdot \\ \cdot \\ Y_n \end{bmatrix}, \quad \text{and} \quad \mathbf{X} = \begin{bmatrix} X_{11} & X_{21} & \cdots & X_{p1} \\ X_{12} & X_{22} & \cdots & X_{p2} \\ \cdot & \cdot & & \cdot \\ \cdot & \cdot & & \cdot \\ \cdot & \cdot & & \cdot \\ X_{1n} & X_{2n} & \cdots & X_{pn} \end{bmatrix}$$

Then we can write the principal component as

$$\mathbf{Y} = \mathbf{X}\hat{\boldsymbol{\beta}} \tag{7-16a}$$

For a given $\hat{\boldsymbol{\beta}}$, the sample variance of $\mathbf{Y}$ is given by[10]

$$\widehat{\text{var}}(\mathbf{Y}) = \hat{\boldsymbol{\beta}}'\mathbf{S}\hat{\boldsymbol{\beta}} \tag{7-17}$$

where the caret over var indicates the sample variance.

The first problem in principal component analysis is to find the principal component, $\mathbf{Y}_1$, with the maximum variance. This problem has no solution without further restrictions, for if we take $\hat{\boldsymbol{\beta}}^* = c\hat{\boldsymbol{\beta}}$ for a fixed $\hat{\boldsymbol{\beta}}$ and an arbitrary constant c, we can make the variance of $\mathbf{Y}$ arbitrarily large by taking c to be arbitrarily large. To avoid this indeterminancy, we usually normalize the vector $\hat{\boldsymbol{\beta}}$ such that

$$\hat{\boldsymbol{\beta}}'\hat{\boldsymbol{\beta}} = \hat{\beta}_1^2 + \hat{\beta}_2^2 + \cdots + \hat{\beta}_p^2 = 1 \tag{7-18}$$

This normalization simply requires that the vector defining the linear combination weights be of unit length.

The problem then is

$$\text{Maximize } \hat{\boldsymbol{\beta}}'\mathbf{S}\hat{\boldsymbol{\beta}} \text{ subject to } \hat{\boldsymbol{\beta}}'\hat{\boldsymbol{\beta}} = 1$$

This type of problem was treated in detail in Secs. 1.5 and 1.6, but we shall repeat its solution here. Let

$$\phi = \hat{\boldsymbol{\beta}}'\mathbf{S}\hat{\boldsymbol{\beta}} - \hat{\lambda}(\hat{\boldsymbol{\beta}}'\hat{\boldsymbol{\beta}} - 1)$$

where $\hat{\lambda}$ is a Lagrange multiplier. The vector of partial derivatives is

$$\frac{\partial \phi}{\partial \hat{\boldsymbol{\beta}}} = 2\mathbf{S}\hat{\boldsymbol{\beta}} - 2\hat{\lambda}\hat{\boldsymbol{\beta}}$$

---

[10]Note that $\text{var}(\mathbf{Y}) = \text{var}(\mathbf{X}\hat{\boldsymbol{\beta}} = \hat{\boldsymbol{\beta}}'\text{var}(\mathbf{X})\hat{\boldsymbol{\beta}} = \hat{\boldsymbol{\beta}}'\,\boldsymbol{\Sigma}\,\hat{\boldsymbol{\beta}}$, where $\boldsymbol{\Sigma}$ is the population covariance matrix of $\mathbf{X}$. Replacing $\boldsymbol{\Sigma}$ with $\mathbf{S}$ gives Eq. (7-17), where $\mathbf{S}$ is the sample covariance matrix for $\mathbf{X}$ as defined by Eq. (2-32).

which, upon being set equal to zero, reduces to

$$(\mathbf{S} - \hat{\lambda}\mathbf{I})\hat{\boldsymbol{\beta}} = 0 \tag{7-19}$$

Equation (7-19) is the classic type of equation treated in Sec. 1.6, and it will have nonzero solutions only if the determinant of $(\mathbf{S} - \hat{\lambda}\mathbf{I})$ is equal to zero [i.e., $(\mathbf{S} - \hat{\lambda}\mathbf{I})$ must be singular]. The determinant

$$|\mathbf{S} - \hat{\lambda}\mathbf{I}| = 0 \tag{7-20}$$

is a polynomial equation in $\hat{\lambda}$. Therefore, to solve Eq. (7-20), we must find the p characteristic roots of the covariance matrix $\mathbf{S}$. Call these roots $\hat{\lambda}_1 \geq \hat{\lambda}_2 \geq \cdots \geq \hat{\lambda}_p$. To determine which of these roots to use in finding the characteristic vector which will maximize $\hat{\boldsymbol{\beta}}'\mathbf{S}\hat{\boldsymbol{\beta}}$, we premultiply Eq. (7-19) by $\hat{\boldsymbol{\beta}}'$. Then using the fact that $\hat{\boldsymbol{\beta}}'\hat{\boldsymbol{\beta}} = 1$, we have

$$\hat{\boldsymbol{\beta}}'(\mathbf{S} - \hat{\lambda}\mathbf{I})\hat{\boldsymbol{\beta}} = \hat{\boldsymbol{\beta}}'\mathbf{S}\hat{\boldsymbol{\beta}} - \hat{\lambda} = 0$$

or

$$\hat{\boldsymbol{\beta}}'\mathbf{S}\hat{\boldsymbol{\beta}} = \hat{\lambda}$$

However, from Eq. (7-17), $\hat{\boldsymbol{\beta}}'\mathbf{S}\hat{\boldsymbol{\beta}}$ is the variance of $\mathbf{Y}$. Thus, to maximize the variance of $\mathbf{Y}$, we choose the largest characteristic root of the covariance matrix, $\mathbf{S}$. We call this root $\hat{\lambda}_1$ and we call the normalized characteristic vector associated with this root $\hat{\boldsymbol{\beta}}_1$. The first principal component is given by

$$\mathbf{Y}_1 = \mathbf{X}\hat{\boldsymbol{\beta}}_1$$

with variance equal to $\hat{\lambda}_1$.

If there are only two random variables, $\mathbf{X}_1$ and $\mathbf{X}_2$, with a constant density ellipse such as that shown in Figure 7.2, then it is clear that in the population the first principal component corresponds to the major axis of the ellipse (line I) with a semilength proportional to $\sqrt{\lambda_1}$, where λ_1 is the population characteristic root estimated by $\hat{\lambda}_1$. Since $\mathbf{X}$ has a bivariate normal distribution, there is a minor axis (line II) perpendicular to the major axis. The minor axis is called the second principal component and has length proportional to $\sqrt{\lambda_2}$. Clearly, the major and minor axes are given by $\mathbf{X}\boldsymbol{\beta}_1$ and $\mathbf{X}\boldsymbol{\beta}_2$, where $\boldsymbol{\beta}_1$ and $\boldsymbol{\beta}_2$ are the characteristic vectors corresponding to λ_1 and λ_2, respectively. $\mathbf{X}\boldsymbol{\beta}_1$ is called the first population principal component of $\mathbf{X}$ and $\mathbf{X}\boldsymbol{\beta}_2$ is called the second population principal component of $\mathbf{X}$. In a similar way $\mathbf{Y}_1 = \mathbf{X}\hat{\boldsymbol{\beta}}_1$ and $\mathbf{Y}_2 = \mathbf{X}\hat{\boldsymbol{\beta}}_2$ are the first and second sample principal components.

In general, when there are p variables, the first principal component, $\mathbf{Y}_1$, is a linear combination of the p variables with coefficients equal to the normalized characteristic vector associated with the *largest* characteristic root of $\mathbf{S}$. The *second* principal component, $\mathbf{Y}_2$, is the linear combination of the p variables with coefficients equal to the normalized characteristic vector associated with the *second largest* characteristic root of $\mathbf{S}$, and so forth up to the pth principal component, $\mathbf{Y}_p$. Each principal component has variance equal to its corresponding characteristic root and each component is linearly independent of every other component.[11] Thus each principal component merely defines the p axes of the p-dimensional concentration ellipsoid.

Some Additional Properties. Since principal component analysis is based upon the characteristic roots and vectors of the covariance matrix, $\mathbf{S}$, the following properties are important. These properties have already been discussed in Sec. 1.6.

1. Since the sample covariance matrix, $\mathbf{S}$, is positive definite, the characteristic roots of $\mathbf{S}$ are all positive.

2. The sum of the characteristic roots is equal to the sum of the elements on the main diagonal of $\mathbf{S}$. Thus

$$\operatorname{tr}\mathbf{S} = \sum_{i=1}^{p} \hat{\lambda}_i \tag{7-21}$$

Likewise, the product of the characteristic roots is the determinant of the sample covariance matrix. Thus

$$|\mathbf{S}| = \prod_{i=1}^{p} \hat{\lambda}_i \tag{7-22}$$

The properties associated with Eqs. (7-21) and (7-22) are important in the interpretation of the principal components. We have shown previously that the ith principal component, $\mathbf{Y}_i$, is linearly independent with respect to the jth principal component, $\mathbf{Y}_j$ and that the sample variance of $\mathbf{Y}_i$ is equal to $\hat{\lambda}_i$. Thus from Eq. (7-21) the total variance of the p components (i.e., $\sum_{i=1}^{p} \hat{\lambda}_i$) is equal to the trace of the sample covariance matrix:

$$\operatorname{tr}(\mathbf{S}) = \widehat{\operatorname{var}}(\mathbf{Y}_1) + \widehat{\operatorname{var}}(\mathbf{Y}_2) + \cdots + \widehat{\operatorname{var}}(\mathbf{Y}_p)$$

The relative "importance" of the ith principal component in the description of the system may be measured by

$$\frac{\widehat{\operatorname{var}}(\mathbf{Y}_i)}{\sum_{i=1}^{p} \widehat{\operatorname{var}}(\mathbf{Y}_i)} = \frac{\hat{\lambda}_i}{\operatorname{tr}(\mathbf{S})} = \frac{\hat{\lambda}_i}{\sum_{i=1}^{p} \hat{\lambda}_i} \tag{7-23}$$

[11] To prove each of these contentions neatly, recall Eq. (1-31), which states that if $\mathbf{P}$ is a matrix of characteristic vectors calculated from a symmetric matrix $\mathbf{A}$, then $\mathbf{P}'\mathbf{A}\mathbf{P} = \mathbf{D}$ is a diagonal matrix with the characteristic roots of $\mathbf{A}$ on the main diagonal and, of course, zeros elsewhere. Thus let $\mathbf{Y}$ be a $n \times p$ matrix containing the p principal components. Then the sample covariance matrix of $\mathbf{Y}$ is equal to $\hat{\boldsymbol{\beta}}'\mathbf{S}\hat{\boldsymbol{\beta}}$ from Eq. (7-17). Since $\hat{\boldsymbol{\beta}}$ is the matrix of characteristic vectors calculated from $\mathbf{S}$, the covariance matrix $\hat{\boldsymbol{\beta}}'\mathbf{S}\hat{\boldsymbol{\beta}}$ must be a diagonal matrix. Hence all covariances are zero (which proves linear independence) and all variances (main diagonal elements) are equal to the characteristic roots.

Moreover, since

$$\prod_{i=1}^{p} \hat{\lambda}_i = |\widehat{\text{var}}(\mathbf{Y})| = \begin{vmatrix} \widehat{\text{var}}(\mathbf{Y}_1) & 0 & \cdots & 0 \\ 0 & \widehat{\text{var}}(\mathbf{Y}_2) & & \vdots \\ \vdots & & \ddots & \vdots \\ 0 & \cdots & & \widehat{\text{var}}(\mathbf{Y}_p) \end{vmatrix}$$

from Eq. (7-22)

$$|\mathbf{S}| = |\widehat{\text{var}}(\mathbf{Y})| \tag{7-24}$$

The determinant of a covariance matrix is sometimes called *generalized variance*. Therefore Eq. (7-24) indicates the equality of the generalized variance of the original variables and the generalized variance of the principal components.

3. Usually the characteristic roots are unequal. In the special case where one or more of the roots are equal, we see from Eq. (7-23) that all the equal roots have the same relative importance. In the case of two variables with $\hat{\lambda}_1 = \hat{\lambda}_2$, the ellipse becomes a circle and there is no distinction between the major and the minor axes. In fact, any line passing through the center of the circle can be considered to be a major or a minor axis, and hence any two orthogonal lines can be considered as principal components.

A Remark on Computation. It has been shown that the principal component coefficients are the characteristic vectors of the **S** matrix. However, the **S** matrix is not invariant under a change of scale, and neither are the characteristic roots and vectors of the **S** matrix. Thus, if we change the units of measurement of X_1 and/or X_2, there will be a change in the shape, position, or both of the concentration ellipse shown in Figure 7.2.

Another problem with the analysis is that unless all the X variables are measured in the same units it is difficult to understand the meaning of the linear combination of the X variables. A linear combination of pounds of apples and yards of cloth is at best unclear. Therefore it is usual to standardize the X variables before computing the $\hat{\beta}$ coefficients. Since the standard scores are pure numbers, the problem of forming linear combinations with different measurement units vanishes. Furthermore, since the correlation coefficient is invariant with respect to scale, the first problem vanishes as well. Therefore, before computing we usually convert the X scores to z scores, where

$$z_{ij} = \frac{X_{ij} - \bar{X}_i}{s_i}, \qquad \begin{array}{l} i = 1, 2, \ldots, p \\ j = 1, 2, \ldots, n \end{array}$$

The analysis and interpretation is then carried out in terms of standard scores.

Corresponding to **S** for the original variables is **R**, the correlation matrix, for the z scores.[12] When standard scores are used, the ith principal component becomes

$$Y_{ij}^* = \hat{\beta}_{i1}^* z_{1j} + \hat{\beta}_{i2}^* z_{2j} + \cdots + \hat{\beta}_{ip}^* z_{pj}, \qquad j = 1, 2, \ldots, n \qquad (7\text{-}16b)$$

Notice that the subscript i is used here to denote the ith principal component. More compactly we may write

$$\mathbf{Y}_i^* = \mathbf{z}\hat{\boldsymbol{\beta}}_i^*$$

The coefficients $\hat{\boldsymbol{\beta}}_i^*$ are the elements of the normalized characteristic vectors associated with the $\hat{\lambda}_i^*$ characteristic roots of the correlation matrix, **R**.

The three properties of the principal components cited previously with reference to **S** can be stated with regard to **R**:

1. All characteristic roots, $\hat{\lambda}_i^*$, are positive.
2. Since the correlation matrix **R** has only units as diagonal elements, it follows that

$$\text{tr}\,\mathbf{R} = \sum_{i=1}^{p} \hat{\lambda}_i^* = p$$

and

$$|\mathbf{R}| = \prod_{i=1}^{p} \hat{\lambda}_i^*$$

Correspondingly, the importance of the ith principal component in the system can be measured by

$$\frac{\widehat{\text{var}}(\mathbf{Y}_i^*)}{\sum_{i=1}^{p} \widehat{\text{var}}(\mathbf{Y}_i^*)} = \frac{\hat{\lambda}_i^*}{p}$$

3. If some or all of the characteristic roots, $\hat{\lambda}_i^*$, are equal, the principal axes corresponding to these roots will have equal length. If all of the roots are equal, the ellipsoid will be spherical.

In summary, the use of standardized variables is helpful when the units of measurement for the variables are not the same. To the authors' knowledge, there is no general way to relate the principal components calculated from **S** with those calculated from **R**.

[12] For example, we know that $s_{12} = \sum x_1 x_2/(n-1)$ and in general that $\mathbf{S} = \mathbf{x}'\mathbf{x}/(n-1)$. In a similar way $r_{12} = \sum z_1 z_2/(n-1) = \sum x_1 x_2/[s_1 s_2 (n-1)] = \sum x_1 x_2/(\sqrt{\sum x_1^2}\sqrt{\sum x_2^2})$. In general, if $\mathbf{z}$ is a matrix of standardized scores corresponding to $\mathbf{x}$, then $\mathbf{R} = \mathbf{z}'\mathbf{z}/(n-1)$.

Some Examples. Let us illustrate this technique with a historical example. Table 7.1 shows three "cost-of-living" indices relating to the years 1900–1910 in the United States. Suppose that we are uncertain as to which

Table 7.1 Three "Cost-of-Living" Indices

Year	*F.R.B. NY, X_1*	*Hansen, X_2*	*Burgess, X_3*
1900	80	76	67.7
1901	82	75	70.6
1902	84	78	74.8
1903	88	81	74.8
1904	87	81	76.1
1905	87	81	76.0
1906	90	85	78.2
1907	95	90	82.0
1908	91	87	84.4
1909	91	91	88.6
1910	96	94	93.1

Source: United States Department of Commerce, *Historical Statistics of the United States* (Washington, D.C.: U.S. Government Printing Office, 1949), p. 235.

of these indices to adopt in a larger economic study. Under these conditions one approach would be to form a linear combination of the indices and use the result to represent them all. One kind of linear combination is a simple average; another is a principal component.

Since the variables are all index numbers, they are all expressed in the same measurement units (percent). Thus we shall calculate the principal components from the covariance matrix, which is

$$\mathbf{S} = \begin{bmatrix} 25.218 & 30.336 & 34.144 \\ 30.336 & 40.073 & 46.187 \\ 34.144 & 46.187 & 58.005 \end{bmatrix}$$

The characteristic roots and vectors of this matrix are

Root	*Corresponding Vector*		
118.44	0.4398	0.5762	0.6889
4.00	−0.7170	−0.2365	0.6557
0.85	0.5408	−0.7823	0.3091

The sum of the characteristic roots (also the trace of **S**) is 123.29. Thus the first principal component accounts for (118.44/123.29) about 96 percent of the generalized variance of the variables. The first principal component is given by

$$Y_{1j} = 0.4398X_{1j} + 0.5762_{2j} + 0.6889X_{3j}, \qquad j = 1, 2, \ldots, 11$$

If we wish to calculate the component, its first value is

$$Y_{11} = 0.4398(80) + 0.5762(76) + 0.6889(67.7) = 125.61$$

Other values are shown by the computer program in the Appendix to this chapter. The first principal component might now be used to represent the variation in the cost of living during this historical period.

In the particular example just discussed, the meaning of the first principal component is clear. The three price indices are highly correlated so that it makes sense to combine them into one index. The first principal component accounts for virtually all the generalized variance in the three series, and each of the three series is given positive weights in that component. Except for the fact that the weights on the three index numbers are not 1/3, we could have accomplished almost the same result by taking a simple average of the three index numbers.

Great difficulties arise in principal component analysis when we attempt to impart a nonstatistical meaning to the principal components. The following example will illustrate this point. It was concocted purely for enjoyment by one of the authors and it is not intended as a contribution to a controversial literature.[13]

For some time now there has been an interesting subfield in monetary economics dealing with the problem of a monetary indicator. According to two authorities, a monetary indicator is "a scale that is invariant up to a monotone transformation and that provides a logical foundation for statements comparing the thrust of monetary policy."[14] To put it in a layman's terms, the problem is to form some index that will tell us whether monetary policy is "tighter" or "easier" now than, say, it was 6 months ago. Some economists believe that a measure of the stock of money is a good indicator of monetary conditions. Thus if the rate of change in a measure of the stock of money increases, it must mean that monetary conditions are easing up. The problem with using this variable as an indicator of policy, says another group, is that the rate of change in any measure of the stock of money is endogenous to the economic system. That is, the measure depends on things other than monetary policy. For example, during economic expansions the demand for money increases and the fractional reserve banking system in

[13] We wish to thank John Pilgrim for helping with this project.

[14] Brunner and Meltzer (1969), p. 2.

the United States is able to increase the supply of money without any change in monetary policy. Thus "the business cycle generates procyclical movements in the money stock. This implies that those who use the money stock as an indicator of Federal Reserve policy actions are biased toward an unfavorable assessment of monetary policy."[15]

Suppose that we begin with the theory that all monetary series (free reserves, interest rates, the money stock, and so forth) have a *monetary component* and a *real component* which are independent. In this case, an indicator of monetary policy that would be free of the influence of nonmonetary phenomena would be the principal component associated with *monetary* behavior. We choose three series: the percentage change in free reserves of all member banks (X_1), the percentage change in the prime commercial interest rate (X_2), and the percentage change in a narrow definition of the stock of money (X_3).[16] For purposes of illustration we shall calculate the components from the correlation matrix even though there is no measurement problem. The characteristic roots and vectors appear as follows:

Root	*Corresponding Vector*		
1.76	0.6059	−0.6391	0.4737
0.79	−0.4516	0.2139	0.8662
0.45	0.6549	0.7388	0.1590

Now, we are interested in the monetary principal component. The first component explains most of the variation ($1.76/3 = 58.7$ percent) and in addition has interesting signs on the coefficients. For example, if we observe an increase in free reserves, an increase in the money stock, and a *decrease* in interest rates, we would think that monetary conditions have eased. Thus a negative sign on the interest rate variable with positive signs on the other two variables leads us to interpret this component as being monetary. Recall that if $\hat{\boldsymbol{\beta}}$ is a normalized characteristic vector of a matrix, so is $-\hat{\boldsymbol{\beta}}$. Thus our interpretation is based upon the *pattern* of the signs, not the signs themselves. The second component has signs on its coefficients which might lead us to call it *real.* An increase in the real demand for credit would be expected to be associated with both an increase in the money stock and interest rates while *decreasing* free reserves. Hence the sign on free reserves which is opposite to that on the other variables might lead us to call this component real.

Attempts to interpret principal components as *primitive* or *latent* characteristics of variables is extremely dangerous. First, there may be wide dis-

[15]Hendershott (1968), p. 92.

[16]Data run from 1918–1939, monthly.

agreement among qualified persons as to what the pattern of signs means. Second, the analysis assumes that there are primitive characteristics of the data set and that they are linearly independent. Neither one of these assumptions may be true, for there may be no primitive characteristics, and if there are, they may not be independent. Third, principal component analysis is often used as a "fishing" technique. There is a bad joke among statisticians which goes: "when all else fails, try principal component analysis or its extension factor analysis."[17] It should be remembered that classic statistical inference does not allow hypotheses to be developed from the data and that fishing the data for hypotheses nearly always leads eventually to grief.

In short, it is best to keep in mind that principal component analysis is nothing more than the determination of the axes of a concentration ellipsoid. It may or may not be a method for detecting hidden factors in the data.

Hypothesis Tests. It has been shown that principal component analysis is equivalent to the analysis of the sample covariance matrix, **S**, or correlation matrix, **R**. The principal component coefficients are the characteristic vectors of **S** or **R**, and the importance of the components is determined by the characteristic roots. Thus tests concerning principal components are really tests concerning characteristic roots and vectors.

Let us clarify the kind of hypotheses we wish to test. Suppose that the characteristic roots of **S** or **R** are all equal, that is, the relative importance of each component as measured by $\hat{\lambda}_i/\sum \hat{\lambda}_i$ or $\hat{\lambda}_i^*/p$ is equal, and so are the variances of the components. Suppose that the p characteristic roots of the population covariance matrix $\boldsymbol{\Sigma}$ (the population correlation matrix $\boldsymbol{\rho}$) are given by $\lambda_1 \geq \lambda_2 \geq \cdots \geq \lambda_p$ ($\lambda_1^* \geq \lambda_2^* \geq \cdots \geq \lambda_p^*$). As previously mentioned, in the bivariate case, if $\lambda_1 = \lambda_2$ in Figure 7.2, the constant density ellipse will become a circle, and the original axes X_1 and X_2 are just as good as any other orthogonal axes for the purpose of representing the distribution. The sort of transformation which principal component analysis tries to perform is superfluous. Thus one set of hypotheses is

$$H_0: \lambda_1 = \lambda_2 = \cdots = \lambda_p$$
$$H_1: \text{Not all characteristic roots are equal}$$

Under the null hypothesis the constant density ellipsoid becomes a constant density sphere. This test is therefore described as a *sphericity test.*

The test statistic is based upon the likelihood ratio described by Eq. (3-29) of Sec. 3.5. It can be shown that the *modified* likelihood ratio is

$$L^* = \left[\frac{|\mathbf{S}|}{[(1/p)\,\mathrm{tr}\,\mathbf{S}]^p}\right]^{(n-1)/2}$$

[17] Equally deplorable is the use of stepwise regression as a fishing device.

The implication of the null hypothesis is that the constant density ellipsoid is spherical, so that the original variables **X** are independent and have equal variances. Let $\bar{s} = (s_{11} + \cdots + s_{pp})/p$ be the pooled variance of all the X variables. The generalized variance under the null hypothesis is therefore equal to

$$\begin{vmatrix} \bar{s} & 0 & \cdots & 0 \\ 0 & \bar{s} & & \cdot \\ \cdot & & \cdot & \cdot \\ \cdot & & & \cdot \\ 0 & & \cdots & \bar{s} \end{vmatrix} = \left(\frac{s_{11} + \cdots + s_{pp}}{p}\right)^p = \left(\frac{1}{p}\operatorname{tr}\mathbf{S}\right)^p$$

which is the denominator inside the brackets of L^* above. Thus the likelihood ratio is the ratio of two generalized variances: One is $|\mathbf{S}|$ and the other is the generalized variance under the null hypothesis.

For large samples, the statistic $-2 \ln L^*$ is distributed approximately as chi square,

$$\chi^2 = -(n-1)\left[\ln|\mathbf{S}| - p \ln \frac{\operatorname{tr}\mathbf{S}}{p}\right] \tag{7-25}$$

with $\frac{1}{2}p(p+1) - 1$ degrees of freedom. From the properties of characteristic roots, we know that $|\mathbf{S}| = \prod_{i=1}^{p} \hat{\lambda}_i$ and $\operatorname{tr}\mathbf{S} = \sum_{i=1}^{p} \hat{\lambda}_i$. Thus Eq. (7-25) may be written as

$$\chi^2 = -(n-1)\left[\sum_{i=1}^{p} \ln \hat{\lambda}_i - p \ln \frac{\sum_{i=1}^{p} \hat{\lambda}_i}{p}\right] \tag{7-25a}$$

If we extract the components from the correlation matrix, **R**, a test for independence is

$$\begin{aligned} H_0&: \ \lambda_1^* = \lambda_2^* = \cdots = \lambda_p^* \\ H_1&: \ \text{Not all roots are equal} \end{aligned}$$

The likelihood ratio becomes

$$L^* = |R|^{(n-1)/2}$$

since $\operatorname{tr}\mathbf{R} = p$. Thus for large samples we have, approximately,

$$\chi^2 = -(n-1)\ln|R|$$

with $\frac{1}{2}p(p-1)$ degrees of freedom.

We may, however, legitimately ask a second question. Suppose that the first K characteristic roots are large and hence that the first K components

$\mathbf{Y}_1, \mathbf{Y}_2, \ldots, \mathbf{Y}_K$ explain a substantial fraction (say 95 percent) of the generalized variance of **X**. We wish to ask, do the remaining roots differ significantly among themselves? If they do not, the use of the remaining $q = p - K$ components is superfluous. Thus the hypotheses are

$$H_0: \ \lambda_{K+1} = \lambda_{K+2} = \cdots = \lambda_p$$
$$H_1: \ \text{Not all roots are equal}$$

The test statistic is also based upon a likelihood ratio. Under the hypothesis, the pooled variance for $\mathbf{Y}_{K+1}, \ldots, \mathbf{Y}_p$ is $\bar{\lambda} = (\hat{\lambda}_{K+1} + \cdots + \hat{\lambda}_p)/q$. Thus the likelihood ratio is equal to

$$L^* = \left[\frac{\sum_{i=1}^{p} \hat{\lambda}_i}{\bar{\lambda}^q \sum_{i=1}^{K} \hat{\lambda}_i} \right]^{(n-1)/2} \tag{7-26}$$

For large samples, the statistic $-2 \ln L^*$ is distributed approximately as chi square,

$$\chi^2 = -(n-1)\left[\sum_{j=K+1}^{p} \ln \hat{\lambda}_j - q \ln \frac{\sum_{j=K+1}^{p} \hat{\lambda}_j}{q} \right] \tag{7-27}$$

with $\frac{1}{2}q(q+1) - 1$ degrees of freedom.

Similarly, if we extract the components from the correlation matrix, **R**, a test for independence is

$$H_0: \ \lambda^*_{K+1} = \lambda^*_{K+2} = \cdots = \lambda^*_p$$
$$H_1: \ \text{Not all roots are equal}$$

The test statistic is the same as Eq. (7-27) except that $\hat{\lambda}^*_j$ replaces $\hat{\lambda}_j$.

We warn that these hypothesis tests are sensitive to the normality assumptions. They are especially dubious when economic time series are being studied since independence can rarely be assumed with time series data.

7.3 canonical correlations

The simple correlation between two random variables X and Y is defined [see Eq. (2-11)] as the ratio $\rho = \operatorname{cov}(Y, X)/\sqrt{\operatorname{var}(Y) \operatorname{var}(X)}$. In Sec. 5.1 we extended simple correlation to multiple correlation. There, the variable Y was correlated with several X variables. One view of multiple correlation is that it is the simple correlation between the variable Y and the variable $\hat{Y}$, where $\hat{Y}$ is a linear combination of several X variables, $\mathbf{X} = [\mathbf{X}_1 \quad \mathbf{X}_2 \quad \cdots$

$\mathbf{X}_p$]. If we define

$$\hat{Y}_j = \hat{\beta}_1 X_{1j} + \hat{\beta}_2 X_{2j} + \cdots + \hat{\beta}_p X_{pj}, \qquad j = 1, 2, \ldots, n$$

where the $\hat{\beta}$s are the slopes of a multiple regression equation which has been estimated by ordinary least-squares, then the multiple correlation coefficient is given by the simple correlation of Y and $\hat{Y}$. As an exercise, the reader should prove this contention.

Just as multiple correlation extends simple correlation to the case where there are several X variables, so *canonical correlation* extends simple correlation to the case where there are several X variables *and* several Y variables. Thus canonical correlation is the correlation between a linear combination of several X variables and a linear combination of several Y variables which are denoted as $\mathbf{Y} = [\mathbf{Y}_1 \quad \mathbf{Y}_2 \cdots \mathbf{Y}_q]$. We define the linear combinations of the q Y-variables and the p X-variables for the population as

$$\begin{aligned} \mathbf{X}^* &= \alpha_1 \mathbf{X}_1 + \alpha_2 \mathbf{X}_2 + \cdots + \alpha_p \mathbf{X}_p \\ \mathbf{Y}^* &= \beta_1 \mathbf{Y}_1 + \beta_2 \mathbf{Y}_2 + \cdots + \beta_q \mathbf{Y}_q \end{aligned} \tag{7-28}$$

One problem in canonical correlation will be to determine the unknown α_i and β_i coefficients such that the correlation between $\mathbf{Y}^*$ and $\mathbf{X}^*$ will be a maximum. The canonical correlation coefficient, ρ_c, between the two resulting variables will then be defined as

$$\rho_c = \frac{\text{cov}(X^*, Y^*)}{\sqrt{\text{var}(X^*)\,\text{var}(Y^*)}} \tag{7-29}$$

Let us denote the maximum canonical correlation coefficient as $\rho_c^{(1)}$, and call it the *first* canonical correlation coefficient between the two variable sets. We call the associated linear combinations $\mathbf{X}_1^*$ and $\mathbf{Y}_1^*$ the *first* canonical variables. The second canonical correlation will be determined by a linear combination $\mathbf{X}_2^*$ and $\mathbf{Y}_2^*$ such that of all linear combinations uncorrelated with $\mathbf{X}_1^*$ and $\mathbf{Y}_1^*$ these second canonical variables will produce the second largest canonical correlation $\rho_c^{(2)}$. The third canonical correlation will be determined by a linear combination $\mathbf{X}_3^*$ and $\mathbf{Y}_3^*$ such that of all linear combinations uncorrelated with the first two canonical variables these third canonical variables will produce the third largest canonical correlation. Each successive set of canonical variables is determined in a similar way, and if $q \leq p$, there will be q canonical correlations and q sets of canonical variables. Let us now formalize these ideas assuming, without loss of generality, that $q \leq p$.

Assume that the variables $\mathbf{X}$ and $\mathbf{Y}$ are distributed, respectively, as $N_p(\boldsymbol{\mu}_1, \boldsymbol{\Sigma}_{11})$ and $N_q(\boldsymbol{\mu}_2, \boldsymbol{\Sigma}_{22})$. Also, let

$$\boldsymbol{\Sigma} = \left[\begin{array}{c|c} \boldsymbol{\Sigma}_{11} & \boldsymbol{\Sigma}_{12} \\ \hline \boldsymbol{\Sigma}_{21} & \boldsymbol{\Sigma}_{22} \end{array}\right]$$

Here $\boldsymbol{\Sigma}_{11}$ is $p \times p$ and is the covariance matrix of $\mathbf{X}$, while $\boldsymbol{\Sigma}_{22}$ is $q \times q$ and is the covariance matrix of $\mathbf{Y}$. $\boldsymbol{\Sigma}_{12}$ is $p \times q$ and is the covariance matrix of $\mathbf{X}$ and $\mathbf{Y}$, while $\boldsymbol{\Sigma}_{21}$ is the transpose of $\boldsymbol{\Sigma}_{12}$. Explicitly written, we have

$$\boldsymbol{\Sigma} = \left[\begin{array}{ccc|ccc} \sigma_{X_1X_1} & \cdots & \sigma_{X_1X_p} & \sigma_{X_1Y_1} & \cdots & \sigma_{X_1Y_q} \\ \vdots & & \vdots & \vdots & & \vdots \\ \sigma_{X_pX_1} & \cdots & \sigma_{X_pX_p} & \sigma_{X_pY_1} & \cdots & \sigma_{X_pY_q} \\ \hline \sigma_{X_1Y_1} & \cdots & \sigma_{X_pY_1} & \sigma_{Y_1Y_1} & \cdots & \sigma_{Y_1Y_q} \\ \vdots & & \vdots & \vdots & & \vdots \\ \sigma_{X_1Y_q} & \cdots & \sigma_{X_pY_q} & \sigma_{Y_qY_1} & \cdots & \sigma_{Y_qY_q} \end{array}\right]$$

For convenience let us express the canonical variables in matrix notation. For the population, let

$$\begin{aligned} \mathbf{X}^* &= \mathbf{X}\boldsymbol{\alpha} \\ \mathbf{Y}^* &= \mathbf{Y}\boldsymbol{\beta} \end{aligned} \qquad \text{(7-28a)}$$

where $\mathbf{X}^*$ and $\mathbf{Y}^*$ are both vectors of canonical variables, $\mathbf{X}$ and $\mathbf{Y}$ are matrices of the original random variables, and $\boldsymbol{\alpha}$ and $\boldsymbol{\beta}$ are, respectively, $p \times 1$ and $q \times 1$ coefficient vectors. Then, using Eq. (7-29), we can write a canonical correlation coefficient for the population as

$$\rho_c = \frac{\text{cov}(\mathbf{X}\boldsymbol{\alpha}, \mathbf{Y}\boldsymbol{\beta})}{\sqrt{\text{var}(\mathbf{X}\boldsymbol{\alpha})\,\text{var}(\mathbf{Y}\boldsymbol{\beta})}}$$

or

$$\rho_c = \frac{\boldsymbol{\alpha}'\,\boldsymbol{\Sigma}_{12}\,\boldsymbol{\beta}}{\sqrt{(\boldsymbol{\alpha}'\,\boldsymbol{\Sigma}_{11}\,\boldsymbol{\alpha})(\boldsymbol{\beta}'\,\boldsymbol{\Sigma}_{22}\,\boldsymbol{\beta})}} \qquad \text{(7-29a)}$$

To find ρ_c of Eq. (7-29a), we shall need to maximize ρ_c with respect to $\boldsymbol{\alpha}$ and $\boldsymbol{\beta}$. However, the correlation will be unaffected, and the mathematics simplified, if we require in addition that each of the canonical variables have unit variance. Thus we require that

$$\text{var}\,(\mathbf{X}^*) = \boldsymbol{\alpha}'\boldsymbol{\Sigma}_{11}\boldsymbol{\alpha} = 1 \qquad \text{(7-30)}$$

$$\text{var}\,(\mathbf{Y}^*) = \boldsymbol{\beta}'\boldsymbol{\Sigma}_{22}\boldsymbol{\beta} = 1 \qquad \text{(7-31)}$$

With these restrictions, $\rho_c = \boldsymbol{\alpha}'\boldsymbol{\Sigma}_{12}\boldsymbol{\beta}$ so that we need only to maximize $\boldsymbol{\alpha}'\boldsymbol{\Sigma}_{12}\boldsymbol{\beta}$ given the restrictions of Eqs. (7-30) and (7-31). Upon performing the maximization, we find that the necessary conditions for a maximum are[18]

$$\mathbf{\Sigma}_{12}\boldsymbol{\beta} - \lambda\mathbf{\Sigma}_{11}\boldsymbol{\alpha} = \mathbf{0} \tag{7-32}$$

$$\mathbf{\Sigma}_{21}\boldsymbol{\alpha} - \lambda\mathbf{\Sigma}_{22}\boldsymbol{\beta} = \mathbf{0} \tag{7-33}$$

where λ is a Lagrange multiplier. Multiplying Eq. (7-32) by λ gives $\mathbf{\Sigma}_{12}\lambda\boldsymbol{\beta} = \lambda^2\mathbf{\Sigma}_{11}\boldsymbol{\alpha}$ and multiplying Eq. (7-33) by $\mathbf{\Sigma}_{22}^{-1}$ gives $\mathbf{\Sigma}_{22}^{-1}\mathbf{\Sigma}_{21}\boldsymbol{\alpha} = \lambda\boldsymbol{\beta}$. Upon substitution of the second result into the first, we obtain $\mathbf{\Sigma}_{12}\mathbf{\Sigma}_{22}^{-1}\mathbf{\Sigma}_{21}\boldsymbol{\alpha} = \lambda^2\mathbf{\Sigma}_{11}\boldsymbol{\alpha}$. Finally, multiplication of both sides of this result by $\mathbf{\Sigma}_{11}^{-1}$ and arranging terms gives

$$(\mathbf{\Sigma}_{11}^{-1}\,\mathbf{\Sigma}_{12}\,\mathbf{\Sigma}_{22}^{-1}\,\mathbf{\Sigma}_{21} - \lambda^2\mathbf{I})\boldsymbol{\alpha} = \mathbf{0} \tag{7-34}$$

After some similar manipulation we can also write

$$(\mathbf{\Sigma}_{22}^{-1}\,\mathbf{\Sigma}_{21}\,\mathbf{\Sigma}_{11}^{-1}\,\mathbf{\Sigma}_{12} - \lambda^2\mathbf{I})\boldsymbol{\beta} = \mathbf{0} \tag{7-35}$$

Equations (7-34) and (7-35) are homogeneous equations whose solutions can be found by extracting their characteristic roots and vectors. However, since we have assumed that $q \leq p$, there will be fewer roots and vectors for Eq. (7-35) than for Eq. (7-34) since the degree of the characteristic equation associated with Eq. (7-35) is q. Therefore we shall need to find only the characteristic roots and vectors of Eq. (7-35). Then after finding λ^2 and $\boldsymbol{\beta}$, we can use Eq. (7-32) to find $\boldsymbol{\alpha}$ by noting that

$$\boldsymbol{\alpha} = \frac{\mathbf{\Sigma}_{11}^{-1}\,\mathbf{\Sigma}_{12}\boldsymbol{\beta}}{\lambda} \tag{7-36}$$

Before proceeding to the calculation, let us investigate further the properties of Eqs. (7-32) and (7-33). First, if we multiply Eq. (7-32) by $\boldsymbol{\alpha}'$ and Eq. (7-33) by $\boldsymbol{\beta}'$, we obtain $\boldsymbol{\alpha}'\mathbf{\Sigma}_{12}\boldsymbol{\beta} = \lambda\boldsymbol{\alpha}'\mathbf{\Sigma}_{11}\boldsymbol{\alpha}$ and $\boldsymbol{\beta}'\mathbf{\Sigma}_{21}\boldsymbol{\alpha} = \lambda\boldsymbol{\beta}'\mathbf{\Sigma}_{22}\boldsymbol{\beta}$. Using the restrictions that $\boldsymbol{\alpha}'\mathbf{\Sigma}_{11}\boldsymbol{\alpha} = \boldsymbol{\beta}'\mathbf{\Sigma}_{22}\boldsymbol{\beta} = 1$, we may write

$$\boldsymbol{\alpha}'\mathbf{\Sigma}_{12}\boldsymbol{\beta} = \boldsymbol{\beta}'\mathbf{\Sigma}_{21}\boldsymbol{\alpha} = \lambda \tag{7-37}$$

However, by Eq. (7-29a) the quantity $\boldsymbol{\alpha}'\mathbf{\Sigma}_{12}\boldsymbol{\beta}$ is the canonical correlation under the restrictions of Eqs. (7-30) and (7-31). Therefore the square root of the *largest* characteristic root of Eq. (7-34), or Eq. (7-35), will be the maximum (first) canonical correlation. Notice carefully that the characteristic roots of, say, Eq. (7-35) are represented by λ^2 (not λ). Using Eq. (7-35) to find λ^2 and $\boldsymbol{\beta}$ and then Eq. (7-36) to find $\boldsymbol{\alpha}$, we may state the following proposition. Let $\lambda_1^2 \geq \lambda_2^2 \geq \cdots \geq \lambda_q^2$ be the characteristic roots of Eq. (7-35). Similarly, let $\boldsymbol{\beta}_1, \boldsymbol{\beta}_2, \ldots, \boldsymbol{\beta}_q$ be the characteristic vectors associated with

[18]The mathematics is clearly explained in Anderson (1958), pp. 289ff.

these roots. Finally, let $\boldsymbol{\alpha}_1, \boldsymbol{\alpha}_2, \ldots, \boldsymbol{\alpha}_q$ be the characteristic vectors which can be calculated from Eq. (7-36) for any given $\boldsymbol{\beta}_i$ and λ_i. Then the ith canonical correlation coefficient is given by λ_i and the ith set of canonical variables is given by $\mathbf{X}_i^* = \mathbf{X}\boldsymbol{\alpha}_i$ and $\mathbf{Y}_i^* = \mathbf{Y}\boldsymbol{\beta}_i$.

A second point, which may be puzzling the reader, is that it makes no difference whether λ^2 is calculated from Eq. (7-34) or Eq. (7-35) or, for that matter, whether $\boldsymbol{\alpha}$ is calculated from Eq. (7-34) or Eq. (7-36). What is puzzling is that there will be p roots and sets of $\boldsymbol{\alpha}$ vectors obtainable from Eq. (7-34), whereas there are only $q \leq p$ roots obtainable from Eq. (7-35). The resolution of this apparent conflict rests in the fact that there are $p - q$ roots of Eq. (7-34) which are zero, so that the *nonzero* roots of Eqs. (7-34) and (7-35) are equal. It is in this sense that it makes no difference whether Eq. (7-34) or Eq. (7-35) is used for calculation.[19]

Third, since the canonical correlations are equal to the square root of λ^2, they can be taken to be positive or negative in sign. Sometimes the sign is chosen according to the nature of the problem, but this practice can lead to difficulties. A positive sign might be chosen for the canonical correlation between several price and several production indices on the assumption that higher prices will bring forth greater production. Conversely, a negative sign might be chosen on the assumption that higher prices reduce demand. In truth, the sign on a canonical correlation coefficient is arbitrary.

Finally, canonical correlation coefficients retain many of the same properties as simple correlation coefficients. Namely, the absolute value of a canonical correlation coefficient lies between zero and 1, and canonical correlation coefficients are invariant to changes in units of measurement.

Computation. Since canonical correlation coefficients are invariant to changes in units of measurement, they may be estimated by replacing $\boldsymbol{\Sigma}$ with either

$$\mathbf{S} = \left[\begin{array}{c|c} \mathbf{S}_{11} & \mathbf{S}_{12} \\ \hline \mathbf{S}_{21} & \mathbf{S}_{22} \end{array}\right]$$

where $\mathbf{S}$ is the sample covariance matrix, or

$$\mathbf{R} = \left[\begin{array}{c|c} \mathbf{R}_{11} & \mathbf{R}_{12} \\ \hline \mathbf{R}_{21} & \mathbf{R}_{22} \end{array}\right]$$

where $\mathbf{R}$ is the sample correlation matrix of all the variables. Here, $\mathbf{R}_{11}$ is a

[19]The proof of this contention is rather lengthy. The interested reader may consult Anderson (1958), Chap. 12.

$p \times p$ matrix of the intercorrelations among the p variables $\mathbf{X}$, $\mathbf{R}_{22}$ is a $q \times q$ matrix of the intercorrelations among the q variables $\mathbf{Y}$, $\mathbf{R}_{12}$ is a $p \times q$ matrix of the intercorrelations between $\mathbf{X}$ and $\mathbf{Y}$, and $\mathbf{R}_{21} = \mathbf{R}'_{12}$. Usually, the calculation is done by use of the correlation matrix. However, while the canonical correlation coefficients are the same whether calculated from $\mathbf{S}$ or $\mathbf{R}$, the coefficients $\hat{\boldsymbol{\alpha}}_i$ and $\hat{\boldsymbol{\beta}}_i$ are not. As in principal component analysis, if $\hat{\boldsymbol{\alpha}}_i$ and $\hat{\boldsymbol{\beta}}_i$ are calculated from $\mathbf{S}$, they apply to the original variables; if they are calculated from $\mathbf{R}$, they apply to the *standardized* original variables. The computer program in the Appendix to this chapter will accept either $\mathbf{S}$ or $\mathbf{R}$, but the same measurement advantages that were mentioned for principal component analysis apply here to the use of $\mathbf{R}$. Unlike in principal component analysis, one is able to pass directly from coefficients calculated from $\mathbf{R}$ to those calculated from $\mathbf{S}$. For example, notice that

$$\begin{bmatrix} \frac{1}{\sqrt{s_{11}}} & 0 \\ 0 & \frac{1}{\sqrt{s_{22}}} \end{bmatrix} \begin{bmatrix} s_{11} & s_{12} \\ s_{21} & s_{22} \end{bmatrix} \begin{bmatrix} \frac{1}{\sqrt{s_{11}}} & 0 \\ 0 & \frac{1}{\sqrt{s_{22}}} \end{bmatrix} = \begin{bmatrix} 1 & r_{12} \\ r_{21} & 1 \end{bmatrix}$$

Thus if we pre- and postmultiply a covariance matrix by a diagonal matrix which contains reciprocals of standard deviations, we get a correlation matrix. Therefore, if

$$\hat{\alpha}^*_{i1}, \hat{\alpha}^*_{i2}, \ldots, \hat{\alpha}^*_{ip} \quad \text{and} \quad \hat{\beta}^*_{i1}, \hat{\beta}^*_{i2}, \ldots, \hat{\beta}^*_{iq}, \qquad i = 1, 2, \ldots, q$$

have been calculated from a correlation matrix, they may be caused to apply to the original variables (be the same as if they had been calculated from the $\mathbf{S}$ matrix) by dividing each coefficient by the standard deviation of the original variable associated with the coefficient. Thus we need only calculate $\hat{\alpha}^*_{i1}/\sqrt{s_{X_{11}}}$, $\hat{\alpha}^*_{i2}/\sqrt{s_{X_{22}}}, \ldots, \hat{\alpha}^*_{ip}/\sqrt{s_{X_{pp}}}$ and $\hat{\beta}^*_{i1}/\sqrt{s_{Y_{11}}}$, $\hat{\beta}^*_{i2}/\sqrt{s_{Y_{22}}}, \ldots, \hat{\beta}^*_{iq}/\sqrt{s_{Y_{qq}}}$.

We shall illustrate the calculation with an example given by Tintner.[20] The following production and price indices for the years 1919–1939 are used in the analysis:

X_1	durable goods production
X_2	nondurable goods production
X_3	mineral production
X_4	agricultural production
Y_1	farm prices
Y_2	food prices
Y_3	other prices

[20]Tintner (1946), p. 487ff.

The correlation matrix is as follows:

	X_1	X_2	X_3	X_4	Y_1	Y_2	Y_3
X_1	1.000000	0.495951	0.872836	0.481240	−0.436385	−0.427250	−0.203390
X_2	0.495951	1.000000	0.768279	0.709807	0.425782	0.429576	0.584220
X_3	0.872836	0.768279	1.000000	0.712358	−0.038273	−0.043762	0.138680
X_4	0.481240	0.709807	0.712358	1.000000	0.261010	0.267098	0.378452
Y_1	−0.436385	0.425728	−0.038273	0.261010	1.000000	0.987285	0.904598
Y_2	−0.427250	0.429576	−0.043762	0.267098	0.987285	1.000000	0.914394
Y_3	−0.203390	0.584220	0.138680	0.378542	0.904598	0.914394	1.000000

The question that Tintner asks is the following: What linear combination of the various production indices is most successful in predicting a general price "index," and at the same time what linear combination of the price indices is the most successful in predicting a general production "index"?

The correlation matrix above has been partitioned in such a way that the price indices (the Ys) and the production indices (the Xs) are separated. Let us now use Eq. (7-35) to find the $\hat{\lambda}$s and $\hat{\beta}$s with the elements of $\mathbf{R}$ replacing the elements of $\mathbf{\Sigma}$.

Following Eq. (7-35) and replacing $\mathbf{\Sigma}$s with $\mathbf{R}$s, we calculate

$$\mathbf{R}_{22}^{-1} = \begin{bmatrix} 39.60727 & -38.69914 & -0.44239 \\ -38.69914 & 43.91374 & -5.14729 \\ -\ 0.44239 & -\ 5.14729 & 6.10683 \end{bmatrix}$$

and

$$\mathbf{R}_{21}\mathbf{R}_{11}^{-1}\mathbf{R}_{12} = \begin{bmatrix} 0.77450 & 0.76211 & 0.68112 \\ 0.76211 & 0.75227 & 0.67831 \\ 0.68112 & 0.67831 & 0.66885 \end{bmatrix}$$

Then we have

$$\mathbf{R}_{22}^{-1}\mathbf{R}_{21}\mathbf{R}_{11}^{-1}\mathbf{R}_{12} = \begin{bmatrix} 0.88147 & 0.77281 & 0.43159 \\ -0.01126 & 0.05057 & -0.01459 \\ -0.10595 & -0.06701 & 0.29178 \end{bmatrix}$$

The last matrix is the one from which we desire to calculate the characteristic roots and vectors. Notice that this matrix is *not* symmetric, so that subroutine CHAR of Chap. 1 is *not* suitable for carrying out the calculation. However, subroutine CHAR can be used in conjunction with another subroutine (subroutine CHAR1) to extract the roots and vectors. This additional subroutine is given in the Appendix to this chapter and it takes advantage of the fact that $\mathbf{R}_{22}^{-1}\mathbf{R}_{21}\mathbf{R}_{11}^{-1}\mathbf{R}_{12}$ is the product of two symmetric matrices, namely $\mathbf{R}_{22}^{-1}$ and $\mathbf{R}_{21}\mathbf{R}_{11}^{-1}\mathbf{R}_{12}$. The characteristic roots and vectors then are

Roots	*Corresponding Vector*		
0.7799	1.2510	−0.0139	−0.2697
0.3818	−1.9602	−0.0363	2.3344
0.0622	−5.8481	6.6267	−0.7645

Therefore the estimated canonical correlation coefficients may be obtained by taking the square root of the characteristic roots: $\hat{\lambda}_1 = 0.8831$, $\hat{\lambda}_2 = 0.6179$, and $\hat{\lambda}_3 = 0.2494$. The coefficients associated with the first canonical correlation are $\hat{\beta}_{11} = 1.2510$, $\hat{\beta}_{12} = -0.0139$, and $\hat{\beta}_{13} = -0.2697$. Then, following Eq. (7-36), the $\hat{\boldsymbol{\alpha}}_1$ vector is given by

$$\hat{\boldsymbol{\alpha}}_1 = \frac{\mathbf{R}_{11}^{-1}\mathbf{R}_{12}\hat{\boldsymbol{\beta}}_1}{\hat{\lambda}_1}$$

or, for this example we get[21]

$$\begin{bmatrix} \hat{\alpha}_{11} \\ \hat{\alpha}_{12} \\ \hat{\alpha}_{13} \\ \hat{\alpha}_{14} \end{bmatrix} = \begin{bmatrix} -1.5515 \\ 0.5266 \\ 0.8326 \\ 0.0297 \end{bmatrix}$$

As a check on the calculation we recall that

$$\mathbf{R}_{21}\hat{\boldsymbol{\alpha}}_1 = \hat{\lambda}_1\mathbf{R}_{22}\hat{\boldsymbol{\beta}}_1$$

If the reader carries out this calculation, he will find agreement to at least three decimal places.

A final step may be needed to ensure that $\hat{\boldsymbol{\beta}}_1'\mathbf{R}_{22}\hat{\boldsymbol{\beta}}_1 = 1$ and that $\hat{\boldsymbol{\alpha}}_1'\mathbf{R}_{11}\hat{\boldsymbol{\alpha}}_1 = 1$. The computer program in the Appendix to this chapter automatically scales the coefficients so that these relations hold for a calculation done from a correlation matrix, while $\hat{\boldsymbol{\beta}}_1'\mathbf{S}_{22}\hat{\boldsymbol{\beta}}_1 = 1$ and $\hat{\boldsymbol{\alpha}}_1'\mathbf{S}_{11}\hat{\boldsymbol{\alpha}}_1 = 1$ if the calculation is done from a covariance matrix. If after calculating a characteristic vector we find that, say, $\hat{\boldsymbol{\beta}}_1'\mathbf{R}_{22}\hat{\boldsymbol{\beta}}_1 = k$, one need only divide each element of $\hat{\boldsymbol{\beta}}_1$ by

[21]Direct multiplication gives these results only approximately. The calculation here has been adjusted slightly to agree with the results of the computer program in the Appendix to this chapter.

$\sqrt{k}$ to ensure that $\hat{\boldsymbol{\beta}}_1' \mathbf{R}_{22} \hat{\boldsymbol{\beta}}_1 = 1$. A similar procedure may be followed for each vector.

Since the calculation was done from the correlation matrix, our final results apply to standard scores. Therefore the general production index, X^*, which best predicts the price index is[22]

$$\hat{X}_1^* = -1.5515 z_{X_1} + 0.5266 z_{X_2} + 0.8326 z_{X_3} + 0.0297 z_{X_4}$$

and the general price index, Y^*, which best predicts the general production index is

$$\hat{Y}_1^* = 1.2510 z_{Y_1} - 0.0139 z_{Y_2} - 0.2697 z_{Y_3}$$

The two indices are related with canonical correlation of 0.8831.

Tintner (correctly) interprets these results with caution. He notes that in the general production index the largest weight is given to durable goods, with a large coefficient of opposite sign given to minerals, and that there is a negligible weight given to agricultural products. Tintner points out that it appears to be the weighted difference between production of durable goods and minerals which is decisive. He believes that this result agrees with certain business cycle theories which stress the different behavior of producers' and consumers' goods over the cycle. In a similar way Tintner notes that it appears that the difference between farm prices and other prices is decisive in the general price index. This observation also agrees with certain business cycle theories, Tintner notes. As an exercise, which is not trivial, the reader may wish to give some economic meaning to the other two sets of canonical variables (see the Appendix to this chapter).

Hypothesis Tests. The canonical correlations measure the degree of correlation between two sets of variables Y and X. If the two sets of variables are independent, the covariance matrix $\boldsymbol{\Sigma}_{12}$ and the correlation matrix $\boldsymbol{\rho}_{12}$ both contain only zeros, and all the canonical correlations will be zero. Thus, to test the significance of the q canonical correlation coefficients, Bartlett has proposed a large sample test analogous to the test for principal components.[23] Define $\boldsymbol{\Lambda}_0$ (lambda) as the product of the q factors $(1 - \hat{\lambda}_i^2)$, that is,

$$\boldsymbol{\Lambda}_0 = \prod_{i=1}^{q} (1 - \hat{\lambda}_i^2)$$

---

[22]Our coefficients are scalar multiples of Tintner's. This difference arises because Tintner's scaling procedure is different from ours. Some authors scale the canonical coefficients such that they sum to unity, or 100. Some scale so that the first coefficient on each variable is 1. Of course such scaling generally violates the unit variance condition on the canonical variables.

[23]See Cooley and Lohnes (1962), Chap. 3.

For large samples, a test of the null hypothesis that the q Y variables are uncorrelated with the p X variables can be conducted by use of the chi-square distribution where

$$\chi^2 = -[n - 1 - \tfrac{1}{2}(p + q + 1)] \ln \Lambda_0$$

with pq degrees of freedom. If this value of χ^2 exceeds $\chi^2_{\alpha;pq}$, then at least the first canonical correlation coefficient is significantly different from zero at the stated level of significance. To test the significance of the remaining $q - 1$ coefficients, we form

$$\Lambda_1 = \prod_{i=2}^{q} (1 - \hat{\lambda}_i^2)$$

which is distributed for large samples as

$$\chi^2 = -[n - 2 - \tfrac{1}{2}(p + q + 1)] \ln \Lambda_1$$

with $(p - 1)(q - 1)$ degrees of freedom. In general, if K canonical correlations have been rejected as equal to zero, the criterion for testing that the other $q - K$ coefficients are zero is

$$\Lambda_K = \prod_{i=K+1}^{q} (1 - \hat{\lambda}_i^2)$$

which is distributed for large samples as

$$\chi^2 = -[n - 1 - K - \tfrac{1}{2}(p + q + 1)] \ln \Lambda_K$$

with $(p - K)(q - K)$ degrees of freedom. Again, these tests are sensitive to the normality assumptions and are of dubious value when applied to economic time series.

Questions and Problems

Sec. 7.1

1. Tintner (1946), using data he collected for an earlier study, wishes to discriminate between producer's and consumer's goods on the basis of the cyclical behavior of their prices. Using English wholesale prices for the period 1860–1913, he reports the following data:

	Consumer's Goods					Producer's Goods			
	X_1	X_2	X_3	X_4		X_1	X_2	X_3	X_4
Rice	72.0	50.0	8.0	0.5	Gasoline	57.0	57.0	12.5	0.9
Tea	66.5	48.0	15.0	1.0	Lead	100.0	54.0	17.0	0.5
Sugar	54.0	57.0	14.0	1.0	Pig Iron	100.0	32.0	16.5	0.7
Flour	67.0	60.0	15.0	0.9	Copper	96.5	65.0	20.5	0.9
Coffee	44.0	57.0	14.0	0.3	Zinc	79.0	51.0	18.0	0.9
Potatoes	41.0	52.0	18.0	1.9	Tin	78.5	53.0	18.0	1.2
Butter	34.5	50.0	4.0	0.5	Rubber	48.0	50.0	21.0	1.6
Cheese	34.5	46.0	8.5	1.0	Quicksilver	155.0	44.0	20.5	1.4
Beef	24.0	54.0	3.0	1.2	Copper Sheets	84.0	64.0	13.0	0.8
					Iron Bars	105.0	35.0	17.0	1.8

The variables are

X_1 median length of cycle in months
X_2 median percentage of duration of rising prices relative to the entire cycle length
X_3 median cyclical amplitude as a percentage of the trend
X_4 mean monthly rate of change in the cycle

Do a discriminant analysis with these data. Comment on any possible problems with hypothesis tests.

Sec. 7.2

1. Using the data of Table 4.3, combine the manual-dexterity and the finger-dexterity scores into one score by use of principal component analysis. Then regress Y (rivet-setting ability) on the new variable and comment on the results. Explain how principal component analysis may help with the problem of multicollinearity in regression.

Sec. 7.3

1. Show that the multiple correlation coefficient is the same as the simple correlation between $\mathbf{Y}$ and $\hat{\mathbf{Y}}$.
2. Waugh (1942) reports the following $\mathbf{R}$ matrix of wheat and flour characteristics. The problem is to predict characteristics of the finished product, flour, from characteristics of the intermediate product, wheat. Use the first canonical correlation coefficient and associated canonical variables for this purpose. Explain your results in words.

	Wheat Characteristics					Flour Characteristics			
	X_1	X_2	X_3	X_4	X_5	X_6	X_7	X_8	X_9
X_1, kernel texture	1.00000	0.75409	−0.69048	−0.44578	0.69173	−0.60463	−0.47881	0.77978	−0.15205
X_2, test weight	—	1.00000	−0.71235	−0.51483	0.41184	−0.72236	−0.41878	0.54245	−0.10236
X_3, damaged kernels	—	—	1.00000	0.32326	−0.44393	0.73742	0.36132	−0.54624	0.17224
X_4, foreign material	—	—	—	1.00000	−0.33439	0.52744	0.46092	−0.39266	−0.01873
X_5, crude protein in wheat	—	—	—	—	1.00000	−0.38310	−0.50494	0.73666	−0.14848
X_6, wheat/bbl of flour	—	—	—	—	—	1.00000	0.25056	−0.48993	0.24955
X_7, ash in flour	—	—	—	—	—	—	1.00000	−0.43361	−0.07851
X_8, crude protein in flour	—	—	—	—	—	—	—	1.00000	−0.16276
X_9, gluten quality index	—	—	—	—	—	—	—	—	1.00000

APPENDIX

A7.1 main for discriminant analysis

A. Description. This main program calculates the coefficients of the discriminant function of Eq. (7-1a). It also calculates the critical value Y^* of Eq. (7-13), the actual value of T^2 [see Eq. (3-23)], and the critical value of T^2 [see Eq. (3-24)]. In addition, the program evaluates the discriminant function minus Y^* as given by Eq. (7-15). This evaluation is called "Value of Function" by the program. If "Value of Function" is greater than zero, the item is classified as belonging to group 1. If it is less than or equal to zero, the item is classified as belonging to group 2. Finally, the program summarizes the number of correct and incorrect classifications for each group.

B. Limitations. Same as for "Main for Test that Two Vectors of Means Are Equal" (see Sec. A3.2). Notice, however, that in Sec. A3.2 we read in the scores for the unsuccessful students first. In this program we read in the scores for the successful students first. It makes no difference which group is read in first, but which group is chosen to be read in first will be taken to be group 1 by the program.

C. Use. Prepare the data cards in the same way as they were prepared in Sec. A3.2. The data setup is shown below.

1	2	3	4	5	6	7	8	9	10	11	12	13	14	15	16	17	18	19	20	21	22	23	24	25	26	27	28	29	30
		2		1	3		1	0			5	.	8	5															
						7	5	0	.							3	6	0	.							7	2	0	.
						6	2	0	.							6	9	0	.							6	1	0	.
						5	9	0	.							6	0	0	.							7	5	0	.
						7	3	0	.							8	4	0	.							6	8	0	.
						7	4	0	.							6	7	0	.							5	6	0	.
						5	4	0	.							5	0	0	.										
						6	8	0	.							6	0	0	.							5	5	0	.
						6	3	0	.							6	0	0	.										

The output for the current example is shown below, followed by the main program.

```
GROUP 1
ITEM     VALUE OF  CLASSIFICATION
NUMBER   FUNCTION
   1  -.2535995E+01        2
   2  -.5628176E+00        2
   3   .2220978E+01        1
   4   .1841324E+01        1
   5   .1422577E+01        1
   6   .7708282E+00        1
   7   .3939163E+01        1
   8   .9907990E+00        1
   9   .2074326E+01        1
  10   .4953445E+01        1
  11   .6714325E+00        1
  12   .2376587E+01        1
  13   .2782303E+01        1
GROUP 2
ITEM     VALUE OF  CLASSIFICATION
NUMBER   FUNCTION
   1   .1101074E+00        1
   2  -.1901351E+01        2
   3  -.2872455E+01        2
   4  -.3653743E+01        2
   5  -.3291205E+01        2
   6   .1336227E+01        1
   7  -.1037782E+01        2
   8  -.3161670E+01        2
   9  -.4723854E+00        2
  10  -.1167316E+01        2
SUMMARY OF ERRORS
         CORRECT  INCORRECT
GROUP 1    11.        2.
GROUP 2     8.        2.
COEFFICIENTS
VARIABLE
   SLOPE
    1       -.4317850E-02
    2        .2892153E-01
  Y-STAR     .1636130E+02
T-SQR.ACTUAL=   .1821304E+02    T-SQR.CRITICAL=   .1228500E+02
```

```
C       MEAN,COVAR,AND INVS NEEDED
        DIMENSION X(10,100),X1(10,100),A(10,10),S(10,10),XBAR(10),
       1XBD(10),XBP(10)
        READ(5,5) M,N1,N2,F
 5      FORMAT(3I3,F10.0)
        DO 10 I=1,M
 10     READ(5,15)(X(I,J),J=1,N1)
 15     FORMAT(8F10.0)
        DO 20 I=1,M
 20     READ(5,15)(X1(I,J),J=1,N2)
        CALL MEAN(M,N1,X,XBAR)
        CALL COVAR(M,N1,X,S,XBAR)
        DO 30 I=1,M
        XBD(I)=XBAR(I)
        XBP(I)=XBAR(I)
        DO 30 J=1,M
 30     A(I,J)=S(I,J)
        CALL MEAN (M,N2,X1,XBAR)
        CALL COVAR(M,N2,X1,S,XBAR)
        DO 40 I=1,M
        XBD(I)=XBD(I)-XBAR(I)
        XBP(I)=XBP(I)+XBAR(I)
        DO 40 J=1,M
```

```
40    A(I,J)=(A(I,J)+S(I,J))/FLOAT(N1+N2-2)
      CALL INVS(A,M)
      DO 60 I=1,M
      XBAR(I)=0.0
      DO 60 J=1,M
60    XBAR(I)=XBAR(I)+A(I,J)*XBD(J)
      CON=0.0
      D =0.0
      DO 70 I=1,M
      D=D+XBD(I)*XBAR(I)
70    CON=CON+XBP(I)*XBAR(I)
      CON=0.5*CON
      I1=1
      WRITE(6,75)I1
75    FORMAT(' GROUP',I2,/,' ITEM    VALUE OF  CLASSIFICATION',/,' NUMBE
     1R  FUNCTION')
      S1=0.0
      DO 90 I=1,N1
      CL=0.0
      DO 78 J=1,M
78    CL=CL+XBAR(J)*X(J,I)
      CL=CL-CON
      IF(CL)80,85,85
80    NC=2
      GO TO 90
85    NC=1
      S1=S1+1
90    WRITE(6,95)I,CL,NC
95    FORMAT(I5,E15.7,7X,I1)
      I1=2
      WRITE(6,75)I1
      S2=0.0
      DO 110 I=1,N2
      CL=0.0
      DO 98 J=1,M
98    CL=CL+XBAR(J)*X1(J,I)
      CL=CL-CON
      IF(CL)100,105,105
100   NC=2
      S2=S2+1
      GO TO 110
105   NC=1
110   WRITE(6,95)I,CL,NC
      SS1=N1-S1
      SS2=N2-S2
      WRITE(6,115)S1,SS1,S2,SS2
115   FORMAT(' SUMMARY OF ERRORS',/, 9X,'CORRECT  INCORRECT',/,' GROUP 1
     1',3X,F4.0,6X,F4.0,/,' GROUP 2',3X,F4.0,6X,F4.0)
      WRITE(6,120)
120   FORMAT(' COEFFICIENTS',/,' VARIABLE',/,4X,'SLOPE')
      DO 125 I=1,M
125   WRITE(6,130) I,XBAR(I)
130   FORMAT(4X,I2,3X,E15.7)
      WRITE(6,135) CON
135   FORMAT(3X,'Y-STAR',E15.7)
      XBAR(1)=D*FLOAT(N1*N2)/FLOAT(N1+N2)
      XBAR(2)=F*FLOAT((N1 +N2 -2)*M)/FLOAT(N1+N2-M-1)
      WRITE(6,140) XBAR(1),XBAR(2)
140   FORMAT(' T-SQR.ACTUAL=',E15.7,3X,'T-SQR.CRITICAL=',E15.7)
      CALL EXIT
      END
```

A7.2 main for principal component analysis

If one has already calculated a correlation or covariance matrix, then the use of "Main for CHAR" in the Appendix to Chap. 1 will calculate the characteristic roots and vectors of **R** or S. If one wishes to start from the raw data, the following program may be useful.

A. Description. This main program calculates the covariance matrix S, or the correlation matrix **R**, from a set of data **X**. The characteristic roots and vectors are then calculated from S (or **R**) and the principal components are evaluated according to Eq. (7-16a) or (7-16b). Subroutines MEAN, COVAR, CORR, and CHAR are called by the main program.

B. Limitations. Not more than 10 variables with 100 observations on each variable may be used. This limit may be altered by changing the DIMENSION statement.

C. Use. On the first card, punch the number of variables, M, in columns 1–3. On the same card, punch the number of observations on each variable, N, in columns 4–6. Finally, in column 9 punch 1 if the components are to be calculated from the correlation matrix, **R**, and punch zero (or leave blank) if the components are to be calculated from the covariance matrix, S. M and N are punched without a decimal point as far to the right in their fields as possible.

Starting on a fresh card, punch X_1 in eight fields of width 10 with the decimal point until X_1 is exhausted. Starting on a fresh card, punch X_2 in the same way. This is the usual format for the observations.

The output for the index number problem of this chapter is shown below. Note that "Value of Component" is read from left to right across the page. The roots are not sorted in order of magnitude by the program. Thus the first and subsequent principal components must be determined by visual inspection.

```
COMPONENTS EXTRACTED FROM COVARIANCE MATRIX


CHARACTERISTIC ROOT=      .8509
CHARACTERISTIC VECTOR
  .5408  -.7823   .3091
VALUE OF COMPONENT
  .4734940E+01   .7495316E+01   .7528336E+01   .7344467E+01   .7205582E+01
  .7174667E+01   .6347855E+01   .6314911E+01   .7240738E+01   .5409943E+01
  .7157990E+01
```

```
CHARACTERISTIC ROOT=  118.4407
CHARACTERISTIC VECTOR
  .4398   .5762   .6889
VALUE OF COMPONENT
  .1256142E+03   .1279153E+03   .1334168E+03   .1369047E+03   .1373604E+03
  .1372915E+03   .1424314E+03   .1501293E+03   .1482945E+03   .1534927E+03
  .1605203E+03


CHARACTERISTIC ROOT=    4.0037
CHARACTERISTIC VECTOR
 -.7170  -.2365   .6557
VALUE OF COMPONENT
 -.3095029E+02  -.3024635E+02  -.2963620E+02  -.3321394E+02  -.3164452E+02
 -.3171010E+02  -.3336487E+02  -.3564116E+02  -.3048979E+02  -.2868208E+02
 -.3002634E+02
```

```
C     MEAN,COVAR,CORR,AND CHAR NEEDED
      DIMENSION X(10,100),XBAR(10),S(10,10),R(10,10),
     1VEC(10,10),ROOT(10),Y(100)
      READ(5,5) M,N,MOD
 5    FORMAT(3I3)
      DO 10 I=1,M
 10   READ(5,15)(X(I,J),J=1,N)
 15   FORMAT(8F10.0)
      CALL MEAN(M,N,X,XBAR)
      CALL COVAR(M,N,X,S,XBAR)
      IF(MOD)25,25,20
 20   DO 22 I=1,M
      DO 22 J=1,N
 22   X(I,J)=(X(I,J)-XBAR(I))/SQRT(S(I,I)/FLOAT(N-1))
      CALL CORR(M,S,R)
      WRITE(6,24)
 24   FORMAT(' COMPONENTS EXTRACTED FROM CORRELATION MATRIX')
      GO TO 40
 25   DO 35 I=1,M
      DO 35 J=I,M
      R(I,J)=S(I,J)/FLOAT(N-1)
 35   R(J,I)=R(I,J)
      WRITE(6,37)
 37   FORMAT(' COMPONENTS EXTRACTED FROM COVARIANCE MATRIX')
 40   CALL CHAR(R,M,VEC,ROOT)
      DO 70 K=1,M
      DO 50 I=1,N
      Y(I)=0.0
      DO 50 J=1,M
 50   Y(I)=Y(I)+X(J,I)*VEC(J,K)
      WRITE(6,55) ROOT(K)
 55   FORMAT(/,/,' CHARACTERISTIC ROOT=',F10.4,/,' CHARACTERISTIC VECTOR
     1')
      WRITE(6,65)(VEC(J,K),J=1,M)
 65   FORMAT(10F8.4)
      WRITE(6,68)
 68   FORMAT(' VALUE OF COMPONENT')
 70   WRITE(6,75)(Y(I),I=1,N)
 75   FORMAT(5E15.7)
      CALL EXIT
      END
```

A7.3 subroutine CHAR1

A. Description. Recall that subroutine CHAR is capable of finding the characteristic roots and vectors of symmetric matrices only. Subroutine CHAR1, when used in conjunction with subroutine CHAR, allows one to find the characteristic roots and vectors of a class of nonsymmetric matrices which arise in canonical correlation analysis and in discriminant analysis where there are more than two groups.[1] Since we do not treat the case of discriminant analysis with more than two groups, our use of CHAR1 in this book will be restricted to canonical correlation analysis.

The class of matrices that can be handled by CHAR1 are of the form **A**, where

$$\mathbf{A} = (\mathbf{S1})^{-1}(\mathbf{S2})$$

The matrix **A** is nonsymmetric but the matrices $(\mathbf{S1})^{-1}$ and **S2** are both symmetric and of the same dimensions. In canonical correlation analysis, $\mathbf{S1}^{-1} = \mathbf{R}_{22}^{-1}$, $\mathbf{S2} = \mathbf{R}_{21}\mathbf{R}_{11}^{-1}\mathbf{R}_{12}$, and their product is $\mathbf{R}_{22}^{-1}\mathbf{R}_{21}\mathbf{R}_{11}^{-1}\mathbf{R}_{12}$, which is the matrix whose characteristic roots and vectors are desired.

B. Limitations. **S1** and **S2** must both be symmetric and of the same order and cannot exceed 10×10. The maximum size of each matrix can be altered by changing the DIMENSION statement. Note that the characteristic vectors which emerge from CHAR1 are not normalized.

C. Use. Assuming that **S1** [not $(\mathbf{S1})^{-1}$], **S2**, and NQ are in storage, the statement

CALL CHAR1 (S1, S2, NQ, VEC, ROOT)

will cause entry into the subroutine. NQ gives the dimension of **S1** and **S2**. Thus, if **S1** and **S2** are 4×4, then NQ = 4. The matrix **VEC** contains the characteristic vectors of **A** (columnwise) and the vector **ROOT** contains the characteristic roots of **A**. Both **S1** and **S2** are destroyed by the subroutine. The subroutine calls CHAR and is shown below.

```
      SUBROUTINE CHAR1(S1,S2,NQ,VEC,ROOT)
C     CHAR NEEDED
      DIMENSION S1(10,10),S2(10,10),VEC(10,10),ROOT(10),ST(10,10)
      CALL CHAR(S1,NQ,VEC,ROOT)
      DO 5 J=1,NQ
      DO 5 I=1,NQ
5     S1(I,J)=VEC(I,J)/SQRT(ROOT(J))
      DO 10 I=1,NQ
      DO 10 J=1,NQ
      ST(I,J)=0.0
      DO 10 K=1,NQ
10    ST(I,J)=ST(I,J)+S2(I,K)*S1(K,J)
      DO 15 I=1,NQ
```

[1]See Cooley and Lohnes (1962), pp. 188–189, for an explanation of the mathematics behind this subroutine. The use of this subroutine in discriminant analysis is also explained in this reference.

```
      DO 15 J=1,NQ
      S2(I,J)=0.0
      DO 15 K=1,NQ
15    S2(I,J)=S2(I,J)+S1(K,I)*ST(K,J)
      CALL CHAR(S2,NQ,ST,ROOT)
      DO 20 I=1,NQ
      DO 20 J=1,NQ
      VEC(I,J)=0.0
      DO 20 K=1,NQ
20    VEC(I,J)=VEC(I,J)+S1(I,K)*ST(K,J)
      RETURN
      END
```

A7.4 main for canonical correlation

A. Description. This program calculates the canonical correlation coefficients and the coefficients for the canonical variables from a correlation matrix, **R**. The matrix **R** is assumed to be of the form

$$\mathbf{R} = \left[\begin{array}{c|c} \mathbf{R}_{11} & \mathbf{R}_{12} \\ \hline \mathbf{R}_{21} & \mathbf{R}_{22} \end{array}\right]$$

where $\mathbf{R}_{21} = \mathbf{R}'_{12}$, $\mathbf{R}_{11}$ is NP $\times$ NP, $\mathbf{R}_{12}$ is NP $\times$ NQ, $\mathbf{R}_{21}$ is NQ $\times$ NP, and $\mathbf{R}_{22}$ is NQ $\times$ NQ. The matrix **S** may be used instead of **R**. However, remember that if **S** is used, the coefficients apply to the original variables. If **R** is used, the coefficients apply to the standardized variables unless they are divided by the standard deviation of the original variables. The program does not perform this division.

B. Limitations. The maximum dimension of **R**, or S, is 10 $\times$ 10. This maximum size may be altered by altering the DIMENSION statement. The program assumes that NP is greater than or equal to NQ. Thus the correlation or covariance matrix should be so arranged. This limitation is purely for computational convenience, and, of course, such an arrangement will not alter the canonical results.

C. Use. If the original data (but not **R** or S) are at hand, the first step will be to use the "Main for MEAN, COVAR, and CORR" given in the Appendix to Chap. 2 to obtain the full correlation matrix, **R**, or covariance matrix, S. Then, on the first card, punch NP (the dimension of $\mathbf{R}_{11}$) in columns 1–3. On the same card, punch NQ (the dimension of $\mathbf{R}_{22}$) in columns 4–6. Punch both numbers right-adjusted and without the decimal point.

Starting on a fresh card, punch the first row of **R** in eight fields of width 10 (with the decimal point) until the first row of **R** is exhausted. Starting on a fresh card, punch the second row of **R** in a similar manner.

Continue from card to card until all rows of **R** are punched. Whichever set is entered first (corresponds to $\mathbf{R}_{11}$), it will be called "First Set" by the program. Notice that the canonical correlations are not necessarily given in order of magnitude.

The output for Tintner's example is shown below followed by the main program.

```
            CANONICAL CORRELATION  1=  .8831
                    COEFFICIENTS
            FIRST  SET            SECOND  SET
     1        -1.5515      1           1.2510
     2          .5266      2           -.0139
     3          .8326      3           -.2697
     4          .0297
            CANONICAL CORRELATION  2=  .2494
                    COEFFICIENTS
            FIRST  SET            SECOND  SET
     1         1.4447      1          -5.8480
     2          .7144      2           6.6266
     3        -3.0718      3           -.7645
     4          .8024
            CANONICAL CORRELATION  3=  .6179
                    COEFFICIENTS
            FIRST  SET            SECOND  SET
     1         1.4253      1          -1.9602
     2         1.3347      2           -.0363
     3        -1.7776      3           2.3344
     4          .2191
```

```
C     INVS AND CHAR1 NEEDED
C     NP ASSUMED GREATER THAN OR EQUAL TO NQ
      DIMENSION R(10,10),S(10,10),A(10,10),VEC(10,10),
     1S1(10,10),S2(10,10),ROOT(10),SS(10)
      READ(5,5)NP,NQ
 5    FORMAT(2I3)
      M=NP+NQ
      DO 20 I=1,M
 20   READ(5,25)(R(I,J),J=1,M)
 25   FORMAT(8F10.0)
      DO 30 I=1,NP
      DO 30 J=1,NP
 30   A(I,J)=R(I,J)
      CALL INVS(A,NP)
      DO 35 I=1,NP
      DO 35 J=1,NQ
      S(I,J)=0.0
      DO 35 K=1,NP
 35   S(I,J)=S(I,J)+A(I,K)*R(K,J+NP)
      DO 40 I=1,NQ
      DO 40 J=1,NQ
 40   S1(I,J)=R(I+NP,J+NP)
      DO 42 I=1,NQ
      DO 42 J=1,NQ
      S2(I,J)=0.0
      DO 42 K=1,NP
 42   S2(I,J)=S2(I,J)+R(I+NP,K)*S(K,J)
      CALL CHAR1(S1,S2,NQ,VEC,ROOT)
      DO 100 K=1,NQ
      ROOT(K)=SQRT(ROOT(K))
      DO 45 I=1,NP
      S2(I,1)=0.0
      DO 45 J=1,NQ
```

```
45    S2(I,1)=S2(I,1)+S(I,J)*VEC(J,K)/ROOT(K)
      DO 48 I=1,NQ
      SS(I)=0.0
      DO 48 J=1,NQ
48    SS(I)=SS(I)+VEC(J,K)*R(J+NP,I+NP)
      S1(1,1)=0.0
      DO 50 I=1,NQ
50    S1(1,1)=S1(1,1)+SS(I)*VEC(I,K)
      DO 53 I=1,NQ
53    VEC(I,K)=VEC(I,K)/SQRT(S1(1,1))
      DO 59 I=1,NP
      SS(I)=0.0
      DO 59 J=1,NP
59    SS(I)=SS(I)+S2(J,1)*R(J,I)
      S1(1,1)=0.0
      DO 60 I=1,NP
60    S1(1,1)=S1(1,1)+SS(I)*S2(I,1)
      DO 62 I=1,NP
62    S2(I,1)=S2(I,1)/SQRT(S1(1,1))
      WRITE(6,64) K,ROOT(K)
64    FORMAT(6X,'CANONICAL CORRELATION',I3,'=',F7.4,/,14X,
     1'COEFFICIENTS',/,6X,'FIRST  SET',8X,'SECOND  SET')
      DO 65 I=1,NQ
65    WRITE(6,70)I,S2(I,1),I,VEC(I,K)
70    FORMAT(I3,5X,F8.4,3X,I3,5X,F8.4)
      IF(NP-NQ) 100,100,80
80    NQ1=NQ+1
      DO 85 I=NQ1,NP
85    WRITE(6,90) I,S2(I,1)
90    FORMAT(I3,5X,F8.4)
100   CONTINUE
      CALL EXIT
      END
```

chapter 8

Spectral Analysis[1]

A time series is a set of observations which are associated with time. In economic time series analysis we generally assume that the observations are ordered in time and spaced at equal intervals. The consumption and income series of Table 4.2 offer examples since there is one observation for each of these series for each year. The reader has undoubtedly studied some methods of time series analysis. Typically, elementary statistics courses concentrate on the *time domain* properties of time series. That is, they usually present models which decompose the time series into component parts, each of which is a function of time. One traditional model contends that economic time series are composed of four elements: a trend, a cyclical movement, a seasonal movement, and a random (or residual) element. Time series analysis under this model consists of the operations of measuring and identifying the elements of the series and, in some cases, of removing certain elements from the series. The process of seasonal adjustment consists of the identification of the seasonal element and then the removal (or *filtering*) of the seasonal element from the other elements of the time series.

In this chapter, rather than concentrating on the time domain characteristics of the time series, we shall concentrate on the *frequency domain*

[1]Some authors prefer to use the term *spectrum* analysis. They contend, with some justification, that the procedure is already ghostly enough without calling it by a ghostly name.

characteristics of the time series. Thus we shall study the decomposition of the time series into component parts which are associated with frequency rather than with time.

8.1 stationary time series and periodic functions

Figure 8.1 shows a hypothetical time series. Such a time series would never be observed in practice by an economist, but the series gives us a starting point in our discussion. The values of the series (Y_1, Y_2, Y_3, . . .) fluctuate about a mean value μ. Moreover, the fluctuations repeat themselves. In other words, the time series is *periodic*. The length of time necessary for the time series to repeat itself is called the *period*, denoted by P. If the time series in Figure 8.1 is observed monthly, 10 months elapse between peaks (high points) or troughs (low points). Hence the period of the cycle is 10 months (per cycle). Period is measured in terms of time units per cycle and is not unique since if a time series has period P it also has period $2P$, $3P$, ... In general, a series Y_t is periodic if $Y_t = Y_{t+cP}$, where $c = 1, 2, \ldots$.

The reciprocal of the period is called the *frequency* of the series and is denoted as $f = 1/P$. Frequency tells us the number of repetitions of the cycle per unit of time and is therefore measured in terms of cycles per unit of time. The frequency of the series shown in Figure 8.1 is 1/10 cycle per month. Obviously we can use the terms period and frequency interchangeably. It is also important to notice that a constant time series, $Y_t = k$, may be regarded as a special case of a periodic time series with zero frequency. That is, the period of Y_t is infinite (the time series never cycles).

The *amplitude* of a periodic series is the distance from the mean value of the series to a peak or trough. Amplitude is denoted as A and is shown in Figure 8.1. Finally, the *phase*, ϕ, is the distance between the nearest peak and the origin in time (which is the point when $t = 0$).

Periodic time series may oscillate about a rising (or falling) mean value

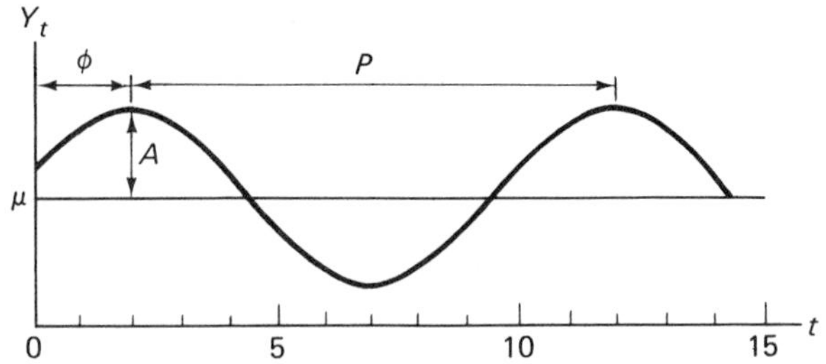

Figure 8.1. Periodic time series.

such as the series shown on the left side of Figure 8.2. Such a series is said to have a *trend in the mean.* Also, the amplitude of a periodic series may increase (or decrease) with time. Such a series is said to have a *trend in the variance*, and an example is shown on the right-hand side in Figure 8.2. Roughly speaking, a series with no trend in the mean and variance is called *stationary.*[2] Most economic time series show a trend in the mean. If the trend is linear, it can be removed (in principle) by fitting a least-squares regression $\hat{Y}_t = \hat{\beta}_0 + \hat{\beta}_1 t$ to the series. The variable t represents time and consists of integers running from $t = 1$ to $t = N$, where N is the number of observations for the variable Y_t. The residual, $Y_t - \hat{Y}_t$, will then be free of the trend. Alternatively, we can use the first differences of Y_t, where the first differences are defined as $\Delta Y_t = Y_t - Y_{t-1}$. A disadvantage of using first differences is that one observation is lost. We shall discuss this problem further in Sec. 8.5. In any event, to attempt to meet the assumption of stationarity, one should attempt to remove the trend prior to conducting a spectral analysis of a series.

As previously discussed, a periodic time series which is stationary can be described by four parameters: the period (or frequency), the amplitude, the phase, and the mean value. Therefore a stationary periodic time series can be expressed as

$$Y_t = \mu + A \cos 2\pi f(t - \phi)$$

which is known as a *harmonic representation.* The reader may verify that the time series in Figure 8.1 can be expressed by this model. For example, when $t = \phi$, then $Y_t = \mu + A$ since $\cos 0 = 1$.

As a matter of convenience it is useful to express the periodic function in terms of *angular frequency*, ω. Angular frequency is measured in terms of

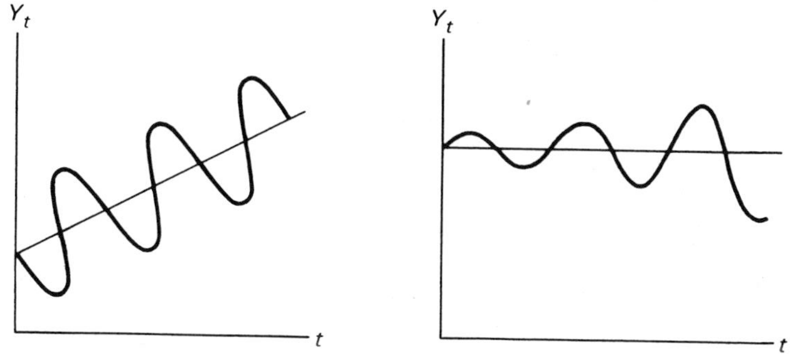

Figure 8.2. Two nonstationary time series.

[2]There are various formal mathematical definitions of stationarity. Our definition corresponds to the term *wide stationarity*, which implies that the mean and variance are independent of time.

radians per time unit and is

$$\omega = 2\pi f, \qquad 0 \leq \omega \leq 2\pi$$

Thus we may write the periodic time series model as

$$Y_t = \mu + A\cos(\omega t - \theta) \tag{8-1}$$

where $\theta = 2\pi f\phi$. This will be our fundamental representation of a periodic function. We shall use the term *frequency* to mean f, *angular frequency* to mean ω, and *phase* to mean θ.

Equation (8-1) is often expressed in terms of sines and cosines, eliminating any explicit reference to phase. Thus

$$Y_t = \mu + \alpha\cos\omega t + \beta\sin\omega t \tag{8-1a}$$

is the same representation as Eq. (8-1) provided that we define $\alpha = A\cos\theta$ and $\beta = A\sin\theta$. To show this equality, recall the trigonometric identity: $\cos(x - y) = \cos x\cos y + \sin x\sin y$. Then

$$\begin{aligned} A\cos(\omega t - \theta) &= A(\cos\omega t\cos\theta + \sin\omega t\sin\theta) \\ &= \alpha\cos\omega t + \beta\sin\omega t \end{aligned} \tag{8-2}$$

where

$$\alpha = A\cos\theta \tag{8-3}$$

$$\beta = A\sin\theta \tag{8-4}$$

Furthermore, we can pass easily between the coefficients of Eqs. (8-1) and (8-1a). Recalling the trigonometric identity $\cos^2 x + \sin^2 x = 1$, from Eqs. (8-3) and (8-4) we have

$$A^2(\cos^2\theta + \sin^2\theta) = \alpha^2 + \beta^2$$

or

$$A^2 = \alpha^2 + \beta^2 \tag{8-5}$$

Also, recalling that $\tan x = \sin x/\cos x$ and that $\arctan(x) = \tan^{-1}(x)$, we have, after dividing Eq. (8-4) by Eq. (8-3),

$$\tan\theta = \frac{\sin\theta}{\cos\theta} = \frac{\beta}{\alpha}$$

or

$$\arctan\frac{\beta}{\alpha} = \theta \tag{8-6}$$

A time series may be composed of the sum of many periodic time series.

One model of a stationary time series states that it is composed of the sum of infinitely many periodic series plus a mean. The model is

$$Y_t = \mu + \sum_{i=1}^{\infty} A_i \cos(\omega_i t - \theta_i) \tag{8-7}$$

where $0 \leq \omega_i \leq 2\pi$. Alternatively, using Eq. (8-1a), we may write the model as

$$Y_t = \mu + \sum_{i=1}^{\infty} \alpha_i \cos \omega_i t + \sum_{i=1}^{\infty} \beta_i \sin \omega_i t \tag{8-8}$$

Theoretically, it can be shown that any stationary time series can be approximated by an infinite series of sine and cosine curves.[3] Such a decomposition is called a *Fourier series* representation. As before, each of the A_i coefficients of Eq. (8-7) is related to the α_i and β_i coefficients of Eq. (8-8) as

$$A_i^2 = \alpha_i^2 + \beta_i^2, \qquad i = 1, 2, \ldots \tag{8-5a}$$

Similarly,

$$\theta_i = \arctan \frac{\beta_i}{\alpha_i}, \qquad i = 1, 2, \ldots \tag{8-6a}$$

We now turn to the decomposition of a time series by use of a Fourier series.

8.2 the fourier series and the correlogram

Continue to suppose that a time series is representable by the model of Eq. (8-7) or Eq. (8-8). Each component is a periodic function and each can be identified with its angular frequency, ω_i. Obviously, if we have an actual time series it will be of finite length, and we cannot estimate from it the infinite number of parameters of Eq. (8-7) or (8-8). Suppose, therefore, that there are N observations for the series, $Y_1, Y_2, \ldots, Y_N$. Then we can estimate at most N parameters for, say, Eq. (8-7). However, before formulating the model with a finite number of components, we need to mention two points. First, if the sampling interval is some constant unit of time, say a month, then, apart from the mean value, the greatest period (sometimes called slowest) cosine curve that can be observed is one with a period of N months, or angular frequency of $2\pi/N$. Clearly, this fact must be true because such a wave can repeat itself only once in N observations if it has a period of N. The smallest period (sometimes called fastest) cosine curve that can be observed is one of a period of 2 months, since it takes at least 2 months for a

[3] For formal properties, see Taylor (1955), Chap. 22, or Lighthill (1958).

cosine curve to complete a cycle.[4] Thus the fastest curve that can be observed has angular frequency of $2\pi/2 = \pi$ radians per month. Second, suppose that N is an even number, that is, $N = 2n$. Then the angular frequency of the ith wave is

$$\omega_i = \frac{2\pi i}{N}, \qquad i = 0, 1, 2, \ldots, n \tag{8-9}$$

When $i = 1$, $\omega_1 = 2\pi/N$, which is the slowest wave that can be observed. When $i = n$, $\omega_n = 2\pi n/N = \pi$ (since $n = N/2$), which is the fastest curve that can be observed. When $i = 0$, $\omega_0 = 0$, and the zero frequency wave represents the mean value. Thus Eq. (8-8) can be written as $Y_t = \sum_{i=0}^{\infty} \alpha_i \cos \omega_i t + \sum_{i=0}^{\infty} \beta_i \sin \omega_i t$ with $\omega_0 = 0$. Since $\sin \omega_0 t = 0$ and $\cos \omega_0 t = 1$, the mean value $\mu = \alpha_0 \cos \omega_0 t = \alpha_0$ is identified with the amplitude of the cosine wave associated with the zero angular frequency, ω_0.

With these considerations in mind, we may write a finite analogy of Eq. (8-7) as

$$Y_t = \mu + \sum_{i=1}^{n} A_i \cos(\omega_i t - \theta_i) \tag{8-10}$$

where $n = N/2$. Corresponding to Eq. (8-8), we may write the finite analogy as

$$Y_t = \mu + \sum_{i=1}^{n} \alpha_i \cos \omega_i t + \sum_{i=1}^{n} \beta_i \sin \omega_i t \tag{8-11}$$

The parameters in Eq. (8-10) are μ, A_i, and θ_i; in Eq. (8-11) they are μ, α_i, and β_i. It appears that we have one more parameter in each of these equations than we have observations, N. However, we shall show that the actual effective number of parameters in both equations is only N rather than $N + 1$. Since $\omega_n = \pi$, the nth component in Eq. (8-10) is

$$A_n \cos(\omega_n t - \theta_n) = \alpha_n \cos \omega_n t$$

from Eq. (8-2) and the fact that $\sin \omega_n t = \sin \pi t = 0$ for any integer t. Thus we may write Eq. (8-10) as

$$Y_t = \mu + \sum_{i=1}^{n-1} A_i \cos(\omega_i t - \theta_i) + \alpha_n \cos \omega_n t$$

or, equivalently, we may write Eq. (8-11) as

$$Y_t = \mu + \sum_{i=1}^{n} \alpha_i \cos \omega_i t + \sum_{i=1}^{n-1} \beta_i \sin \omega_i t$$

since $\sin \omega_n t = \sin \pi t = 0$.

Now, there are N parameters in both equations: one with μ, $(n - 1)A$s, $(n - 1)\theta$s, and α_n; the other has μ, n αs, and $(n - 1)\beta$s.

---

[4]This smallest frequency, $f = 1/2$, is sometimes called the Nyquist frequency.

As a matter of convenience in presentation, we often explicitly include the zero frequency wave in Eq. (8-10). Suppose that we *define* the phase $\theta_0 = 0$. Then Eq. (8-10) can be written as

$$Y_t = \sum_{i=0}^{n} A_i \cos(\omega_i t - \theta_i) \tag{8-10a}$$

where the mean μ is equal to the amplitude, A_0, of the zero frequency wave: $\mu = A_0$. Corresponding to Eq. (8-11), we have

$$Y_t = \sum_{i=0}^{n} \alpha_i \cos \omega_i t + \sum_{i=0}^{n} \beta_i \sin \omega_i t \tag{8-11a}$$

where, again, the mean value $\mu = \alpha_0$. Remember that $\sin \omega_n t = 0$, so that the very last term of Eq. (8-11a) is equal to zero.

By making use of some orthogonality properties of the sine and cosine functions, the estimates of the parameters of Eq. (8-11a) are[5]

$$\hat{\alpha}_i = \begin{cases} \dfrac{2}{N}\sum_{t=1}^{N} Y_t \cos \omega_i t, & \text{for } i = 1, 2, \ldots, n-1 \\ \dfrac{1}{N}\sum_{t=1}^{N} Y_t \cos \omega_i t, & \text{for } i = 0, n \end{cases}$$

$$\hat{\beta}_i = \begin{cases} \dfrac{2}{N}\sum_{t=1}^{N} Y_t \sin \omega_i t, & \text{for } i = 1, 2, \ldots, n-1 \\ \dfrac{1}{N}\sum_{t=1}^{N} Y_t \sin \omega_i t, & \text{for } i = 0, n \end{cases} \tag{8-12}$$

where the angular frequency is $\omega_i = 2\pi i/N$.

---

[5]See Jenkins and Watts (1968), pp. 17ff. The orthogonality properties of the sine and cosine functions are

$$\sum_{t=1}^{N} \sin \omega_i t \sin \omega_j t = \begin{cases} 0, & i \neq j \\ \dfrac{N}{2}, & i = j \neq 0, n \\ 0, & i = j = 0, n \end{cases}$$

$$\sum_{t=1}^{N} \cos \omega_i t \cos \omega_j t = \begin{cases} 0, & i \neq j \\ \dfrac{N}{2}, & i = j \neq 0, n \\ N, & i = j = 0, n \end{cases}$$

$$\sum_{t=1}^{N} \sin \omega_i t \cos \omega_j t = 0, \qquad i, j = 0, 1, 2, \ldots, n$$

$$\sum_{t=1}^{N} \sin \omega_i t = 0, \qquad i = 0, 1, 2, \ldots, n$$

$$\sum_{t=1}^{N} \cos \omega_i t = 0, \qquad i = 1, 2, \ldots, n$$

Thus the estimates, Eq. (8-12), can be obtained by multiplying both sides of Eq. (8-11a) by $\cos \omega_j t$ (or $\sin \omega_j t$) and summing over N. Then make use of the orthogonality properties. This procedure is nothing more than forming the normal equations of least-squares. Thus Eqs. (8-12) are really the least-squares estimates. Also, see Parzen (1963), pp. 198ff.

The estimates of A_i and θ_i can be obtained directly from the relationships of Eqs. (8-5a) and (8-6a). Thus for Eq. (8-10a) we have

$$\begin{aligned} \hat{A}_i &= \sqrt{\hat{\alpha}_i^2 + \hat{\beta}_i^2} \\ \hat{\theta}_i &= \arctan \frac{\hat{\beta}_i}{\hat{\alpha}_i}, \qquad i = 0, 1, 2, \ldots, n \end{aligned} \tag{8-13}$$

There are two points we should observe carefully in Eqs. (8-12) and (8-13). First, when $i = 0$, $\hat{\alpha}_0 = \hat{\mu}$ and

$$\hat{\alpha}_0 = \frac{1}{N} \sum_{t=1}^{N} Y_t = \bar{Y}, \qquad \hat{\beta}_0 = 0$$

since $\cos \omega_0 t = 1$ and $\sin \omega_0 t = 0$. The amplitude and phase of the zero frequency wave are therefore

$$\hat{A}_0 = \hat{\alpha}_0 \quad \text{and} \quad \hat{\theta}_0 = 0$$

Second, when $i = n$, the estimates are

$$\hat{\alpha}_n = \frac{1}{N} \sum_{t=1}^{N} (-1)^{t-1} Y_t, \qquad \hat{\beta}_n = 0$$

since $\cos \omega_n t = (-1)^{t-1}$ and $\sin \omega_n t = 0$. The amplitude and phase of the nth frequency wave are therefore

$$\hat{A}_n = \hat{\alpha}_n \quad \text{and} \quad \hat{\theta}_n = 0$$

By moving the mean μ to the left-hand side of Eq. (8-11), we obtain a model in deviation form:

$$Y_t - \mu = \sum_{i=1}^{n} \alpha_i \cos \omega_i t + \sum_{i=1}^{n} \beta_i \sin \omega_i t \tag{8-14}$$

The estimation of the α_i and β_i coefficients then can also be carried out by use of deviations about the mean. Thus

$$\begin{aligned} \hat{\alpha}_i &= \begin{cases} \dfrac{2}{N} \displaystyle\sum_{t=1}^{N} (Y_t - \bar{Y}) \cos \omega_i t, & \text{for } i = 1, 2, \ldots, n-1 \\ \dfrac{1}{N} \displaystyle\sum_{t=1}^{N} (Y_t - \bar{Y}) \cos \omega_i t, & \text{for } i = n \end{cases} \\ \hat{\beta}_i &= \begin{cases} \dfrac{2}{N} \displaystyle\sum_{t=1}^{N} (Y_t - \bar{Y}) \sin \omega_i t, & \text{for } i = 1, 2, \ldots, n-1 \\ \dfrac{1}{N} \displaystyle\sum_{t=1}^{N} (Y_t - \bar{Y}) \sin \omega_i t, & \text{for } i = n \end{cases} \end{aligned} \tag{8-15}$$

We pause to illustrate the calculation. Suppose that a time series, Y_t, consists of a mean value $\mu = 5$ and four periodic components C_{1t}, C_{2t}, C_{3t}, and C_{4t}. That is,

$$Y_t = \mu + \sum_{i=1}^{4} C_{it}$$

where each periodic component is a harmonic function of the type given by Eq. (8-1):

$$\begin{aligned} C_{it} &= A_i \cos(\omega_i t - \theta_i) \\ &= \alpha_i \cos \omega_i t + \beta_i \sin \omega_i t \end{aligned}$$

Table 8.1 shows the parametric values of these harmonic functions. Note also that the mean is presented as a harmonic cosine function with zero frequency.

Table 8.1 Parametric Values of Harmonic Functions

Periodic Component	*Amplitude*			*Period or Frequency*			*Phase*	
	A_i	α_i	β_i	P_i	f_i	ω_i	ϕ_i	θ_i
$C_0 = \mu$	5	5.000	0.000	∞	0	0.000	0	0.000
C_1	10	5.002	8.659	20	$\frac{1}{20}$	0.314	$\frac{10}{3}$	1.047
C_2	9	6.366	6.361	10	$\frac{1}{10}$	0.628	$\frac{5}{4}$	0.785
C_3	8	8.000	0.000	4	$\frac{1}{4}$	1.571	0	0.000
C_4	5	5.000	0.000	2	$\frac{1}{2}$	3.142	0	0.000

For example, the first periodic component $C_{1t} = 10 \cos(0.314t - 1.047) = (5.002) \cos(0.314t) + (8.659) \sin(0.314t)$. The angular frequency of 0.314 is equal to the frequency of $f_1 = 1/20$ cycles per unit of time (i.e., $f_i = \omega_i/2\pi$ with $\pi \doteq 3.14159$); the wave repeats itself with a period of 20 units of time (say, 20 months). The reader may verify the consistency of Table 8.1, e.g., $A_i^2 = \alpha_i^2 + \beta_i^2$, for all i.

Table 8.2 shows values of the four periodic time series, Y_t, with $N = 20$ observations $t = 1, \ldots, 20$ generated from the parametric values. The series are plotted in Figure 8.3 and their sum plus a mean of 5, Y_t, is also plotted in the figure. Notice that the series Y_t could bear some resemblance to the type of series that might be encountered in economic research. An interesting point to note from Table 8.2 is that the two series C_{1t} and C_{2t} never quite reach their amplitudes of 10.0 and 9.0, respectively. This fact occurs because we are evaluating these series only at discrete time intervals rather than continuously. If we evaluated the time series continuously, the problem

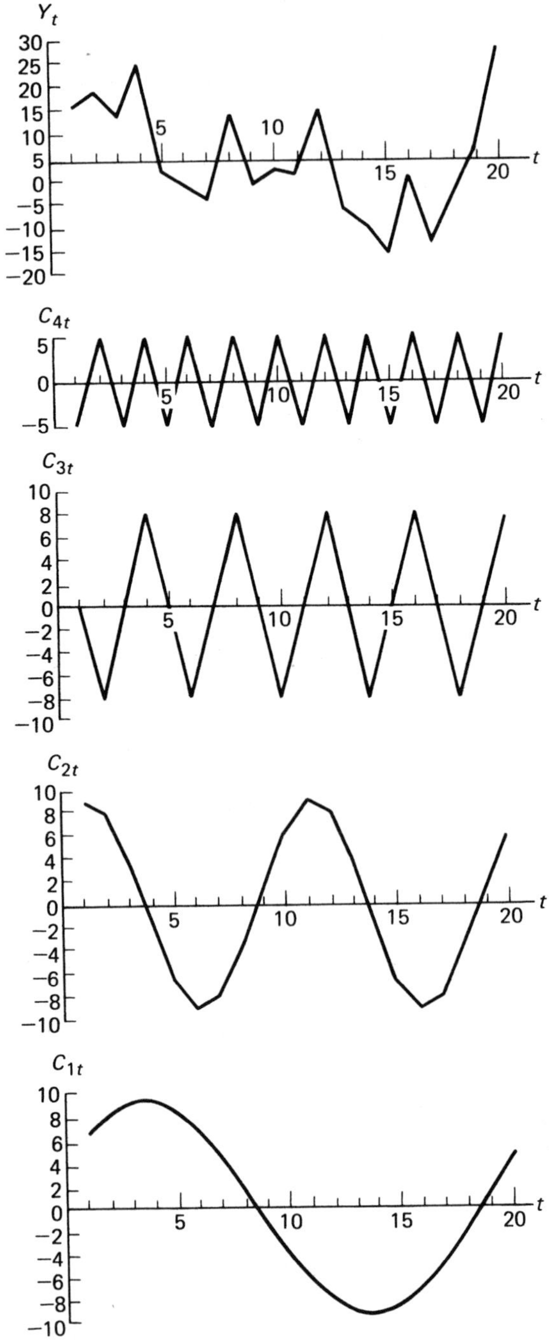

Figure 8.3. Four periodic series and Y_t ($Y_t = 5.0 + C_{1t} + C_{2t} + C_{3t} + C_{4t}$).

Table 8.2 Four Periodic Series and Their Sum ($Y_t = 5.0 + C_{1t} + C_{2t} + C_{3t} + C_{4t}$)

t	C_{1t}	C_{2t}	C_{3t}	C_{4t}	Y_t
1	7.43	8.89	0.00	−5.00	16.32
2	9.14	8.02	−8.00	5.00	19.15
3	9.95	4.09	0.00	−5.00	14.03
4	9.78	−1.41	8.00	5.00	26.37
5	8.66	−6.36	0.00	−5.00	2.30
6	6.69	−2.89	−8.00	5.00	−0.20
7	4.07	−8.02	0.00	−5.00	−3.95
8	1.05	−4.09	8.00	5.00	14.96
9	−2.08	1.41	0.00	−5.00	−0.67
10	−5.00	6.36	−8.00	5.00	3.36
11	−7.43	8.89	0.00	−5.00	1.46
12	−9.14	8.02	8.00	5.00	16.88
13	−9.95	4.09	0.00	−5.00	−5.86
14	−9.78	−1.41	−8.00	5.00	−9.19
15	−8.66	−6.36	0.00	−5.00	−15.02
16	−6.69	−8.89	8.00	5.00	2.42
17	−4.07	−8.02	0.00	−5.00	−12.09
18	−1.05	−4.09	−8.00	5.00	−3.13
19	2.08	1.41	0.00	−5.00	3.49
20	5.00	6.36	8.00	5.00	29.36

would vanish. Series C_{1t} reaches its first peak at $t = 3.33\ldots$, and series C_{2t} reaches its first peak at $t = 1.25$. That is, since $C_{1t} = 10 \cos(0.314t - 0.785)$, the first series reaches its first peak when $0.314t - 1.047 = 0$, that is, $t = 3.33\ldots$, with amplitude of 10, while the second series reaches its first peak when $0.628t - 0.785 = 0$, that is, $t = 1.25$, with amplitude of 9.

Given the 20 observations for Y_t in Table 8.2, let us now try to recover the four periodic components from this time series. With 20 observations, the model of Eq. (8-11a) becomes

$$Y_t = \sum_{i=0}^{10} \alpha_i \cos \omega_i t + \sum_{i=0}^{10} \beta_i \sin \omega_i t$$

with $\omega_i = 2\pi i/20$. The estimates of the parameters can be obtained from Eq. (8-12). For example,

$$\hat{\alpha}_1 = \frac{2}{20}\left\{16.32 \cos\left(\frac{2\pi}{20}\right) + 19.15 \cos\left(\frac{4\pi}{20}\right) + \cdots + 29.36 \cos\left(\frac{40\pi}{20}\right)\right\} = 5.0$$

Similarly, $\hat{\beta}_1$ is

$$\hat{\beta}_1 = \frac{2}{20}\left\{16.32 \sin\left(\frac{2\pi}{20}\right) + 19.15 \sin\left(\frac{4\pi}{20}\right) + \cdots + 29.36 \sin\left(\frac{40\pi}{20}\right)\right\} = 8.6602$$

Table 8.3 gives the other values for $\hat{\alpha}$ and $\hat{\beta}$. Also shown in Table 8.3 are $\hat{A} = \sqrt{\hat{\alpha}^2 + \hat{\beta}^2}$ and $\hat{\theta} = \arctan(\hat{\beta}/\hat{\alpha})$.

Table 8.3 Fourier Coefficients and Phase Measures

i	$\omega_i = 2\pi i/N$	$f_i = \omega_i/2\pi$	$\hat{\alpha}_i$	$\hat{\beta}_i$	$\hat{A}_i$	$\hat{\theta}_i$
0	$0.000 = 0$	0.00	5.0	0.0	5.0	0.0
1	$0.314 = 2\pi/20$	0.05	5.0	8.6602	9.95	1.047
2	$0.628 = 4\pi/20$	0.10	6.3639	6.3639	8.89	0.785
3	$0.942 = 6\pi/20$	0.15	0.0	0.0	0.0	—
4	$1.257 = 8\pi/20$	0.20	0.0	0.0	0.0	—
5	$1.571 = 10\pi/20$	0.25	8.0	0.0	8.0	0.0
6	$1.885 = 12\pi/20$	0.30	0.0	0.0	0.0	—
7	$2.199 = 14\pi/20$	0.35	0.0	0.0	0.0	—
8	$2.513 = 16\pi/20$	0.40	0.0	0.0	0.0	—
9	$2.827 = 18\pi/20$	0.45	0.0	0.0	0.0	—
10	$3.142 = 20\pi/20$	0.50	5.0	0.0	5.0	0.0

A comparison of Tables 8.1 and 8.3 reveals that, except for rounding errors and errors induced by finite sampling, the parameters of the periodic functions have been recovered.

A remarkable property of a Fourier series is that the Fourier coefficients $\hat{A}_i$ bear a strict relationship to the sample variance of the composite series, Y_t. By utilizing the orthogonality properties of the sine and cosine functions (see footnote 5), it can be shown that the mean of the sum of squares of the N observations can be written as[6]

$$\sum_{t=1}^{N} \frac{Y_t^2}{N} = \hat{\alpha}_0^2 + \frac{1}{2} \sum_{i=1}^{n-1} (\hat{\alpha}_i^2 + \hat{\beta}_i^2) + \hat{\alpha}_n^2$$

$$= \hat{A}_0^2 + \frac{1}{2} \sum_{i=1}^{n-1} \hat{A}_i^2 + \hat{A}_n^2 \tag{8-16}$$

since $\hat{\alpha}_0 = \hat{A}_0$ and $\hat{\alpha}_n = \hat{A}_n$. The quantity on the left-hand side of Eq. (8-16) is called *average power* or *mean square*. Alternatively, we can write Eq. (8-16) as

$$SD^2 = \sum_{t=1}^{N} \frac{(Y_t - \bar{Y})^2}{N} = \frac{1}{2} \sum_{i=1}^{n-1} \hat{A}_i^2 + \hat{A}_n^2 \tag{8-17}$$

since $\hat{A}_0 = \bar{Y}$. Notice that the only difference between Eqs. (8-16) and (8-17) concerns the mean. If the mean of the series is zero, then the variance SD^2 is equal to the average power. Thus in Eq. (8-17) we are working with

[6]This is a special case of Parseval's theorem.

deviations about the mean (the mean of these deviations is zero) and there is no contribution to average power at the zero frequency.

We have shown that the total variance of the time series may be decomposed into a linear combination of the squares of the amplitudes of each periodic component. Fourier analysis therefore lets us do a kind of analysis of variance of the sample time series. It makes good sense that there should be some relationship between the amplitudes of the periodic series of which Y_t is composed and the variance of Y_t, since the greater the amplitude of a series which makes up Y_t, the greater will be the amplitude of fluctuation of Y_t. In the physical sciences the words *power* and *variance* are used interchangeably. Thus the quantities $\frac{1}{2}\hat{A}_i^2$, $\hat{A}_0^2$, and $\hat{A}_n^2$ are sometimes called the estimated average *power* at angular frequency ω_i. The plot of these coefficients against frequency is a version of what is called the *Fourier line spectrum*.[7]

Figure 8.4 shows the Fourier line spectrum for the artificial series Y_t and gives us a visual analysis of variance of the series. It tells us at a glance, for instance, that apart from the mean all the average power of the Y_t series can be explained by four harmonic curves located at frequencies 0.05, 0.10, 0.25 and 0.50.

The Correlogram. The *autocovariance function* for a stationary time series with mean μ and variance σ^2 is usually defined for the population as

$$\gamma(K) = E[(Y_t - \mu)(Y_{t+K} - \mu)] \tag{8-18}$$

where K takes on values 0, 1, 2, ... Note that $\gamma(K)$ is defined to be a func-

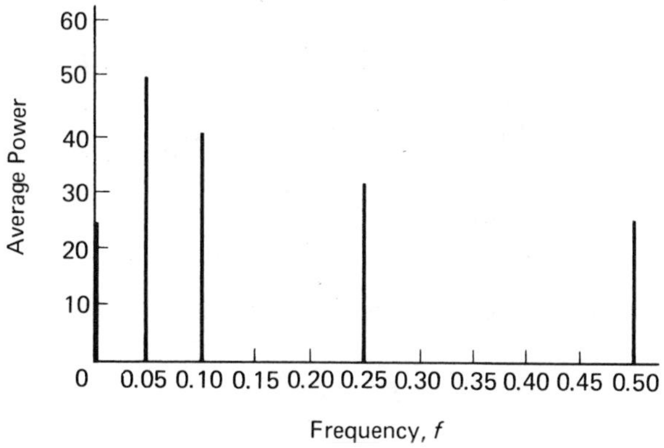

Figure 8.4. Fourier line spectrum.

[7]The word *line* is used because the harmonic contributions are considered to be at discrete frequencies. We shall drop this assumption later in the chapter. The representation is also sometimes called the *periodogram*.

tion of the *lag* K alone and not of the point in time, t, upon which Y_t is dependent.

Since the variance, σ^2, of Y_t is equal to $\gamma(0)$, the *autocorrelation function* is defined as

$$\rho(K) = \frac{\gamma(K)}{\gamma(0)}, \qquad K = 0, 1, 2, \ldots \tag{8-19}$$

Notice that this definition corresponds to our usual definition of the simple correlation between Y_t and Y_{t+K} if the series is stationary. Thus our usual definition of simple correlation is

$$\rho(K) = \frac{\text{cov}(Y_t, Y_{t+K})}{\sqrt{\text{var}(Y_t)\,\text{var}(Y_{t+K})}}$$

However, if the series is stationary, $\text{var}(Y_t) = \text{var}(Y_{t+K}) = \gamma(0)$. Since $\text{cov}(Y_t, Y_{t+K}) = \gamma(K)$, the usual definition corresponds to Eq. (8-19).

For the sample, the autocovariance function is usually defined as[8]

$$C(K) = \frac{1}{N}\sum_{t=1}^{N-K}(Y_t - \bar{Y})(Y_{t+K} - \bar{Y}) \tag{8-20}$$

Notice that when $K = 0$, $C(0) = SD^2$, so that $C(0)$ is a biased estimate of σ^2. The divisor N is used rather than $N - K - 1$ because in many cases the former divisor gives us an estimate with a smaller mean square error and is convenient in relation to Parseval's theorem.

We shall illustrate the calculation with a short artificial series: $Y_1 = 4$, $Y_2 = 2$, $Y_3 = 3$. Then $\bar{Y} = 3$, and

$$C(0) = \frac{(1)(1) + (-1)(-1) + (0)(0)}{3} = 0.667$$

$$C(1) = \frac{(1)(-1) + (-1)(0)}{3} = -0.333$$

$$C(2) = \frac{(1)(0)}{3} = 0.0$$

The sample autocorrelation function is defined as

$$r(K) = \frac{C(K)}{C(0)} \tag{8-21}$$

For the artificial series $r(0) = 1.0$, $r(1) = -0.5$, and $r(2) = 0.0$.

---

[8] Parzen (1963) makes a good case for not subtracting the mean $\bar{Y}$ from each value of Y_t. However, we shall not change our definitions, which do subtract the mean because of their wide usage.

The sample autocorrelation function is sometimes useful in interpreting time series. If the sample time series is strictly periodic, the sample autocorrelation function will also be periodic and of the same period as the series from which it was calculated. For the series C_{3t} above (see Table 8.2), the reader should convince himself that the sample autocorrelation function will take on the values $+1, 0, -1, 0, +1, \ldots$, so that it will have a period of 4, just as C_{3t} does. The autocorrelation function is also useful because it gives us a picture of the way in which the temporal dependency in economic time series usually "fades away" with increasing lags. However, the autocorrelation function is generally difficult to interpret because the neighboring values of the autocorrelation coefficients are not independent. It is likely that a positive autocorrelation coefficient will be followed by another positive autocorrelation coefficient for most economic time series. It is this problem that leads us to prefer to study strictly periodic time series on the frequency domain since the Fourier coefficients are orthogonal (independent). The sample autocorrelation function for the artificial series Y_t of Table 8.2 is shown in Figure 8.5; this kind of chart is known as a *correlogram*. The reader should spend a few moments attempting to interpret this figure and should convince himself that the Fourier line spectrum in Figure 8.4 is much easier to understand.[9]

Another problem with the sample autocorrelation function concerns the number of lags, K, to use in the calculation. There are $N - 1$ possible autocorrelation coefficients that can be calculated [excluding $r(0)$, which is always 1] from a series of length N. As K approaches $N - 1$, the number of observa-

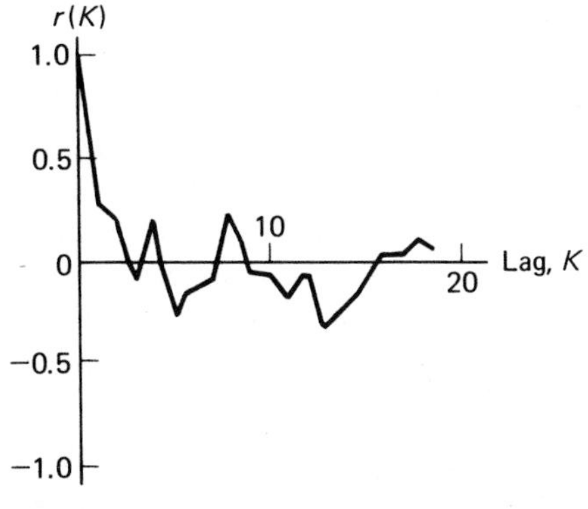

Figure 8.5. Correlogram for Y_t of Table 8.1.

---

[9]Since the autocorrelation coefficients are calculated only for discrete lags, it is theoretically preferable to plot the coefficients using vertical lines, as was done for the line spectrum. However, in practice, the diagram is nearly always drawn as it is in Figure 8.5, where the points are connected with straight lines.

tions upon which each autocorrelation coefficient is based declines, and each coefficient becomes less comparable to coefficients based upon fewer lags. Furthermore, each coefficient is less reliable than the one before it. A rule of thumb is that the maximum value for K should not exceed one fifth of the value of N. Thus, contrary to the calculation shown in Figure 8.5, we should not calculate all possible autocorrelation coefficients and expect to make much sense out of the last 80 percent of them. We shall have more to say about this problem in the next section.

If autocorrelation coefficients are calculated from a random series (sometimes called *white noise*),[10] the sample autocorrelation coefficients will tend to be zero [except, of course, $r(0) = 1.0$]. Under the hypothesis that the population autocorrelation coefficients are zero, it can be shown that the distribution of $r(K)$ tends toward the normal form with mean zero and variance $1/N$ for large samples.[11] Thus a $100(1-\alpha)$ percent confidence interval for a particular $\rho(K)$ is given by

$$r(K) \pm \frac{z_{\alpha/2}}{\sqrt{N}}$$

where $z_{\alpha/2}$ is the normal deviate. Figure 8.6 shows 10 autocorrelation coefficients calculated from a series of 100 observations which were obtained by use of a computer random number generator. For a 90 percent confidence interval $z_{0.05}/\sqrt{N} = 1.645/\sqrt{100} = 0.1645$. When this interval is set about any of the coefficients shown in Figure 8.6, it covers zero. Thus we cannot reject the hypothesis that any of the coefficients differ from zero. Sometimes the confidence limits are connected by straight lines as in Figure 8.6 to produce a *confidence band.* Care must be taken to realize that these bands are made up of univariate intervals and are, therefore, not simultaneous. We shall illustrate an alternative test for white noise in a later section.

A remarkable fact about the autocorrelation function is that it is directly related to the Fourier line spectrum. In fact, as we shall now show, the time domain representation of a time series by use of the autocorrelation function is equivalent to the frequency domain representation by use of the Fourier line spectrum. The difference is only that once we become accustomed to thinking in terms of frequency, the Fourier representation is easier to interpret for a time series that is made up of strictly periodic series.

---

[10]The reader should attempt to overcome being discouraged by our repeated introduction of exotic terms. Much of the battle of trying to understand spectral analysis is won when this jargon of physics qua statistics qua economics is mastered. White light contains all the colors of the spectrum. A random series may be thought of as containing cosine curves of every frequency, each with the same amplitude as every other. Thus no frequency stands out as dominant.

[11]See Jenkins and Watts (1968), p. 187.

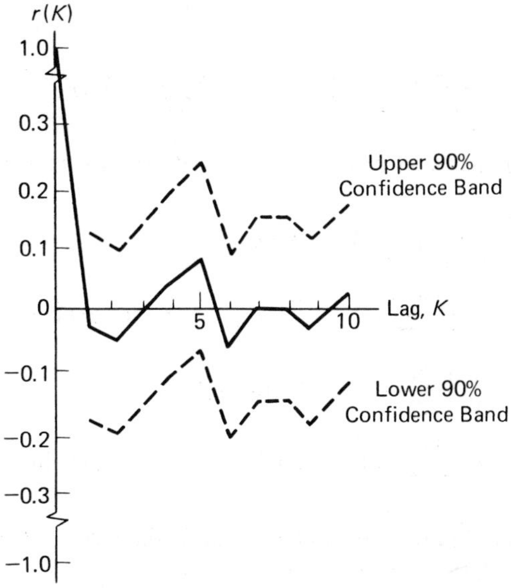

Figure 8.6. Autocorrelation coefficients calculated from white noise.

Consider the following expression for average power of the ith frequency wave:

$$\frac{1}{2}\hat{A}_i^2 = \frac{1}{2}(\hat{\alpha}_i^2 + \hat{\beta}_i^2)$$

$$= \frac{1}{2}\left[\frac{2}{N}\sum_{t=1}^{N}(Y_t - \bar{Y})\cos\omega_i t\right]^2 + \frac{1}{2}\left[\frac{2}{N}\sum_{t=1}^{N}(Y_t - \bar{Y})\sin\omega_i t\right]^2,$$

$$i = 1, 2, \ldots, n-1 \qquad (8\text{-}22)$$

This expression follows from Eqs. (8-5a) and (8-15). The average power of nth frequency wave can be written as

$$\hat{A}_n^2 = (\hat{\alpha}_n^2 + \hat{\beta}_n^2)$$

$$= \left[\frac{1}{N}\sum_{t=1}^{N}(Y_t - \bar{Y})\cos\omega_n t\right]^2 + \left[\frac{1}{N}\sum_{t=1}^{N}(Y_t - \bar{Y})\sin\omega_n t\right]^2 \qquad (8\text{-}23)$$

since $\hat{\beta}_n = 0$ and $\sin\omega_n t = 0$.

Let us rearrange Eq. (8-22) by making use of a well-known mathematical expansion. Notice that for, say,

$$\left(\sum_{t=1}^{3} X_t\right)^2 = X_1^2 + X_2^2 + X_3^2 + 2X_1X_2 + 2X_1X_3 + 2X_2X_3$$

we can write

$$(\sum_{t=1}^{3} X_t)^2 = \sum_{t=1}^{3} X_t^2 + 2\sum_{K=1}^{2}\sum_{t=1}^{3-K} X_t X_{t+K}.$$

Keeping this expansion in mind, we can write the expression as

$$\frac{1}{2}\hat{A}_i^2 = \frac{2}{N^2}\Big[\sum_{t=1}^{N}(Y_t - \bar{Y})^2 \cos^2 \omega_i t$$

$$+ 2\sum_{K=1}^{N-1}\sum_{t=1}^{N-K}(Y_t - \bar{Y})(Y_{t+K} - \bar{Y}) \cos \omega_i t \cos \omega_i(t+K)\Big]$$

$$+ \frac{2}{N^2}\Big[\sum_{t=1}^{N}(Y_t - \bar{Y})^2 \sin^2 \omega_i t$$

$$+ 2\sum_{K=1}^{N-1}\sum_{t=1}^{N-K}(Y_t - \bar{Y})(Y_{t+K} - \bar{Y}) \sin \omega_i t \sin \omega_i(t+K)\Big]$$

Recalling that $\cos^2 \omega_i t + \sin^2 \omega_i t = 1$ and that $\cos \omega_i t \cos \omega_i(t+K) + \sin \omega_i t \sin \omega_i(t+K) = \cos \omega_i K$, we can write

$$\frac{1}{2}\hat{A}_i^2 = \frac{2}{N^2}\Big[\sum_{t=1}^{N}(Y_t - \bar{Y})^2 + 2\sum_{K=1}^{N-1}\sum_{t=1}^{N-K}(Y_t - \bar{Y})(Y_{t+K} - \bar{Y}) \cos \omega_i K\Big]$$

The first term in the brackets of this expression, if divided by N, is equal to the sample variance $C(0)$ of Eq. (8-20). Furthermore, we may use Eq. (8-20) to write the second term in brackets as $2N\sum_{K=1}^{N-1} C(K) \cos \omega_i K$. Thus the final form of the expression becomes

$$\frac{1}{2}\hat{A}_i^2 = \frac{2}{N}\Big[C(0) + 2\sum_{K=1}^{N-1} C(K) \cos \omega_i K\Big], \qquad i = 1, 2, \ldots, n-1 \qquad (8\text{-}24)$$

Similarly, the average power of the nth wave, Eq. (8-23), can be expressed as

$$\hat{A}_n^2 = \frac{1}{N}\Big[C(0) + 2\sum_{K=1}^{N-1} C(K) \cos \omega_n K\Big] \qquad (8\text{-}25)$$

These final forms of the expansion tell us that the Fourier coefficients, and hence average power $\hat{A}_i^2/2$ and $\hat{A}_n$, can be calculated from the autocovariance coefficients $C(K)$. Technically, the Fourier coefficients are the result of a *Fourier transformation* of the autocovariance function. Thus a knowledge of the autocovariance function implies a knowledge of the Fourier coefficients and vice versa.

We also noted earlier in Eq.(8-17) that the sum of the $\hat{A}_i^2/2$ and $\hat{A}_n^2$ coefficients is equal to the sample variance of the series. Thus the *relative average power* at each frequency is given by $\hat{A}_i^2/2C(0)$ and $\hat{A}_n^2/C(0)$. Therefore, if we divide both sides of Eqs. (8-24) and (8-25) by $C(0)$ and keep in mind the definition of the sample autocorrelation function of Eq. (8-21), we have

$$\frac{1}{2}\frac{\hat{A}_i^2}{C(0)} = \frac{2}{N}\left[1 + 2\sum_{K=1}^{N-1} r(K)\cos\omega_i K\right], \qquad i = 1, 2, \ldots, n-1 \qquad \text{(8-26)}$$

$$\frac{\hat{A}_n^2}{C(0)} = \frac{1}{N}\left[1 + 2\sum_{K=1}^{N-1} r(K)\cos\omega_n K\right]$$

Thus the estimated relative average power at any frequency is the Fourier transformation of the sample autocorrelation function. A knowledge of one is equivalent to a knowledge of the other. It is in this sense that there is a direct relationship between the autocorrelation function and the Fourier line spectrum. The advantage of using relative average power over average power is that the former is independent of measurement units since it is implicitly calculated from correlation coefficients. We now turn to the problem of estimating the Fourier line spectrum from the autocorrelation function or the autocovariance function, whichever is convenient.

8.3 spectral estimation

We have assumed up to now that the power is concentrated at distinct angular frequencies ω_i. Also, as noted before, with a finite number of observations we can estimate the power only at a finite number of angular frequencies. Suppose, however, that there are periodic components of a time series whose angular frequencies do not correspond to any of our estimated angular frequencies. Suppose that there is a periodic curve whose angular frequency lies between, say, ω_1 and ω_2. One way to estimate the power at these frequencies is to convert the line spectrum into a histogram. That is, one estimation technique is to distribute the power at the frequencies over a *band* about the frequencies.

For the spectrum expressed as a function of angular frequency, the distance between any two estimates is $2\pi/N$ (see Table 8.3). Thus if we divide each of the power ordinates by $2\pi/N$, we shall convert each of the ordinates into a bar of a histogram such that the area under the bar is the same as the height of the original ordinate.[12] Each bar in the histogram is now an estimate of the power that is found over the band between $\omega_i + \pi/N$ and $\omega_i - \pi/N$. Dividing both sides of Eq. (8-26) by $2\pi/N$, we have a new estimate of relative

[12]Recall that in plotting a frequency distribution of a continuous variable we construct bars whose width is equal to the class interval. The height of each bar is given by the frequency of observations falling in that class. Such a diagram is called a *histogram* or a *bar chart*. Sometimes the presentation is accomplished by connecting the top centers of each bar with a straight line. The resulting curve is called a *frequency polygon*.

power as a *function of angular frequency*:

$$p(\omega_i) = \frac{1}{\pi}\left[1 + 2\sum_{K=1}^{N-1} r(K)\cos(\omega_i K)\right], \qquad i = 1, 2, \ldots, n \qquad (8\text{-}27)$$

Note that we use the same formula to measure $p(\omega_n)$ for the nth frequency. Actually this value should be divided by 2, but this *end-point* correction is often neglected in practice.

In a similar way the distance between any two estimates expressed in terms of frequency is $1/N$ (see Table 8.3). Thus if we divide the power of Eq. (8-26) at each frequency by $1/N$ and recall that $\omega_i = 2\pi f_i$, we have an estimate of relative power as a *function of frequency*:

$$p(f_i) = 2\left[1 + 2\sum_{K=1}^{N-1} r(K)\cos(2\pi f_i K)\right], \qquad i = 1, 2, \ldots, n \qquad (8\text{-}28)$$

The spectrum is now assumed to be continuous so that it may be represented by a frequency polygon (see also Figure 8.7). Such a curve is the usual graphic representation of a spectrum. Finally, if the original series (rather than $Y_t - \bar{Y}$) is used to calculate the spectrum, then there will be an estimate at the zero frequency.

Equations (8-27) and (8-28) serve the same purpose. We have gone to the trouble to illustrate them both because the reader will find both expressions in the literature of spectral analysis, and it is initially difficult to understand at first why some authors have a $1/\pi$ in their formulas and others have a 2. In theoretical work the spectrum is generally expressed as a function of angular frequency. In practical work (especially in economics) the spectrum is expressed in terms of frequency, f. Also notice that we could have used Eqs. (8-24) and (8-25) rather than Eq. (8-26) to arrive at the histogram representation if we had been interested in power rather than in relative power. Thus using Eq. (8-24), we can write

$$P(\omega_i) = \frac{1}{\pi}\left[C(0) + 2\sum_{K=1}^{N-1} C(K)\cos(\omega_i K)\right], \qquad i = 1, 2, \ldots, n \qquad (8\text{-}29)$$

or

$$P(f_i) = 2\left[C(0) + 2\sum_{K=1}^{N-1} C(K)\cos(2\pi f_i K)\right], \qquad i = 1, 2, \ldots, n \qquad (8\text{-}30)$$

Let us take Eq. (8-28) as the fundamental representation of what is called the sample *power density spectrum*.[13] Modern spectral analysis takes the view that at any point in time, t, the sample time series value Y_t is but a sample of size 1 from an *ensemble* of Y_t values that exist at that point in time. Thus

---

[13] Equation (8-30) may be taken as the fundamental representation of the sample *power spectrum*.

there is a distribution of Y_t values for any t. If we replace X with t in Figure 4.1, we can get a visual impression of this view. Now the time series that we sample may be made up of any combination of the points in each of the distributions associated with t. However, if the series is stationary (in that the mean and variance of each of these distributions is independent of the mean and variance of any other distribution), it can be shown that as the sample size increases (at any point in time), the mean and variance for any particular sample time series (*realization*) drawn at random from this ensemble will approach the mean and variance of the ensemble at any point in time. Thus under the assumption of stationarity we can say something about the entire ensemble by use of a single *realized* series.

Under these conditions it is possible to define a theoretical population spectral density function, and Bartlett (1948) has shown that Eq. (8-28) is not a consistent estimator of the population spectral density function. In fact, the variance of Eq. (8-28) does not decrease as the sample size increases. Thus Eq. (8-28) tends to show peaks indicating high relative power when in the population high relative power does not exist. Thus modern spectral analysis makes several modifications of Eq. (8-28) which are designed to improve upon the estimate of the population function. We now introduce these changes, which will lead us to our final computing formula.

First, *not all* $N - 1$ autocorrelation coefficients are used in the estimation. We have already noted that not all autocorrelation coefficients were calculated for the correlogram. Theoretically, there is an indeterminancy problem in the choice of how many autocorrelation coefficients to use. Suppose that we use a maximum of $L \leq N - 1$ autocorrelation coefficients, where L is called the *truncation point*. As L increases, the variance of the sample spectral density can be shown to increase. However, as L is reduced, the bias of the sample spectral density as an estimator of the true spectral density can be shown to increase. This conflict of being able to achieve less bias only at the expense of greater variance is central to spectral estimation, and about all that one can do in practice is to choose several different values for L and compute the spectral estimates for each of these values. In practice, one starts with a value of L which is small (perhaps 10 percent of N) and then increases L. This procedure is sometimes called *window closing*. The spectral density with a small L will give estimates with a high order of bias (*low fidelity*). As the size of L is increased, one is often able to obtain an insight into the details of the spectrum while (as the variance increases) running the risk of finding peaks where no peaks should be. Rules of thumb vary, but most of them state that L should be from 10 to 40 percent of N with 20 percent being a popular number. Estimation therefore requires considerable experience, and critics of spectral analysis are quick to point out that one is hard-pressed to know the meaning of a particular estimated spectral density function because it can be changed in appearance by changing L. In truth, a similar problem crops up in many places in statistics. To give but one example, the choice of the number of class intervals in forming any frequency distribution involves a trade-off between information loss and generality. If one class interval is chosen, we have a frequency distribution which says that "all the data lie between *a* and *b*." If as many classes are chosen as there are data points, the frequency distribution is the same as the raw data and there is no point in having a frequency distribution at all. With a frequency distribution we try to present a picture that gives us some meaningful insight into the properties of the data through a judicious choice of class intervals. This is precisely what we attempt to do in spectral analysis through a judicious choice of L.

The analogy between forming a "good" frequency distribution and forming a "good" set of spectral estimates is made even clearer if we remember that in both cases we generally use a priori information as an aid in the analysis. For a frequency distribution we often postulate that it is, say, unimodal and approaches zero as we move toward the right tail. Such a set of beliefs aid us in determining the proper number of class intervals. In spectral analysis we sometimes postulate, say, a 20-year building cycle or a 4-year durable goods cycle. We then increase L until such cycles appear, or fail to appear.

Second, modern spectral analysis smooths the spectral estimates by use of a *lag window*. These lag windows are used in an attempt to reduce the variance of the sample spectral density function. Many windows have been proposed,[14] but the authors have obtained good results with a window proposed by Parzen:

$$g(K) = \begin{cases} 1 - \dfrac{6K^2}{L^2}\left(1 - \dfrac{K}{L}\right), & 0 \le K \le \dfrac{L}{2} \\ 2\left(1 - \dfrac{K}{L}\right)^3 & \dfrac{L}{2} \le K \le L \end{cases} \tag{8-31}$$

We shall illustrate shortly how this window is used.

Let us now incorporate these two modifications into Eq. (8-28):

$$p(f_i) = 2\left[1 + 2\sum_{K=1}^{L-1} g(K)r(K)\cos(2\pi f_i K)\right], \qquad i = 0, 1, \ldots, L \tag{8-32}$$

In comparing Eqs. (8-32) and (8-28), notice that the upper limit of summation is changed from $N - 1$ to $L - 1$ and that the lag window $g(K)$ is also inserted. Finally, notice that the index i starts from zero rather than 1.

---

[14]John Tukey has called the problem of the design of these windows *window carpentry*. Many of the windows are given in Parzen (1963). A comparison of several windows is given in Blackman and Tukey (1959). In particular, the Parzen window will never lead to negative estimates of the spectral density function.

This inclusion of an estimate at the zero frequency enables us to check whether or not we have successfully removed the trend prior to spectral estimation. A linear trend, like the mean value, may be considered to be a harmonic of zero frequency. Thus a spectral peak at the zero frequency may mean that there is a trend left in the sample series.[15]

A final modification of Eq. (8-32) is currently made. In that equation $f_i = i/2L$. Hence, $f_i = 0, 1/2L, 2/2L, \ldots, 1/2$. If $L = 2$, say, then $f_i = 0$, 1/4, 1/2, which gives as many frequency estimates (excluding $f = 0$) as there are lags. This frequency spacing has been found to be too wide for good results, and it is generally recommended that the spectrum be calculated at from two to four times this many frequencies. Since $0 \leq f_i \leq 1/2$, this narrower spacing can be accomplished by dividing f_i by some constant, c. If the constant is chosen to be 2, then $f_i^* = f_i/2$ and $f_i^* = 0, 1/4L, 2/4L, \ldots$, 1/2. Thus if $L = 2$, we have $f_i^* = 0$, 1/8, 2/8, 3/8, 1/2, which gives twice as many estimates as before. While the constant need not be 2, it will be chosen to be 2 in this book to avoid presenting the student with another variable to worry about other than L. The final computing formula, then, is

$$p(f_i^*) = 2\left[1 + 2\sum_{K=1}^{L-1} g(K)r(K)\cos\left(\frac{\pi i K}{2L}\right)\right], \qquad i = 0, 1, \ldots, 2L \qquad (8\text{-}33)$$

Let us illustrate the calculations with an artificial series. Table 8.4 shows the series, and it appears to have an upward trend. For simplicity we have taken the first differences of the series in hopes of removing the trend. In

Table 8.4 Artificial Series and First Differences

t	Y_t	$\Delta Y_t = Y_t - Y_{t-1}$
1	92.94	9.70
2	102.64	7.46
3	110.10	−10.94
4	99.16	−1.56
5	97.60	10.20
6	107.80	5.74
7	113.54	−13.52
8	100.02	1.44
9	101.46	13.32
10	114.78	0.22
11	115.00	−12.60
12	102.40	

[15]It may also mean that there is "leakage" from low-frequency cycles. Leakage is discussed in the next section. Of course, if we follow Parzen's advice (see footnote 8), the zero frequency estimate may only represent the mean value.

so doing, one observation will be lost so that $N = 11$. As an exercise the reader may wish to repeat the calculation using the residuals about a least-squares trend line. Let us choose a maximum lag of $L = 4$.

The autocorrelation coefficients are calculated by Eq. (8-21) to be

K	$r(K)$
0	1.0000
1	−0.0419
2	−0.7822
3	0.0923
$L = 4$	0.6399

The Parzen window by Eq. (8-31) is

K	*Calculation*	$g(K)$
0	$1 - \frac{6(0)^2}{(4)^2}\left(1 - \frac{0}{4}\right)$	1.00000
1	$1 - \frac{6(1)^2}{(4)^2}\left(1 - \frac{1}{4}\right)$	0.71875
$\frac{L}{2} = 2$	$1 - \frac{6(2)^2}{(4)^2}\left(1 - \frac{2}{4}\right)$	0.25000
3	$2\left(1 - \frac{3}{4}\right)^3$	0.03125
$L = 4$	$2(1 - 1)^3$	0.00000

Note that when $K = L/2$, the window calculation could have been done using $g(K) = 2(1 - K/L)^3 = 2(1 - 0.5)^3 = 0.250$.

We now calculate $\cos(\pi iK/2L)$ for $K = 1, 2, 3$ and $i = 0, 1, 2, \ldots, 8$. We shall illustrate the calculation for $i = 0$ and 2 and shall leave the other points for the reader to calculate. For $i = 0$ and 2, the cosines are

K \ i	0	2
1	1.0000	0.0000
2	1.0000	−1.0000
3	1.0000	0.0000

We can now find $\sum_{K=1}^{L-1} g(K)r(K)\cos(\pi iK/2L)$ for $i = 0, 1, \ldots, 4$. Let us first multiply $r(K)$ by $g(K)$ for $K = 1$ to 3:

K	$r(K)$	$g(K)$	$g(K)r(K)$
1	−0.0419	0.71875	−0.03012
2	−0.7822	0.25000	−0.19555
3	0.0923	0.03125	0.00288

We now complete the summation using the cosines previously calculated:

K \ i	0	2
1	(−0.03012)(1.0)	(−0.03012)(0.0)
2	(−0.19555)(1.0)	(−0.19555)(1.0)
3	(0.00288)(1.0)	(0.00288)(0.0)
Sum	−0.22279	0.19555

We can now complete the calculation by multiplying these sums by 2, adding 1, and multiplying the result by 2 according to Eq. (8-33). Thus the two-power density spectrum estimates are

i	$f_i^* = i/2L$	*Calculation*	*Power Density Spectrum*
0	0	2[1 + 2(−0.22279)]	1.109
2	$\frac{1}{4}$	2[1 + 2(0.19555)]	2.782

The remainder of points are given by the computer program in the Appendix to this chapter.

The power density spectrum is usually plotted by connecting straight lines between the estimates. A chart with a vertical logarithmic scale is generally used. The logarithmic scale has the advantage of enlarging the estimates over regions where they are small and hence facilitates interpretation of the spectrum. The logarithmic scale also has an advantage in hypothesis testing. Figure 8.7 shows the power density spectrum for which we have just calculated two points.

Hypothesis Testing. Suppose that the theoretical spectrum, $\phi(f_i^*)$, is reasonably smooth. Then, for large samples, an approximate $100(1-\alpha)$ percent confidence interval for $\phi(f_i^*)$ is bounded by

$$\frac{p(f_i^*)}{(\chi^2_{\alpha/2;\,\nu})/\nu} < \phi(f_i^*) < \frac{p(f_i^*)}{(\chi^2_{1-\alpha/2;\,\nu})/\nu} \qquad \text{(8-34)}$$

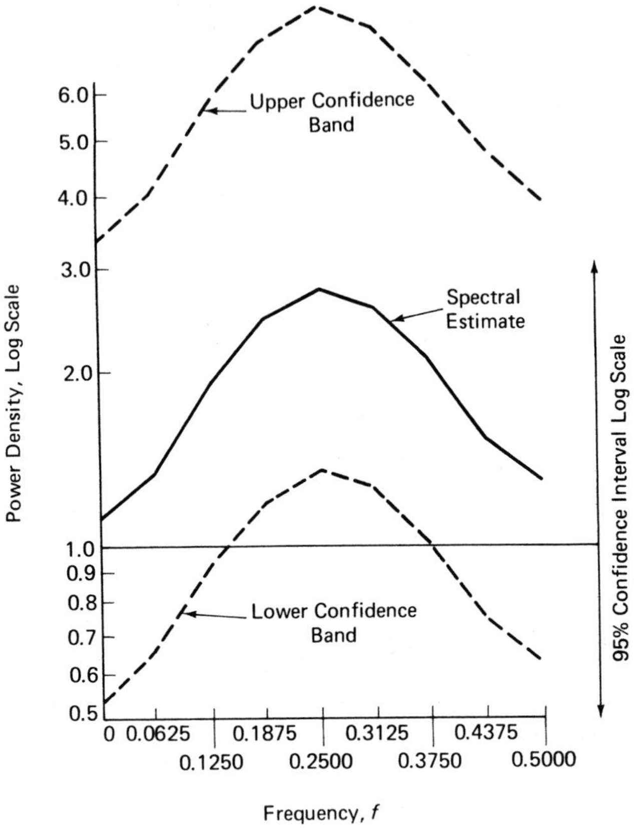

Figure 8.7. Estimated power density spectrum of first differences of the artificial series of Table 8.4.

For the Parzen window, the *equivalent* number of degrees of freedom, ν, is $(3.71)N/L$, which for our example is $(3.71)(11)/4 \doteq 10$. Also, for our example $\chi^2_{0.025;\,10} = 20.483$ and $\chi^2_{0.975;\,10} = 3.247$, so that a 95 percent confidence interval for, say, $\phi(f_2^*)$ is bounded by

$$\frac{p(f_2^*)}{(\chi^2_{0.025;\,10})/\nu} = \frac{2.782}{20.483/10} = 1.358$$

and

$$\frac{p(f_2^*)}{(\chi^2_{0.975;\,10})/\nu} = \frac{2.782}{3.247/10} = 8.567$$

However, by taking the logarithm of Eq. (8-34), the confidence interval for

$\log \phi(f_i^*)$ is bounded by

$$\log p(f_i^*) + \log (\nu/\chi^2_{\alpha/2;\nu}) < \log \phi(f_i^*) < \log p(f_i^*) + \log (\nu/\chi^2_{1-\alpha/2;\nu})$$

Thus, if logarithmic paper is used (and it should be), the interval is set by taking 1.0 to be the center of a line whose lower value is $\nu/\chi^2_{\alpha/2;\nu}$ and whose upper value is given by $\nu/\chi^2_{1-\alpha/2;\nu}$. For our example the lower value of the line is thus $10/20.483 = 0.49$ and the upper value is $10/3.247 = 3.08$. This line is plotted in Figure 8.7, and if the point 1.0 is placed upon a spectral estimate, the upper and lower confidence limits are given by the end points of the line. If such limits are plotted for each estimate and connected by straight lines, the result produces upper and lower confidence bands. These bands are shown by the dashed lines in Figure 8.7. The reader is warned that this technique does *not* give the simultaneous confidence interval for the entire spectrum.

In Figure 8.7 the estimated spectral density peaks at $f^* = 1/4$. It can be shown that the true spectrum of a random series (white noise) is a horizontal line. That is, the power is evenly distributed over all frequencies. One way to test that a series is white noise is to attempt to draw a horizontal line between the confidence bands without touching either the upper or the lower bands. If this is possible, we cannot reject the hypothesis that the true spectrum differs from the spectrum of white noise. It is clearly possible to draw such a horizontal line in Figure 8.7, and we conclude that the peak in the spectrum at $f^* = 1/4$ *does not* signal a significant peak in power in the neighborhood of this frequency.

Another way to establish the confidence bands is to fit a curve (perhaps by least-squares) through the spectral estimates.[16] The line representing the confidence limits is then set about this curve. Any peaks in the spectral estimates which go outside the resulting confidence bands are then considered to be significant.

8.4 uses of spectral analysis

In this section we shall review three uses of spectral analysis found in the literature of economics and business.

Analysis of Stock Prices. Many techniques of security price analysis rely on the study of the past behavior of security prices. One may attempt to find *patterns* in stock price behavior in hopes of using these patterns to predict

[16]The independent variable is $X = 0, 1, 2, \ldots, 2L$. The curve may be a polynomial. More often, it is fit by inspection, since the confidence bands are not precise.

the future.[17] A considerable controversy has arisen over whether the past behavior of stock prices is of any help in analysis. As one leading student of this question puts it, "it is hard to find a practitioner, no matter how sophisticated, who does not believe that by looking at the past history of [stock] prices one can learn something about their prospective behavior, while it is almost as difficult to find an academician who believes that such a backward look is of any *substantial* value."[18]

Many statistical techniques have been used to investigate this question and among them is spectral analysis. Granger and Morgenstern (1963) proposed that stock prices were a random walk. That is, if Y_t is a stock price, then

$$\Delta Y_t = Y_t - Y_{t-1} = \epsilon_t$$

where ϵ_t is a random variable with mean zero and ϵ_t is uncorrelated with ϵ_{t+K} for $K \neq 0$.

Thus the first differences of the series (which, as we know, eliminates a linear trend) is white noise. Granger and Morgenstern computed the spectrum of a number of price series first differences and found them to be approximately flat: "In every case the resulting spectrum was very flat over the whole frequency range, apart from a few exceptional frequencies. . . ."[19] Thus they concluded that "The evidence of 'cycles' obtained in our studies is so weak that 'cyclical investment' is at best only marginally worthwhile."[20]

Long Swings in Economic Activity. A number of writers, following the pioneering work of Simon Kuznets, have found long waves in economic activity which last 10–20 years. The issue developed over the years as to whether or not these swings were different from the shorter period swings associated with typical business cycles, or, even, whether the swings existed at all—perhaps being the result of moving averages applied to the raw data to smooth it prior to analysis.[21] Adelman tested the long swing hypothesis (without prior moving average smoothing) by use of spectral analysis. If long swings exist, they should show up as spectral peaks in the frequency

[17]One interesting point that should be mentioned is the problem of *forecast feedback*. If someone were to discover a pattern in some stock price series which was a perfect forecaster, the pattern would probably vanish if any sizable trading were done on the basis of the forecast. If one found, for example, that the price of XYZ stock always rose in June, he would buy prior to the June rise and sell in June. This action would tend to raise the price prior to June and lower it in June, thus wiping out the pattern. A rule for pattern discoverers is not to be greedy.

[18]Cootner (1964), p. 2. Italics in original.

[19]Granger and Morgenstern (1963), p. 170.

[20]*Ibid.*, p. 178.

[21]It is known that a moving average of certain random series tends to fluctuate. This phenomenon is known as the Yule-Slutsky effect.

range between $f = 1/10$ and $f = 1/20$ cycle per year. Failing to find such peaks, Adelman concluded that "it is likely that the long swings which have been observed in the U.S. economy since 1890 are due in part to the introduction of spurious long cycles by the smoothing process, and in part to the necessity for averaging over a statistically small number of random shocks."[22]

Seasonal Adjustment. A stable seasonal movement is one which repeats itself approximately 1, 2, 3, 4, 5, or 6 times per year, or with a period of 12, 6, 4, 3, 2.4, or 2 months per cycle. "Good" seasonal adjustment might be defined as a filter which removes power in the neighborhood of these seasonal frequencies without removing or adding power elsewhere.

Figure 8.8 shows power density spectra computed for a monthly measure of the money stock in the United States from 1960 to 1970. The spectrum which shows the sharp spikes at the seasonal periods is for the money stock

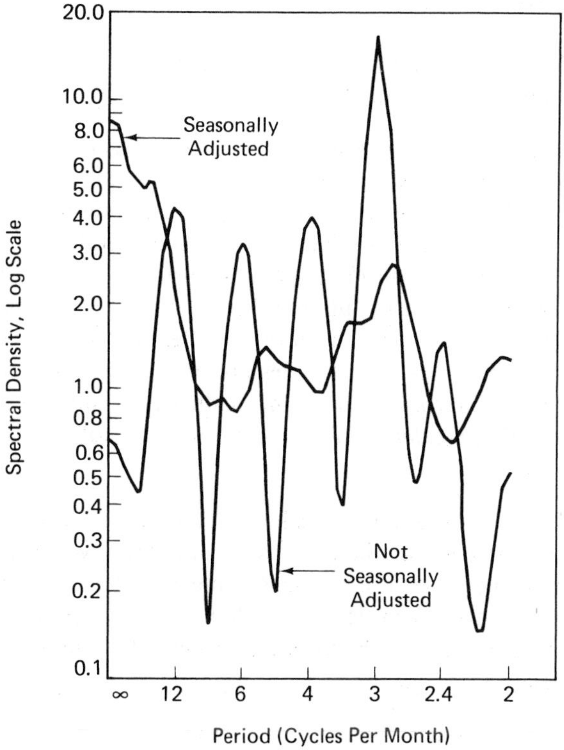

Figure 8.8. Spectrum of seasonally adjusted and not seasonally adjusted money stock series, 1960–1970, monthly.

[22]Adelman (1965), p. 459.

series *not* seasonally adjusted. The other spectrum is for the money stock series after seasonal adjustment. Both series were *detrended* by use of logarithmic first differences.

The seasonal adjustment procedure does seem to remove power at seasonal periods. There is some indication that the strong 3-month cycle (associated with quarterly tax payments) is shifted slightly to the right and not completely removed. Also, low-frequency (high-period) power seems to have been added by seasonal adjustment.

Spectral analysis is of considerable help in the evaluation of alternative seasonal adjustment methods. This type of analysis has been done extensively by the Bureau of the Census under the leadership of Harry Rosenblatt and a report on some alternative seasonal adjustment procedures is provided by Nerlove (1964).

8.5 additional comments

Several additional comments concerning spectral estimation with economic time series need to be made.

Leakage. Because the lag window averages adjacent spectral estimates, if one estimate is large, it will tend to increase the estimates in its immediate neighborhood. Thus a very slow cycle may impart power to, say, the zero frequency, leading us to believe that there is a trend in the series. This problem is called *leakage through the window*. Unwanted peaks in the spectrum can sometimes be removed by filtering the series (just as is done in seasonal adjustment). The estimated spectrum of the filtered series is then calculated, and as a last step the spectral estimates are divided by the *transfer function* of the filter. Such filtering is sometimes called *prewhitening*, and the final adjustment is called *recoloring*. A presentation of recoloring is beyond the scope of this text.[23]

Trend in Mean. If the time series has a trend in the mean, it will produce a large spectral estimate at the zero frequency. However, a large estimate at the zero frequency does not necessarily mean that there is a trend in the mean since it may arise from leakage of low-frequency power. The original data should probably always be inspected visually for possible trends in the mean and variance.

Observations. Regular fluctuations in time series need to be repeated several times in order to be detected by spectral analysis. If sampling is done

[23]Nerlove (1964) recommends the use of *quasi-differences*. Here $\Delta_k Y_t = Y_t - kY_{t-1}$, where k is some number not necessarily 1. In this introductory exposition we have set $k = 1$, which gives ordinary first differences.

at yearly intervals, it takes more observations than are currently available for most economic time series to get an accurate measurement of cycles that last 10–20 years. Some statisticians will not attempt a spectral analysis of economic time series with less than 100 observations.

Cross Spectrum. Spectral analysis can be extended to two or more series. With two series we have *cross spectral analysis*; with many time series, we have *multivariate spectral analysis*. Cross spectral analysis is useful because it gives us an indication of the degree of association across frequencies between two times series by means of a measure called *coherency*. Lead and lag patterns between two series can also be studied by use of *phase lag* measures. We shall not treat cross (or multivariate) spectral analysis in this text.[24]

Questions and Problems

Secs. 8.1–8.4

1. The table below shows the total money stock as defined by the Federal Reserve System for 11 years. Do a spectral analysis of this series. Comment especially on
   a. Trend removal problems.
   b. Seasonal movements in the series.
   c. Cyclical movements (movements of slower than seasonal frequencies) in the data.

Total Money Stock, 1960–1970, Not Seasonally Adjusted

	1960	1961	1962	1963	1964	1965	1966	1967	1968	1969	1970
Jan.	145.7	145.2	149.7	152.5	158.4	165.7	174.2	176.4	189.3	204.2	211.4
Feb.	141.9	142.4	146.4	149.1	154.3	160.8	168.8	171.8	183.3	197.8	202.8
Mar.	140.4	141.6	145.6	148.3	153.6	160.3	168.9	173.1	184.1	198.3	204.7
Apr.	141.3	143.2	147.6	150.5	155.7	162.9	172.6	174.9	187.6	202.0	209.3
May	139.1	141.6	145.0	148.3	153.0	158.9	168.0	172.4	185.0	197.7	205.3
June	139.3	142.1	145.2	149.0	154.2	160.8	170.0	175.7	188.5	200.5	207.8
July	139.8	142.4	145.5	150.2	156.0	162.1	169.3	177.2	190.1	201.5	209.0
Aug.	140.3	142.4	144.8	149.9	156.0	161.5	168.4	177.2	189.8	199.6	208.7
Sept.	141.2	143.9	145.8	151.2	158.1	164.3	170.9	179.8	192.2	201.4	211.4
Oct.	142.0	145.3	147.3	153.1	160.1	166.8	171.7	182.0	194.3	203.2	213.0
Nov.	142.9	147.0	149.0	155.4	161.8	168.5	172.7	183.9	197.7	205.3	215.6
Dec.	145.5	150.1	152.3	157.9	165.3	173.1	176.9	188.6	203.4	209.8	221.1

Source: *Federal Reserve Bulletin*, Dec. 1970 and Jan. 1971

[24]Examples of cross spectral analysis are given in Granger and Hatanaka (1964). See also Bolch (1967).

APPENDIX

A8.1 main for spectral analysis

A. Description. This main program computes the mean, $\bar{Y}$; variance, $C(0)$; power density spectrum, $p(f_i^*)$, of Eq. (8-33); and autocorrelation function, $r(K)$, from a time series. It also computes the frequency, f, associated with each spectral estimate. Parzen weights are used in the spectral estimation [see Eq. (8-31)].

B. Limitations. The time series may have at most 500 observations. A maximum of 100 lags may be specified for the autocorrelation function, which will afford at most 200 spectral estimates. These two restrictions may be altered by changing the DIMENSION statement. The time series is assumed to be stationary (see the next section for trend removal).

C. Use. On the first card, punch the number of observations, N, for the time series in columns 1–3 (right-adjusted and without a decimal point). On the same card, punch in the same way the number of lags desired, L, in columns 4–6.

Starting on a fresh card, punch the time series in eight fields of width 10, with the decimal point. Continue from card to card until the entire series is punched.

The results for the first differences of the artificial series of Table 8.4 are shown below followed by the main program.

```
             MEAN              VARIANCE
        .85999955E+00     .83800037E+02

  LAG          AUTO-         FREQUENCY   SPECTRAL
            CORRELATION                   DENSITY
    0         1.0000           .0000      1.1088
    1         -.0419           .0625      1.3399
    2         -.7822           .1250      1.9066
    3          .0923           .1875      2.4963
    4          .6399           .2500      2.7822
                               .3125      2.6099
                               .3750      2.0934
                               .4375      1.5539
                               .5000      1.3268
```

```
      DIMENSION X(500),R(100),PDS(200),W(100)
      READ (5,5)N,L
    5 FORMAT(2I3)
      L=L+1
      READ(5,10)(X(I),I=1,N)
   10 FORMAT(8F10.0)
      XBAR=0.0
      DO 30 I=1,N
   30 XBAR=XBAR+X(I)
      XBAR=XBAR/FLOAT(N)
      LL=1+L/2
      LL1=LL+1
      L1=L-1
      FL1=L1
      L2=L*2
      ANG=3.14159265/(FL1*2.0)
      DO 40 J=1,L
      RR=0.0
      NJ=N-J+1
      DO 38 I=1,NJ
      IJ=I+J-1
   38 RR=RR+(X(I)-XBAR)*(X(IJ)-XBAR)
   40 R(J)=RR
      DO 45 J=2,L
   45 R(J)=R(J)/ R(1)
      VAR=R(1)/FLOAT(N)
      R(1)=1.0
      DO 50 J=2,LL
      TAU=J-1
   50 W(J)=1.0-(6.0*TAU**2/(FL1**2))+(6.0*TAU**3/(FL1**3))
      DO 55 J=LL1,L1
      TAU=J-1
   55 W(J)=2.0*(1.0-TAU/FL1)**3
      DO 60 I=1,L2
      PDS(I)=0.0
      FI1=I-1
      DO 60 J=2,L1
      FJ1=J-1
   60 PDS(I)=PDS(I)+R(J)*W(J)*COS(ANG*FJ1*FI1)
      WRITE(6,65) XBAR,VAR
   65 FORMAT(6X,'MEAN',9X,'VARIANCE',/2E15.7,/)
      WRITE(6,68)
   68 FORMAT('LAG',6X,'AUTO-'5X,'FREQUENCY',2X,'SPECTRAL',/,7X,
     1'CORRELATION',13X,'DENSITY')
      DO 70 I=1,L
      I1=I-1
      FI1=I1
      PDS(I)=(1.0+2.0*PDS(I))*2.0
      F=FI1/(4.0*FL1)
   70 WRITE(6,75) I1,R(I),F,PDS(I)
   75 FORMAT(I4,F11.4,F12.4,F10.4)
      L1=L+1
      L2=L2-1
      DO 80 I=L1,L2
      PDS(I)=(1.0+2.0*PDS(I))*2.0
      FI1=I-1
      F=FI1/(4.0*FL1)
   80 WRITE(6,85)F,PDS(I)
   85 FORMAT(15X,F12.4,F10.4)
      CALL EXIT
      END
```

A8.2 trend removal

It is a good idea to plot the time series prior to spectral analysis in order to obtain some idea of the type of trend in the series. If the series is linear or polynomial, the series may be regressed on $X = 0, 1, \ldots, N - 1$ and/or X and its powers, as described in Sec. A6.1. The residuals are then subjected to spectral analysis.

In a similar way, loglinear regression may be used for a logarithmic trend. If the trend appears to be a straight line, one can use first differences by inserting the following statements in the main program. This procedure has the advantage of saving computer time. Thus, insert the following immediately after statement 10 in the main program:

```
      N = N - 1
      DO 20 I = 1, N
   20 X(I) = X(I + 1) - X(I)
```

If the trend appears to be linear in the logs (which can be determined by plotting the series on paper with a logarithmic vertical axis), one may use logarithmic first differences if all the sample values of the time series are positive. Thus, insert after statement 10

```
      N = N - 1
      DO 20 I = 1, N
   20 X(I) = ALOG (X(I + 1)) - ALOG (X(I))
```

Appendix Tables

Each of these tables gives right-tail percentage points for the distribution in question. If we call the relevant right-tail areas $Q(\ )$, then, say, for the normal distribution the percentage point is z_Q corresponding to $Q(z)$. The accompanying diagrams illustrate the percentage points for each of the tables to follow:

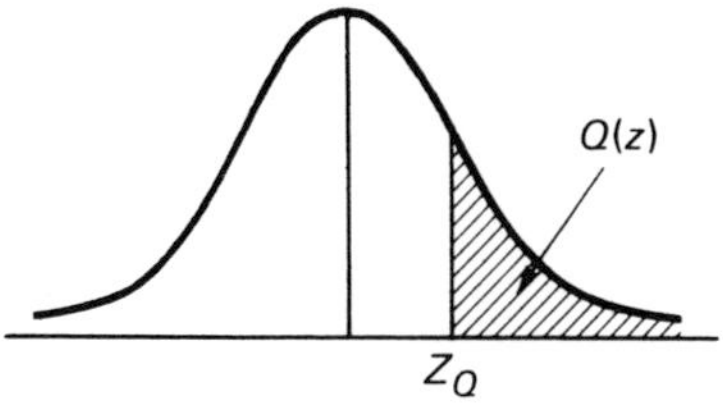

Table 1.

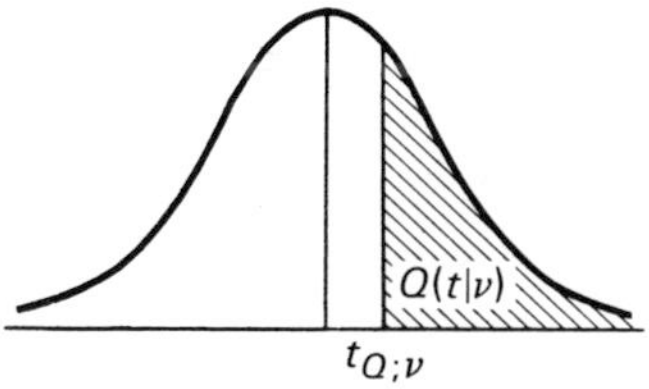

Table 2.

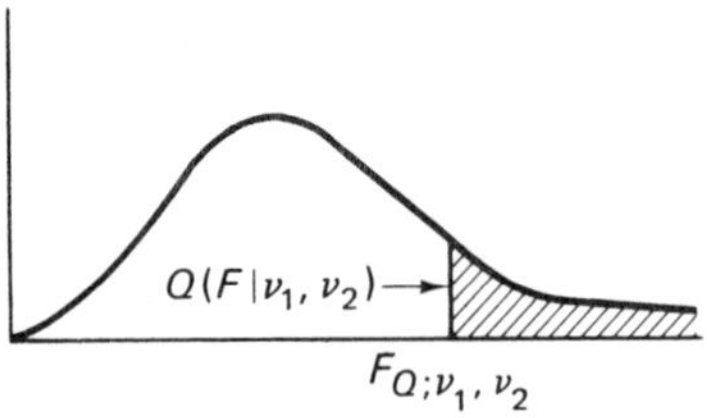

Table 3.

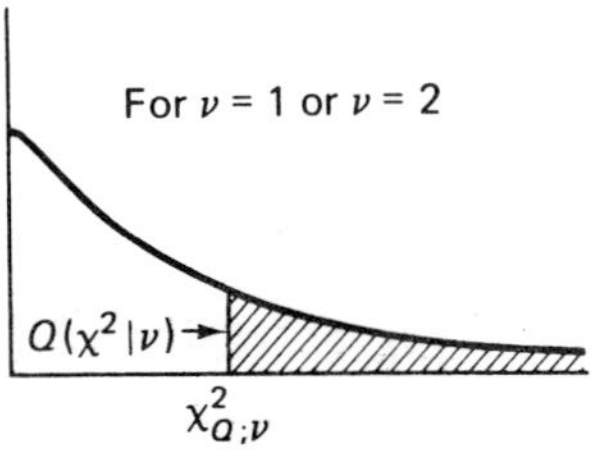

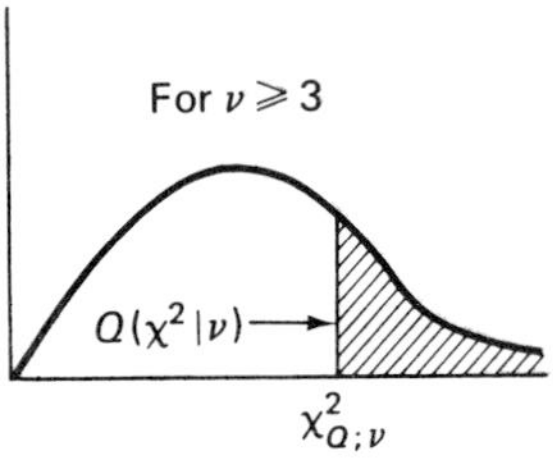

Table 4.

Table 1   Percentage Points for z

z	0.00	0.01	0.02	0.03	0.04	0.05	0.06	0.07	0.08	0.09
0.0	0.50000	0.49601	0.49202	0.48803	0.48405	0.48006	0.47608	0.47210	0.46812	0.46414
0.1	0.46017	0.45620	0.45224	0.44828	0.44433	0.44038	0.43644	0.43251	0.42858	0.42465
0.2	0.42074	0.41683	0.41294	0.40905	0.40517	0.40129	0.39743	0.39358	0.38974	0.38591
0.3	0.38209	0.37828	0.37448	0.37070	0.36693	0.36317	0.35942	0.35569	0.35197	0.34827
0.4	0.34458	0.34090	0.33724	0.33360	0.32997	0.32636	0.32276	0.31918	0.31561	0.31207
0.5	0.30854	0.30503	0.30153	0.29806	0.29460	0.29116	0.28774	0.28434	0.28096	0.27760
0.6	0.27425	0.27093	0.26763	0.26435	0.26109	0.25785	0.25463	0.25143	0.24825	0.24510
0.7	0.24196	0.23885	0.23576	0.23270	0.22965	0.22663	0.22363	0.22065	0.21770	0.21476
0.8	0.21186	0.20897	0.20611	0.20327	0.20045	0.19766	0.19489	0.19215	0.18943	0.18673
0.9	0.18406	0.18141	0.17879	0.17619	0.17361	0.17106	0.16853	0.16602	0.16354	0.16109
1.0	0.15866	0.15625	0.15386	0.15151	0.14917	0.14686	0.14457	0.14231	0.14007	0.13786
1.1	0.13567	0.13350	0.13136	0.12924	0.12714	0.12507	0.12302	0.12100	0.11900	0.11702
1.2	0.11507	0.11314	0.11123	0.10935	0.10749	0.10565	0.10383	0.10204	0.10027	0.09853
1.3	0.09680	0.09510	0.09342	0.09176	0.09012	0.08851	0.08691	0.08534	0.08379	0.08226
1.4	0.08076	0.07927	0.07780	0.07636	0.07493	0.07353	0.07215	0.07078	0.06944	0.06811
1.5	0.06681	0.06552	0.06426	0.06301	0.06178	0.06057	0.05938	0.05821	0.05705	0.05592
1.6	0.05480	0.05370	0.05262	0.05155	0.05050	0.04947	0.04846	0.04746	0.04648	0.04551
[illegible]7	0.04457	0.04363	0.04272	0.04182	0.04093	0.04006	0.03920	0.03836	0.03754	0.03673
[illegible]	0.03593	0.03515	0.03438	0.03362	0.03288	0.03216	0.03144	0.03074	0.03005	0.02938
[illegible]	0.02872	0.02807	0.02743	0.02680	0.02619	0.02559	0.02500	0.02442	0.02385	0.02330
[illegible]	0.02275	0.02216	0.02169	0.02118	0.02068	0.02018	0.01970	0.01923	0.01876	0.01831
[illegible]	0.01786	0.01743	0.01700	0.01659	0.01618	0.01578	0.01539	0.01500	0.01463	0.01426
[illegible]	0.01390	0.01355	0.01321	0.01287	0.01255	0.01222	0.01191	0.01160	0.01130	0.01101
[illegible]	0.01072	0.01044	0.01017	0.00990	0.00964	0.00939	0.00914	0.00889	0.00866	0.00842
[illegible]	0.00820	0.00798	0.00776	0.00755	0.00734	0.00714	0.00695	0.00676	0.00657	0.00639

Table 1 (continued)

z	*0.00*	*0.01*	*0.02*	*0.03*	*0.04*	*0.05*	*0.06*	*0.07*	*0.08*	*0.09*
2.5	0.00621	0.00604	0.00587	0.00570	0.00554	0.00539	0.00523	0.00508	0.00494	0.00480
2.6	0.00466	0.00453	0.00440	0.00427	0.00415	0.00402	0.00391	0.00379	0.00368	0.00357
2.7	0.00347	0.00336	0.00326	0.00317	0.00307	0.00298	0.00289	0.00280	0.00272	0.00264
2.8	0.00256	0.00248	0.00240	0.00233	0.00226	0.00219	0.00212	0.00205	0.00199	0.00193
2.9	0.00187	0.00181	0.00175	0.00169	0.00164	0.00159	0.00154	0.00149	0.00144	0.00139
3.0	0.00135	0.00131	0.00126	0.00122	0.00118	0.00114	0.00111	0.00107	0.00104	0.00100
3.1	0.00097	0.00094	0.00090	0.00087	0.00084	0.00082	0.00079	0.00076	0.00074	0.00071
3.2	0.00069	0.00066	0.00064	0.00062	0.00060	0.00058	0.00056	0.00054	0.00052	0.00050
3.3	0.00048	0.00047	0.00045	0.00043	0.00042	0.00040	0.00039	0.00038	0.00036	0.00035
3.4	0.00034	0.00032	0.00031	0.00030	0.00029	0.00028	0.00027	0.00026	0.00025	0.00024
3.5	0.00023	0.00022	0.00022	0.00021	0.00020	0.00019	0.00019	0.00018	0.00017	0.00017
3.6	0.00016	0.00015	0.00015	0.00014	0.00014	0.00013	0.00013	0.00012	0.00012	0.00011
3.7	0.00011	0.00010	0.00010	0.00010	0.00009	0.00009	0.00008	0.00008	0.00008	0.00008
3.8	0.00007	0.00007	0.00007	0.00006	0.00006	0.00006	0.00006	0.00005	0.00005	0.00005
3.9	0.00005	0.00005	0.00004	0.00004	0.00004	0.00004	0.00004	0.00004	0.00003	0.00003
4.0	0.00003	—	—	—	—	—	—	—	—	—
4.5	0.000003	—	—	—	—	—	—	—	—	—
5.0	0.0000003	—	—	—	—	—	—	—	—	—

t	0.45	0.40	0.35	0.30	0.25	0.20	0.15	0.10	0.05	0.025	0.01	0.005	0.0005
1	0.158	0.325	0.510	0.727	1.000	1.376	1.963	3.078	6.314	12.706	31.821	63.657	636.692
2	0.142	0.289	0.445	0.617	0.816	1.061	1.386	1.886	2.920	4.303	6.965	9.925	31.598
3	0.137	0.277	0.424	0.584	0.765	0.978	1.250	1.638	2.353	3.182	4.541	5.841	12.924
4	0.134	0.271	0.414	0.569	0.741	0.941	1.190	1.533	2.132	2.776	3.747	4.604	8.610
5	0.132	0.267	0.408	0.559	0.727	0.920	1.156	1.476	2.015	2.571	3.365	4.032	6.869
6	0.131	0.265	0.404	0.553	0.718	0.906	1.134	1.440	1.943	2.447	3.143	3.707	5.959
7	0.130	0.263	0.402	0.549	0.711	0.896	1.119	1.415	1.895	2.365	2.998	3.499	5.408
8	0.130	0.262	0.399	0.546	0.706	0.889	1.108	1.397	1.860	2.306	2.896	3.355	5.041
9	0.129	0.261	0.398	0.543	0.703	0.883	1.100	1.383	1.833	2.262	2.821	3.250	4.781
10	0.129	0.260	0.397	0.542	0.700	0.879	1.093	1.372	1.812	2.228	2.764	3.169	4.587
11	0.129	0.260	0.396	0.540	0.697	0.876	1.088	1.363	1.796	2.201	2.718	3.106	4.437
12	0.128	0.259	0.395	0.539	0.695	0.873	1.083	1.356	1.782	2.179	2.681	3.055	4.318
13	0.128	0.259	0.394	0.538	0.694	0.870	1.079	1.350	1.771	2.160	2.650	3.012	4.221
14	0.128	0.258	0.393	0.537	0.692	0.868	1.076	1.345	1.761	2.145	2.624	2.977	4.140
15	0.128	0.258	0.393	0.536	0.691	0.866	1.074	1.341	1.753	2.131	2.602	2.947	4.073
16	0.128	0.258	0.392	0.535	0.690	0.865	1.071	1.337	1.746	2.120	2.583	2.921	4.015
17	0.128	0.257	0.392	0.534	0.689	0.863	1.069	1.333	1.740	2.110	2.567	2.898	3.965
18	0.127	0.257	0.392	0.534	0.688	0.862	1.067	1.330	1.734	2.101	2.552	2.878	3.922
19	0.127	0.257	0.391	0.533	0.688	0.861	1.066	1.328	1.729	2.093	2.539	2.861	3.883
20	0.127	0.257	0.391	0.533	0.687	0.860	1.064	1.325	1.725	2.086	2.528	2.845	3.850
21	0.127	0.257	0.391	0.532	0.686	0.859	1.063	1.323	1.721	2.080	2.518	2.831	3.819
22	0.127	0.256	0.390	0.532	0.686	0.858	1.061	1.321	1.717	2.074	2.508	2.819	3.792
23	0.127	0.256	0.390	0.532	0.685	0.858	1.060	1.319	1.714	2.069	2.500	2.807	3.767
24	0.127	0.256	0.390	0.531	0.685	0.857	1.059	1.318	1.711	2.064	2.492	2.797	3.745
25	0.127	0.256	0.390	0.531	0.684	0.856	1.058	1.316	1.708	2.060	2.485	2.787	3.725
26	0.127	0.256	0.390	0.531	0.684	0.856	1.058	1.315	1.706	2.056	2.479	2.779	3.707
27	0.127	0.256	0.389	0.531	0.684	0.855	1.057	1.314	1.703	2.052	2.473	2.771	3.690
28	0.127	0.256	0.389	0.530	0.683	0.855	1.056	1.313	1.701	2.048	2.467	2.763	3.674
29	0.127	0.256	0.389	0.530	0.683	0.854	1.055	1.311	1.699	2.045	2.462	2.756	3.659
30	0.127	0.256	0.389	0.530	0.683	0.854	1.055	1.310	1.697	2.042	2.457	2.750	3.646
40	0.126	0.255	0.388	0.529	0.681	0.851	1.050	1.303	1.684	2.021	2.423	2.704	3.551
60	0.126	0.254	0.387	0.527	0.679	0.848	1.046	1.296	1.671	2.000	2.390	2.660	3.460
120	0.126	0.254	0.386	0.526	0.677	0.845	1.041	1.289	1.658	1.980	2.358	2.617	3.373
∞	0.126	0.253	0.385	0.524	0.674	0.842	1.036	1.282	1.645	1.960	2.326	2.576	3.291

Source: This table is reprinted from Table III of R. A. Fisher and F. Yates, *Statistical Tables for Biological, Agricultural and Medical Research*, 6th ed. (Edinburgh: Oliver & Boyd Ltd., 1963), by permission.

Table 3 Percentage Points for F

values of F_Q for $Q(F|\nu_1, \nu_2) = 0.05$

ν_2 \ ν_1	1	2	3	4	5	6	8	10	12	20	24	30	40	60	120	∞
1	161.4	199.5	215.7	224.6	230.2	234.0	238.9	241.9	243.9	248.0	249.1	250.1	251.1	252.2	253.3	254.3
2	18.51	19.00	19.16	19.25	19.30	19.33	19.37	19.40	19.41	19.45	19.45	19.47	19.46	19.47	19.49	19.50
3	10.13	9.55	9.28	9.12	9.01	8.94	8.85	8.79	8.74	8.66	8.64	8.62	8.59	8.57	8.55	8.53
4	7.71	6.94	6.59	6.39	6.26	6.16	6.04	5.96	5.91	5.80	5.77	5.72	5.75	5.69	5.66	5.63
5	6.61	5.79	5.41	5.19	5.05	4.95	4.82	4.74	4.68	4.56	4.53	4.50	4.46	4.43	4.40	4.36
6	5.99	5.14	4.76	4.53	4.39	4.28	4.15	4.06	4.00	3.87	3.84	3.81	3.77	3.74	3.70	3.67
8	5.32	4.46	4.07	3.84	3.69	3.58	3.44	3.35	3.28	3.15	3.12	3.08	3.04	3.01	2.97	2.93
10	4.96	4.10	3.71	3.48	3.33	3.22	3.07	2.98	2.91	2.77	2.74	2.70	2.66	2.62	2.58	2.54
12	4.75	3.89	3.49	3.26	3.11	3.00	2.85	2.75	2.69	2.54	2.51	2.47	2.43	2.38	2.34	2.30
20	4.35	3.49	3.10	2.87	2.71	2.60	2.45	2.35	2.28	2.12	2.08	2.04	1.99	1.95	1.90	1.84
24	4.26	3.40	3.01	2.78	2.62	2.51	2.36	2.25	2.18	2.03	1.98	1.94	1.89	1.84	1.79	1.73
30	4.17	3.32	2.92	2.69	2.53	2.42	2.27	2.16	2.09	1.93	1.89	1.84	1.79	1.74	1.68	1.62
40	4.08	3.23	2.84	2.61	2.45	2.34	2.18	2.08	2.00	1.84	1.79	1.74	1.69	1.64	1.58	1.51
60	4.00	3.15	2.76	2.53	2.37	2.25	2.10	1.99	1.92	1.75	1.70	1.65	1.59	1.53	1.47	1.39
120	3.92	3.07	2.68	2.45	2.29	2.17	2.02	1.91	1.83	1.66	1.61	1.55	1.50	1.43	1.35	1.25
∞	3.84	3.00	2.60	2.37	2.21	2.10	1.94	1.83	1.75	1.57	1.52	1.46	1.39	1.32	1.22	1.00

Table 3 (continued)

values of F_Q for $Q(F|\nu_1, \nu_2) = 0.025$

ν_2 \ ν_1	1	2	3	4	5	6	8	10	12	20	24	30	40	60	120	∞
1	647.8	799.5	864.2	899.6	921.8	937.1	956.7	968.6	976.7	993.1	997.2	1001	1006	1010	1014	1018
2	38.51	39.00	39.17	39.25	39.30	39.33	39.37	39.40	39.41	39.45	39.46	39.46	39.47	39.48	39.49	39.50
3	17.44	16.04	15.44	15.10	14.88	14.73	14.54	14.42	14.34	14.17	14.12	14.08	14.04	13.99	13.95	13.90
4	12.22	10.65	9.98	9.60	9.36	9.20	8.98	8.84	8.75	8.56	8.51	8.46	8.41	8.36	8.31	8.26
5	10.01	8.43	7.76	7.39	7.15	6.98	6.76	6.62	6.52	6.33	6.28	6.23	6.18	6.12	6.07	6.02
6	8.81	7.26	6.60	6.23	5.99	5.82	5.60	5.46	5.37	5.17	5.12	5.07	5.01	4.96	4.90	4.85
8	7.57	6.06	5.42	5.05	4.82	4.65	4.43	4.30	4.20	4.00	3.95	3.89	3.84	3.78	3.73	3.67
10	6.94	5.46	4.83	4.47	4.24	4.07	3.85	3.72	3.62	3.42	3.37	3.31	3.26	3.20	3.14	3.08
12	6.55	5.10	4.47	4.12	3.89	3.73	3.51	3.37	3.28	3.07	3.02	2.96	2.91	2.85	2.79	2.72
20	5.87	4.46	3.86	3.51	3.29	3.13	2.91	2.77	2.68	2.46	2.41	2.35	2.29	2.22	2.16	2.09
24	5.72	4.32	3.72	3.38	3.15	2.99	2.78	2.64	2.54	2.33	2.27	2.21	2.15	2.08	2.01	1.94
30	5.57	4.18	3.59	3.25	3.03	2.87	2.65	2.51	2.41	2.20	2.14	2.07	2.01	1.94	1.87	1.79
40	5.42	4.05	3.46	3.13	2.90	2.74	2.53	2.39	2.29	2.07	2.01	1.94	1.88	1.80	1.72	1.64
60	5.29	3.93	3.34	3.01	2.79	2.63	2.41	2.27	2.17	1.94	1.88	1.82	1.74	1.67	1.58	1.48
120	5.15	3.80	3.23	2.89	2.67	2.52	2.30	2.16	2.05	1.82	1.76	1.69	1.61	1.53	1.43	1.31
∞	5.02	3.69	3.12	2.79	2.57	2.41	2.19	2.05	1.94	1.71	1.64	1.57	1.48	1.39	1.27	1.00

Table 3 (continued)

values of F_Q for $Q(F|\nu_1, \nu_2) = 0.01$

ν_2 \ ν_1	*1*	*2*	*3*	*4*	*5*	*6*	*8*	*10*	*12*	*20*	*24*	*30*	*40*	*60*	*120*	∞
1	4052	5000	5403	5625	5764	5859	5982	6056	6106	6209	6235	6261	6287	6313	6339	6366
2	98.50	99.00	99.17	99.25	99.30	99.33	99.37	99.40	99.42	99.45	99.46	99.47	99.47	99.48	99.49	99.50
3	34.12	30.82	29.46	28.71	28.24	27.91	27.49	27.23	27.05	26.69	26.60	26.50	26.41	26.32	26.22	26.13
4	21.20	18.00	16.69	15.98	15.52	15.21	14.80	14.55	14.37	14.02	13.93	13.84	13.75	13.65	13.56	13.46
5	16.26	13.27	12.06	11.39	10.97	10.67	10.29	10.05	9.89	9.55	9.47	9.38	9.29	9.20	9.11	9.02
6	13.75	10.92	9.78	9.15	8.75	8.47	8.10	7.87	7.72	7.40	7.31	7.23	7.14	7.06	6.97	6.88
8	11.26	8.65	7.59	7.01	6.63	6.37	6.03	5.81	5.67	5.36	5.28	5.20	5.12	5.03	4.95	4.86
10	10.04	7.56	6.55	5.99	5.64	5.39	5.06	4.85	4.71	4.41	4.33	4.25	4.17	4.08	4.00	3.91
12	9.33	6.93	5.95	5.41	5.06	4.82	4.50	4.30	4.16	3.86	3.78	3.70	3.62	3.54	3.45	3.36
20	8.10	5.85	4.94	4.43	4.10	3.87	3.56	3.37	3.23	2.94	2.86	2.78	2.69	2.61	2.52	2.42
24	7.82	5.61	4.72	4.22	3.90	3.67	3.36	3.17	3.03	2.74	2.66	2.58	2.49	2.40	2.31	2.21
30	7.56	5.39	4.51	4.02	3.70	3.47	3.17	2.98	2.84	2.55	2.47	2.39	2.30	2.21	2.11	2.01
40	7.31	5.18	4.31	3.83	3.51	3.29	2.99	2.80	2.66	2.37	2.29	2.20	2.11	2.02	1.92	1.80
60	7.08	4.98	4.13	3.65	3.34	3.12	2.82	2.63	2.50	2.20	2.12	2.03	1.94	1.84	1.73	1.60
120	6.85	4.79	3.95	3.48	3.17	2.96	2.66	2.47	2.34	2.03	1.95	1.86	1.76	1.66	1.53	1.38
∞	6.63	4.61	3.78	3.32	3.02	2.80	2.51	2.32	2.18	1.88	1.79	1.70	1.59	1.47	1.32	1.00

Table 3 (continued)

values of F_Q for $Q(F|\nu_1, \nu_2) = 0.001$

ν_2 \ ν_1	1	2	3	4	5	6	8	10	12	20	24	30	40	60	120	∞
1*	405.3	500.0	540.4	562.5	576.4	585.9	598.1	605.6	610.7	620.9	623.5	626.1	628.7	631.3	634.0	636.6
2	998.5	999.0	999.2	999.2	999.3	999.3	999.4	999.4	999.4	999.4	999.5	999.5	999.5	999.5	999.5	999.5
3	167.0	148.5	141.1	137.1	134.6	132.8	130.6	129.2	128.3	126.4	125.9	125.4	125.0	124.5	124.0	123.5
4	74.14	61.25	56.18	53.44	51.71	50.53	49.00	48.05	47.41	46.10	45.77	45.43	45.09	44.75	44.40	44.05
5	47.18	37.12	33.20	31.09	29.75	28.84	27.64	26.92	26.42	25.39	25.14	24.87	24.60	24.33	24.06	23.79
6	35.51	27.00	23.70	21.92	20.81	20.03	19.03	18.41	17.99	17.12	16.89	16.67	16.44	16.21	15.99	15.75
8	25.42	18.49	15.83	14.39	13.49	12.86	12.04	11.54	11.19	10.48	10.30	10.11	9.92	9.73	9.53	9.33
10	21.04	14.91	12.55	11.28	10.48	9.92	9.20	8.75	8.45	7.80	7.64	7.47	7.30	7.12	6.94	6.76
12	18.64	12.97	10.80	9.63	8.89	8.38	7.71	7.29	7.00	6.40	6.25	6.09	5.93	5.76	5.59	5.42
20	14.82	9.95	8.10	7.10	6.46	6.02	5.44	5.08	4.82	4.29	4.15	4.00	3.86	3.70	3.54	3.38
24	14.03	9.34	7.55	6.59	5.98	5.55	4.99	4.64	4.39	3.87	3.74	3.59	3.45	3.29	3.14	2.97
30	13.29	8.77	7.05	6.12	5.53	5.12	4.58	4.24	4.00	3.49	3.36	3.22	3.07	2.92	2.76	2.59
40	12.61	8.25	6.60	5.70	5.13	4.73	4.21	3.87	3.64	3.15	3.01	2.87	2.73	2.57	2.41	2.23
60	11.97	7.76	6.17	5.31	4.76	4.37	3.87	3.54	3.31	2.83	2.69	2.55	2.41	2.25	2.08	1.89
120	11.38	7.32	5.79	4.95	4.42	4.04	3.55	3.24	3.02	2.53	2.40	2.26	2.11	1.95	1.76	1.54
∞	10.83	6.91	5.42	4.62	4.10	3.74	3.27	2.96	2.74	2.27	2.13	1.99	1.84	1.66	1.45	1.00

*Multiply all entries on this line by 1000.

Source: This table is abridged from Table 18 of E. S. Pearson and H. O. Hartley, eds., *Biometrika Tables for Statisticians*, Vol. I (New York: Cambridge University Press, 1966), by permission.

Table 4 Percentage Points for χ^2

values of χ^2_Q for selected values of $Q(\chi^2 \mid \nu)$

ν	0.995	0.99	0.98	0.975	0.95	0.90	0.80	0.75	0.70	0.50
1	0.0^4393	0.0^3157	0.0^3628	0.0^3982	0.00393	0.0158	0.0642	0.102	0.148	0.455
2	0.0100	0.0201	0.0404	0.0506	0.103	0.211	0.446	0.575	0.713	1.386
3	0.0717	0.115	0.185	0.216	0.352	1.584	1.005	1.213	1.424	2.366
4	0.207	0.297	0.429	0.484	0.711	1.064	1.649	1.923	2.195	3.357
5	0.412	0.554	0.752	0.831	1.145	1.610	2.343	2.675	3.000	4.351
6	0.676	0.872	1.134	1.237	1.635	2.204	3.070	3.455	3.828	5.348
7	0.989	1.239	1.564	1.690	2.167	2.833	3.822	4.255	4.671	6.346
8	1.344	1.646	2.032	2.180	2.733	3.490	4.594	5.071	5.527	7.344
9	1.735	2.088	2.532	2.700	3.325	4.168	5.380	5.899	6.393	8.343
10	2.156	2.558	3.059	3.247	3.940	4.865	6.179	6.737	7.267	9.342
11	2.603	3.053	3.609	3.816	4.575	5.578	6.989	7.584	8.148	10.341
12	3.074	3.571	4.178	4.404	5.226	6.304	7.807	8.438	9.034	11.340
13	3.565	4.107	4.765	5.009	5.892	7.042	8.634	9.299	9.926	12.340
14	4.075	4.660	5.368	5.629	6.571	7.790	9.467	10.165	10.821	13.339
15	4.601	5.229	5.985	6.262	7.261	8.547	10.307	11.036	11.721	14.339
16	5.142	5.812	6.614	6.908	7.962	9.312	11.152	11.912	12.624	15.338
17	5.697	6.408	7.255	7.564	8.672	10.085	12.002	12.792	13.531	16.338
18	6.265	7.015	7.906	8.231	9.390	10.865	12.857	13.675	14.440	17.338
19	6.844	7.633	8.567	8.907	10.117	11.651	13.716	14.562	15.352	18.338
20	7.434	8.260	9.237	9.591	10.851	12.443	14.578	15.452	16.266	19.337
21	8.034	8.897	9.915	10.283	11.591	13.240	15.445	16.344	17.182	20.337
22	8.643	9.542	10.600	10.982	12.338	14.041	16.314	17.240	18.101	21.337
23	9.260	10.196	11.293	11.688	13.091	14.848	17.187	18.137	19.021	22.337
24	9.886	10.856	11.992	12.401	13.848	15.659	18.062	19.037	19.943	23.337
25	10.520	11.524	12.697	13.120	14.611	16.473	18.940	19.930	20.867	24.337
26	11.160	12.198	13.409	13.844	15.379	17.292	19.820	20.843	21.792	25.336
27	11.808	12.879	14.125	14.573	16.151	18.114	20.703	21.749	22.719	26.336
28	12.461	13.565	14.847	15.308	16.928	18.939	21.588	22.657	23.647	27.336
29	13.121	14.256	15.574	16.047	17.708	19.768	22.475	23.567	24.577	28.336
30	13.787	14.953	16.306	16.791	18.493	20.599	23.364	24.478	25.508	29.336

Table 4 (continued)

values of χ^2_Q for selected values of $Q(\chi^2 \mid \nu)$

ν	0.30	0.25	0.20	0.10	0.05	0.025	0.02	0.01	0.005	0.001
1	1.074	1.323	1.642	2.706	3.841	5.024	5.412	6.635	7.879	10.827
2	2.408	2.773	3.219	4.605	5.991	7.378	7.824	9.210	10.597	13.815
3	3.665	4.108	4.642	6.251	7.815	9.348	9.837	11.345	12.838	16.266
4	4.878	5.385	5.089	7.779	9.488	11.143	11.668	13.277	14.860	18.467
5	6.064	6.626	7.289	9.236	11.070	12.832	13.388	15.086	16.750	20.515
6	7.231	7.841	8.558	10.645	12.592	14.449	15.033	16.812	18.548	22.457
7	8.383	9.037	9.803	12.017	14.067	16.013	16.622	18.475	20.278	24.322
8	9.524	10.219	11.030	13.362	15.507	17.535	18.168	20.090	21.955	26.125
9	10.656	11.389	12.242	14.684	16.919	19.023	19.679	21.666	23.589	27.877
10	11.781	12.549	13.442	15.987	18.307	20.483	21.161	23.209	25.188	29.588
11	12.899	13.701	14.631	17.275	19.675	21.920	22.618	24.725	26.757	31.264
12	14.011	14.845	15.812	18.549	21.026	23.337	24.054	26.217	28.300	32.909
13	15.119	15.984	16.985	19.812	22.362	24.736	25.472	27.688	29.819	34.528
14	16.222	17.117	18.151	21.064	23.685	26.119	26.873	29.141	31.319	36.123
15	17.322	18.245	19.311	22.307	24.996	27.488	28.259	30.578	32.801	37.697
16	18.418	19.369	20.465	23.542	26.296	28.845	29.633	32.000	34.267	39.252
17	19.511	20.489	21.615	24.769	27.587	30.191	30.995	33.409	35.718	40.790
18	20.601	21.605	22.760	25.989	28.869	31.526	32.346	34.805	37.156	42.312
19	21.689	22.718	23.900	27.204	30.144	32.852	33.687	36.191	38.582	43.820
20	22.775	23.828	25.038	28.412	31.410	34.170	35.020	37.566	39.997	45.315
21	23.858	24.935	26.171	29.615	32.671	35.479	36.343	38.932	41.401	46.797
22	24.939	26.039	27.301	30.813	33.924	36.781	37.659	40.289	42.796	48.268
23	26.018	27.141	28.429	32.007	35.172	38.076	38.968	41.638	44.181	49.728
24	27.096	28.241	29.553	33.196	36.415	39.364	40.270	42.980	45.558	51.179
25	28.172	29.339	30.675	34.382	37.652	40.646	41.566	44.314	46.928	52.620
26	29.246	30.434	31.795	35.563	38.885	41.923	42.856	45.642	48.290	54.052
27	30.319	31.528	32.912	36.741	40.113	43.194	44.140	46.963	49.645	55.476
28	31.391	32.620	34.027	37.916	41.337	44.461	45.419	48.278	50.993	56.893
29	32.461	33.711	35.139	39.087	42.557	45.722	46.693	49.588	52.336	58.302
30	33.530	34.800	36.250	40.256	43.773	46.979	47.962	50.892	53.672	59.703

Source: This table is abridged from Table IV of R. A. Fisher and F. Yates, *Statistical Tables for Biological, Agricultural and Medical Research*, 6th ed. (Edinburgh: Oliver & Boyd Ltd., 1963), and Table 8 of E. S. Pearson and H. O. Hartley, eds., *Biometrika Tables for Statisticians*, Vol. I (New York: Cambridge University Press, 1966), by permission.

Bibliography

ABRAMOWITZ, M., and I. A. STEGUN, eds. (1964), *Handbook of Mathematical Functions with Formulas, Graphs, and Mathematical Tables* (Washington, D.C.: Government Printing Office).

ADELMAN, I. (1965), "Long Cycles—Fact or Artifact," *The American Economic Review*, Vol. LV, pp. 444–463.

ADELMAN, I., and C. T. MORRIS (1968), "Performance Criteria for Evaluating Economic Development Potential: An Operational Approach," *The Quarterly Journal of Economics*, Vol. LXXXII, pp. 260–280.

ALMON, C., Jr. (1967), *Matrix Methods in Economics* (Reading, Mass.: Addison-Wesley Publishing Company, Inc.).

ANDERSON, L. C. (1969), "Money Market Conditions as a Guide for Monetary Management," in *Targets and Indicators of Monetary Policy*, K. Bruner, ed. (San Francisco: Chandler Publishing Company).

ANDERSON, T. W. (1958), *An Introduction to Multivariate Statistical Analysis* (New York: John Wiley & Sons, Inc.).

BANCROFT, T. A. (1968), *Topics in Intermediate Statistical Methods*, Vol. I (Ames: Iowa State University Press).

BARTLETT, M. S. (1937), "Properties of Sufficiency and Statistical Tests," *Proceedings of the Royal Society*, A, Vol. 160, pp. 268–282.

——— (1948), "Smoothing Periodograms from Times Series with Continuous Spectra," *Nature*, Vol. 161, pp. 686–687.

——— (1962), *An Introduction to Stochastic Processes with Special Reference to Methods and Applications* (New York: Cambridge University Press).

BETANCOURT, R. R. (1971), "Normal Income Hypothesis in Chile," *Journal of the American Statistical Association*, Vol. 66, pp. 258–263.

BLACKMAN, R. B., and J. W. TUKEY (1959), *The Measurement of Power Spectra* (New York: Dover Publications, Inc.).

BOLCH, B. W. (1967), "A Spectral Analysis of Interrelationships Between the Money Stock and Certain Asset Markets," *The Journal of Finance*, Vol. XXII, pp. 475–476.

——— (1968), "More on Unbiased Estimation of the Standard Deviation," *American Statistician*, Vol. 22, p. 27.

BOTTENBERG, R. A., and J. H. Ward, JR. (1963), *Applied Multiple Linear Regression* (Washington, D.C.: U.S. Department of Commerce, Office of Technical Services, AD413128).

BOX, G.E.P. (1949), "A General Distribution Theory for a Class of Likelihood Criteria," *Biometrika*, Vol. 36, pp. 317–346.

BOX, G. E. P., and D. R. COX (1964), "An Analysis of Transformations," *Journal of the Royal Statistical Society*, B. Vol. 26, pp. 211–243.

BROWNE, E. T. (1958), *Introduction to the Theory of Determinants and Matrices* (Chapel Hill: The University of North Carolina Press).

BRUNNER, K., and A. MELTZER (1969), "The Nature of the Policy Problem," in *Targets and Indicators of Monetary Policy*, K. Brunner, ed. (San Francisco: Chandler Publishing Company).

CHASE, G. R., and W. G. Bulgren (1971), "Monte Carlo Investigation of the Robustness of T^2," *Journal of the American Statistical Association*, Vol. 66, pp. 499–502.

CHOW, G. C. (1960), "Tests of Equality Between Sets of Coefficients in Two Linear Regressions," *Econometrica*, Vol. 28, pp. 591–605.

CHOW, G. C., and D. K. RAY-CHAUDHURI (1967), "An Alternative Proof of Hannan's Theorem on Canonical Correlation and Multiple Equation Systems," *Econometrica*, Vol. 35, pp. 139–142.

COCHRAN, W. G. (1957), "Analysis of Covariance: Its Nature and Uses," *Biometrics*, Vol. 13, pp. 261–281.

COCHRAN, W. G., and G. M. COX (1957), *Experimental Designs*, 2nd ed. (New York: John Wiley & Sons, Inc.).

COOLEY, W. W., and P. R. LOHNES (1962), *Multivariate Procedures for the Behavorial Sciences* (New York: John Wiley & Sons, Inc.).

COOTNER, P. H., ed. (1964), *The Random Character of Stock Market Prices* (Cambridge, Mass.: The M.I.T. Press).

COWDEN, D. J. (1952), "The Multiple-Partial Correlation Coefficient," *Journal of the American Statistical Association*, Vol. 47, pp. 442–456.

——— (1957), *Statistical Methods in Quality Control* (Englewood Cliffs, N.J.: Prentice-Hall, Inc.).

CROXTON, F. E., D. J. COWDEN, and B. W. BOLCH (1969), *Practical Business Statistics*, 4th ed. (Englewood Cliffs, N. J.: Prentice-Hall, Inc.).

DAVID, F. N. (1938), *Tables of the Ordinates and Probability Integral of the Distribution of the Correlation Coefficient in Small Samples* (New York: Cambridge University Press).

DEMING, W. E. (1938), *Statistical Adjustment of Data* (New York: Dover Publications, Inc.).

DEMPSTER, A. P. (1969), *Elements of Continuous Multivariate Analysis* (Reading, Mass.: Addison-Wesley Publishing Company, Inc.).

DEUTSCH, R. (1965), *Estimation Theory* (Englewood Cliffs, N. J.: Prentice-Hall, Inc.).

DHRYMES, P. J. (1970), *Econometrics: Statistical Foundations and Applications* (New York: Harper & Row, Publishers).

DIXON, W. J., ed. (1967), *BMD, Biomedical Computer Programs*, 2nd ed. (Berkeley: University of California Press).

DRAPER, N. R., and H. SMITH (1966), *Applied Regression Analysis* (New York: John Wiley & Sons, Inc.).

EATON, M. L., and B. EFRON (1970), "Hotelling's T^2 Test Under Symmetry Conditions," *Journal of the American Statistical Association*, Vol. 65, pp. 702–711.

EVANS, M. K., and L. R. KLEIN (1967), *The Wharton Econometric Forecasting Model* (Philadelphia: Economics Research Unit, University of Pennsylvania).

FADDEEVA, V. N. (1959), *Computational Methods of Linear Algebra* (New York: Dover Publications, Inc.).

FOX, K. A. (1968), *Intermediate Economic Statistics* (New York: John Wiley & Sons, Inc.).

GOLDBERGER, A. S. (1964), *Econometric Theory* (New York: John Wiley & Sons, Inc.).

——— (1968), "The Interpretation and Estimation of Cobb-Douglas Production Functions," *Econometrica*, Vol. 35, pp. 464–472.

GRANGER, C. W. J., and M. HATANAKA (1964), *Spectral Analysis of Economic Time Series* (Princeton, N.J.: Princeton University Press).

GRANGER, C. W. J., and O. MORGENSTERN (1963), "Spectral Analysis of New York Stock Market Prices," reprinted from *Kyklos*, Vol. 16, in P. H. Cootner (1964).

GRAYBILL, F. A. (1961), *An Introduction to Linear Statistical Models*, Vol. I (New York: McGraw-Hill Book Company)

GRILICHES, Z. (1967), "Distributed Lags: A Survey," *Econometrica*, Vol. 35, pp. 16–49.

GUJARATI, D. (1970), "Use of Dummy Variables in Testing for Equality Between Sets of Coefficients in Linear Regression: A Generalization," *American Statistician*, Vol. 24, pp. 18–22.

HANNAN, E. J. (1967), "Canonical Correlation and Multiple Equation Systems in Economics," *Econometrica*, Vol. 35, pp. 123–138.

HARMAN, H. H. (1967), *Modern Factor Analysis*, 2nd ed. (Chicago: University of Chicago Press).

HENDERSHOTT, P. H. (1968), *The Neutralized Money Stock—An Unbiased Measure of Federal Reserve Policy Actions* (Homewood, Ill.: Richard D. Irwin, Inc.).

HOTELLING, H. (1931), "The Generalization of Student's Ratio," *Annuals of Mathematical Statistics*, Vol. 2, pp. 360–378.

——— (1933), "Analysis of a Complex of Statistical Variables into Principal Components," *Journal of Educational Psychology*, Vol. 24, pp. 417–441.

——— (1936), "Simplified Calculation of Principal Components," *Psychometrika*, Vol. 1, pp. 27–35.

HUANG, C. J. (1971), "A Test for Multivariate Normality of Disturbance Terms," *Southern Economic Journal*, Vol. XXXVIII, pp. 206–209.

HUANG, D. S. (1970), *Regression and Econometric Methods* (New York: John Wiley & Sons, Inc.).

JENKINS, G. M., and D. G. WATTS (1968), *Spectral Analysis and its Applications* (San Francisco: Holden-Day, Inc.).

JOHNSTON, J. (1963), *Econometric Methods* (New York: McGraw-Hill Book Company).

KEMENTA, J. (1967), "On Estimation of the CES Production Function," *International Economic Review*, Vol. 8, pp. 180–189.

KENDALL, M. G. (1968), *A Course in Multivariate Analysis* (New York: Hafner Publishing Company, Inc.).

KENDALL, M. G., and A. STUART (1968), *The Advanced Theory of Statistics*, Vol. 3, 2nd ed. (New York: Hafner Publishing Company, Inc.).

KLATZKY, S. R., and R. W. HODGE (1971), "A Canonical Correlation Analysis of Occupational Mobility," *Journal of the American Statistical Association*, Vol. 66, pp. 16–22.

Klein, L. R. (1950), *Economic Fluctuations in the United States, 1921–1941* (New York: John Wiley & Sons, Inc.).

Li, C. C. (1964), *Introduction to Experimental Statistics* (New York: McGraw-Hill Book Company).

Lighthill, M. J. (1958), *Introduction to Fourier Analysis and Generalized Functions* (New York: Cambridge University Press).

McCalla, T. R. (1967), *Introduction to Numerical Methods and FORTRAN Programming* (New York: John Wiley & Sons, Inc.).

Mendenhall, W. (1968), *Introduction to Linear Models and the Design and Analysis of Experiments* (Belmont, Calif.: Wadsworth Publishing Company, Inc.).

Morrison, D. F. (1967), *Multivariate Statistical Methods* (New York: McGraw-Hill Book Company).

Nerlove, M. (1964), "Spectral Analysis of Seasonal Adjustment Procedures," *Econometrica*, Vol. 32, pp. 241–286.

Ohlin, G. (1967), *Population Control and Economic Development*, (Paris: Development Centre of the Organisation for Economic Co-operation and Development).

Paige, L. J., and J. D. Swift (1961). *Elements of Linear Algebra* (Boston: Ginn and Company).

Parzen, E. (1961), "Mathematical Considerations in the Estimation of Spectra," *Technometrics*, Vol. 3, pp. 167–190.

——— (1963), "Notes on Fourier Analysis and Spectral Windows," reprinted in E. Parzen (1967), *Time Series Analysis Papers* (San Francisco: Holden-Day, Inc.) pp. 190–250.

Phillips, A. W. (1958), "The Relation Between Unemployment and the Rate of Change of Money Wage Rates in the United Kingdom, 1861–1957," *Economica*, Vol. 25, pp. 283–299.

Pillai, K. C. S., and S. Al-Ani (1970), "Power Comparisons of Tests of Equality of Two Covariance Matrices Based on Individual Characteristic Roots," *Journal of the American Statistical Association*, Vol. 65, pp. 438–446.

Puri, M. L., and P. K. Sen (1971), *Nonparametric Methods in Multivariate Analysis* (New York: John Wiley & Sons, Inc.).

Quade, D. (1967), "Rank Analysis of Covariance," *Journal of the American Statistical Association*, Vol. 62, pp. 1187–1200.

Rao, C. R. (1952), *Advanced Statistical Methods in Biometric Research* (New York: John Wiley & Sons, Inc.).

——— (1965), *Linear Statistical Inference and Its Applications*, (New York: John Wiley & Sons, Inc.).

REKTORYS, K., ed. (1969), *Survey of Applicable Mathematics* (Cambridge, Mass.: The M.I.T. Press).

ROSENBLATT, H. M. (1968), "Spectral Evaluation of BLS and Census Revised Seasonal Adjustment Procedures," *Journal of the American Statistical Association*, Vol. 63, pp. 472–501.

SCHEFFÉ, H. (1959), *The Analysis of Variance* (New York: John Wiley & Sons, Inc.).

SIEGEL, S. (1956), *Nonparametric Statistics for the Behavioral Sciences* (New York: McGraw-Hill Book Company).

SKVARCIUS R., and F. CROMER, (1971), "A Note on the Use of Categorical Vectors in Testing for Equality of Two Regression Equations," *The American Statistician*, Vol. 25, pp. 27–28.

SNEDECOR, G. W., and W. G. COCHRAN (1967), *Statistical Methods*, 6th ed. (Ames: Iowa State University Press).

STUDENT (W. S. GOSSET) (1908), "The Probable Error of a Mean," *Biometrika*, Vol. 6, pp. 1–25.

TAYLOR, A. E. (1955), *Advanced Calculus* (Boston: Ginn and Company).

THEIL, H. (1971), *Principles of Econometrics* (New York: John Wiley & Sons, Inc.).

TINTNER, G. (1946), "Some Applications of Multivariate Analysis to Economic Data," *Journal of the American Statistical Association*, Vol. 41, pp. 472–500.

TOLLES, N. A., and E. MELICHAR (1968), "Studies of the Structure of Economists' Salaries and Income," *American Economic Review Supplement*, Vol. LVIII, No. 5, Part 2.

WALKER, H. M., and J. LEV (1953), *Statistical Inference* (New York: Holt, Rinehart and Winston, Inc.).

WAUGH, F. V. (1942), "Regressions Between Sets of Variates," *Econometrica*, Vol. 10, pp. 290–310.

WILKS, S. S. (1962), *Mathematical Statistics* (New York: John Wiley & Sons, Inc.).

ZAREMBKA, P. (1968), "Functional Form in the Demand for Money," *Journal of the American Statistical Association*, Vol. 63, pp. 502–511.

Index

D

E